The Science of Radio

An American family poses proudly in 1947 Southern California. The new Admiral model 7C73 9-tube AM/FM radio-phonograph console, won in a jingle contest, was more than simply a radio—it was the family entertainment center and king of the livingroom furniture. The shy lad at the left is the author, age seven. Photo courtesy of the author's sister Kaylyn (Nahin) Warner.

Second Edition

The Science of Radio

with MATLAB® and ELECTRONICS WORKBENCH® Demonstrations

Paul J. Nahin

University of New Hampshire

Durham, New Hampshire

Springer

Paul J. Nahin
Department of Electrical & Computer Engineering
University of New Hampshire
Kingsbury Hall
33 College Road
Durham, NH 03824
USA
paul.nahin@unh.edu

Library of Congress Cataloging-in-Publication Data
Nahin, Paul J.
    The science of radio: with MATLAB and electronics workbench demonstrations/Paul
J. Nahin.–2nd ed.
        p.   cm.
    Includes bibliographical references and index.
    ISBN 0-387-95150-4 (softcover: alk. paper)
    1. Radio.   2. MATLAB.   3. Electronics workbench.   I. Title.
TK6550.N16 2001
621.384–dc21   00-062062

Printed on acid-free paper.

Production managed by Frank McGuckin; manufacturing supervised by Joe Quatela.
Typeset by Integre Technical Publishing Co., Inc., Albuquerque, NM.
Printed and bound by Hamilton Printing Co., Rensselaer, NY.
Printed in the United States of America.

9 8 7 6 5 4 3 2 1

ISBN 1-0-387-95150-4   SPIN 10781909

Springer-Verlag   New York Berlin Heidelberg
A member of BertelsmannSpringer Science+Business Media GmbH

ra·di·o (rā′dēō′)—n., of English origin, early 20th century. Abbreviation of *radiotelegram*, itself a variation of *radial* (from the Latin *radius*), *radiation*, and/or *radiative*, i.e., a "radiation" or "radial" telegram (try saying *radial* quickly, several times). In this use one might say, in brevity, "I have just received a radio." This usage was formally endorsed by the 1906 International Radiotelegraphic Conference in Berlin. The word *radio* gradually came to refer to the devices used to transmit the radiotelegram itself, and then to any means of sending information via electromagnetic waves. In England the term used was *wireless*, but today that is reserved for the distinctly different case of two-way, point-to-point private communication (broadcast radio is inherently wide area, public, and one way) devices such as cell phones and remote personal computer links. "Wireless" is a broader technical term than is "radio." *Radio* is used only for devices utilizing propagating fields, while *wireless* can be applied to inductive field devices, as well.

"Canst thou send lightnings, that they may go, and say unto thee, Here we *are*?"   —*Job* 38:35

A top-down, just-in-time first course in electrical engineering, for students who have had freshman calculus and physics, that answers the questions of,

what's inside a kitchen radio?

how did it all get there?

why does the thing work?,

along with MATLAB and Electronics Workbench demonstrations, and lots of history, plus a collection of theoretical discussions and problems to amuse, perplex, enrage, challenge, and otherwise entertain the reader.

"E-mail and other tech talk may be the third, fourth, or nth wave of the future, but old-fashioned radio is true hyperdemocracy."
—*Time*, January 23, 1995

Paul J. Nahin is Professor of Electrical Engineering at the University of New Hampshire. He is the author of *Oliver Heaviside* (IEEE Press 1988), *An Imaginary Tale: the story of* $\sqrt{-1}$ (Princeton 1998), *Time Machines: Time Travel in Physics, Metaphysics, and Science Fiction* (Springer-Verlag 1999), and *The Duelling Idiots and Other Probability Puzzlers* (Princeton 2000). He welcomes comments or suggestions from readers about any of his books; he can be reached on the Internet at paul.nahin@unh.edu.

CONTENTS

ACKNOWLEDGMENTS

While the actual writing of this book has been a lone effort, there are some individuals who aided my writing that I do wish to particularly thank. First is my loving wife Patricia Ann, who never once objected when I would say "I'm going to the computer room to work on the book." Not even when the booming sounds of anti-aircraft artillery leaked through the door as I occasionally played Beachhead 2000™ on the computer, instead of running a MATLAB program. Even when the clanking noise of a sword fight with guards in the medieval castle of Thief™ were so loud they absolutely couldn't be missed, Pat still gave not the slightest hint that she knew I wasn't actually firing-up an Electronics Workbench simulation. It's tolerance like that that helps a marriage reach its fortieth year!

At one time a dear friend of many years, Professor John Molinder at Harvey Mudd College (Claremont, California), and I did talk of doing something like this book together. A separation of thousands of miles (and different professional pressures) made that not practical. Still, John's influence over the last thirty years on my thinking about the topics in this book has been profound. Much of what is in this book came together in my mind when, during a sabbatical at Mudd in the Fall of 1991, John and I team-taught E101, the College's junior year systems engineering course. This is my opportunity to thank him (and our students) for being kindred spirits in "all things convolutional."

After the first edition of this book appeared in 1995, I received a letter from Professor David Rutledge at Caltech, telling me both of his own sophomore EE book on radio then in-progress, as well as how much he and his Caltech EE colleague Professor William Bridges liked my book; enough to direct their students to it. That letter eventually resulted in a visit and a talk to the EE undergraduates at Caltech on the new material I was then adding to Chapter 7 (Van der Pol's non-linear differential equation for negative resistance oscillators), as well as to a sumptuous dinner with Dave and Bill at Caltech's elegant Athenaeum.

To all my University of New Hampshire (UNH) electrical engineering students in the sophomore circuits and electronics, and the junior networks courses (EE541, 548, and the 'old' 645), I owe much. Both for patiently (*most* of the time) listening to me occasionally grope my way to understanding what I was talking about, and for providing me with solutions to a couple of problems I had a hard time doing (and which are now in this book). To that I want to add a special note of appreciation to the students in the Honors section of the EE645 networks class in the Fall of 1992, who were my test subjects for the more advanced technical parts of this book.

Nan Collins, of the Word Processing Center of the College of Engineering and Physical Sciences at UNH, has been my cheerful and always patient typist on previous books, but never more so than with this one. Without her skill at transforming my scrawled, handwritten equations in smeared ink into WordPerfect beauty, I wouldn't have had the strength to finish. The late Professor Hugh Aitken of Amherst College greatly influenced this book through the historical content, and the elegant prose, of his two well-known books on technical radio history. I was fortunate to enjoy a correspondence of several years with Hugh, but I will always regret that I delayed too long in making the short drive from UNH to Amherst. And finally, I greatly appreciated the professionalism of Frank McGuckin, the book's Production Editor at Springer-Verlag/New York for this new, second edition.

The second edition of any book gives the author a welcomed opportunity to eliminate the irritants that slipped through the first time, such as embarrassing misspellings. On a somewhat grander scale, it also allows for the rewriting of a sentence (or even a paragraph or more) that, once in print, no longer seems as appropriate as it first did. Or it may offer an opportunity to add that really neat example discovered the day *after* the absolutely, positively final deadline for making page-proof changes in the first edition had come and gone. There are a fair number of such things in this new *Science of Radio*.

There is also more history, more end-of-chapter problems, and more hints for the problems (a solution manual is available from the publisher, containing detailed analyses for many of the problems). Doing problems, I believe, is a vital part of learning any technical subject, and that is absolutely the case for a mathematical subject like radio (my original working title for the book was, in fact, *The Mathematical Radio*). To quote Ogden Nash, "At last I've found the secret that guarantees success/to err, and err, and err again, but less, and less, and less."

Moving up the scale of grandeur just a bit, I have added a new appendix on differential amplifiers. I do remain convinced that, from a historical view of how radio developed, my original approach of discussing electronic radio circuits in terms of vacuum tubes and *not* transistors is proper (if unusual). The fundamental circuit laws are just as true for vacuum tubes as they are for transistors. Still, I have been convinced by a number of e-mails from EE professors at a large number of schools that differential amplifiers are simply too important to be left out of a sophomore text. With differential amplifiers it is easy to discuss Tellegen's gyrator, and then such amazing circuits as impedance inverters and negative impedance converters with which to build even more complicated circuits that simulate inductors (the one passive element that it is best to avoid because it is so hard to construct to look "ideal"). From *that* it is then an easy step to making filters, useful in radio, that have absolutely astonishing properties. And, curiously, as such things seem to happen, the differential amplifier appendix (which is presented in terms of an "ideal systems model" and so is technology independent) turns out to dovetail nicely with the really big, new feature of the book: the use of two popular computer application programs, MATLAB and Electronics Workbench.

Electronics Workbench (EWB) creates, right on a computer screen, a virtual electronics laboratory complete with endless supplies of components—including triode vacuum tubes!—and a set of virtual electronic instruments. There is a function generator, a multimeter, and an oscilloscope, of course, but most wonderful of all is a Bode-plotter, an instrument that as far as I know doesn't exist (yet), but should (is anybody from Hewlett-Packard or Tektronix reading this?) The Bode-plotter is used, in particular, to demonstrate the performance of an elegant phase-shifting circuit for use in single-sideband radio. With its schematic capture capability, EWB is perfect for preparing neat and precise circuit diagrams, and its SPICE[1]-based simulation engine is both very fast and very accurate in converging to solutions. It allows the study of circuits, right on a student's computer at home, that in the "old days" would have required access to an expensive laboratory facility.

To visualize EWB's use with a classic problem in elementary circuits, consider the calculation of the body-diagonal resistance of a "cube" of 1-ohm resistors. This can be done analytically by the brute-force method of directly applying Kirchhoff's laws, or more elegantly by invoking symmetry and Ohm's law (see Appendix B), with both approaches resulting in the answer of $\frac{5}{6}$ ohm. The EWB solution is shown in Figure 1, where I have used a mouse to draw the cube—folded-out flat, of course—and

---

[1] SPICE, the acronym for "Simulation Program with Integrated Circuit Emphasis," is a circuit simulator originally developed in the mid-1970s by the Electrical and Computer Engineering Department and the Electronics Research Laboratory at the University of California at Berkeley.

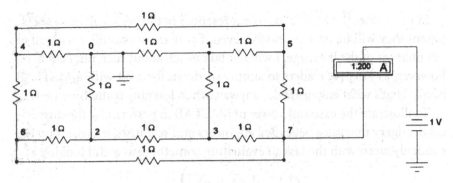

Figure 1 Solving for the body-diagonal resistance of a cube whose edges are 1-ohm resistors. The theoretical answer is $\frac{5}{6}$ ohm, which agrees with an applied 1-volt potential difference causing 1.2 amperes to flow.

have connected a 1-volt battery in series with a dc ammeter to a pair of body-diagonal nodes (node 0, the common ground, and node 7). The meter displays 1.2 A flowing, and so Ohm's law immediately gives us $\frac{5}{6}$ ohm. With EWB it is easy to determine the resistance between *any* two nodes, for *any* values of the edge resistors; the presence of symmetry is irrelevant. (In Appendix D, I discuss how to use EWB to solve the far more interesting case of a *four*-dimensional resistor cube.)

MATLAB (for MATrix LABoratory) allow the easy-to-program calculation *and plotting* of the most complicated, tedious expressions that an electrical engineer—in his or her worst nightmares—could ever imagine encountering. I use that one feature, alone, all through this book to provide graphical illustrations of many of the theoretical calculations e.g., Fourier's theorem and solutions to Van der Pol's nonlinear second order differential equation for non-sinusoidal oscillations in negative resistance circuits, such as the early arc radio transmitters. I also take great advantage of MATLAB's ability to do *symbolic* manipulation, e.g. Fourier and Hilbert transforms, integration and differentiation, and equation solving. MATLAB takes the mathematical entity of the matrix as its fundamental construct. Even a scalar is thought of as a 1×1 matrix. This structure allows extremely compact coding, a virtue (at least most computer scientists seem to think it is a virtue) that I will *not* take advantage of—you will see in this book mostly the "old fashioned" *for, while*, and *if* loops. I am sure most readers will think of many ways to write far more efficient codes than I have, but I make no apologies for that because, in this book, it is not programming but *radio* that is center stage.

In most cases the MATLAB codes presented in the book will be so transparent they will be understandable even if you are not familiar with all of the nuances of the language. I will *not* talk much about such nuances here, however, as I expect readers to simply sit down for a bit with a MATLAB book. That's what engineers do, anyway, when learning something new.

To illustrate the ease and power of MATLAB in performing the numerical drudgery that once bedeviled engineers and scientists, imagine you are suddenly faced with the task of evaluating something as awful looking as

$$\int_0^1 \frac{1}{x} \ln\left\{ \left(\frac{1+x}{1-x} \right)^2 \right\} \, dx.$$

Later in this book, after reading about Fourier series expansions of periodic signals, you'll see it is pretty easy—if you see the trick!—to show that the answer is $\frac{1}{2}\pi^2$ (see Problem 10.6). But with MATLAB you don't need any tricks, just the following two lines of routine[2] code that approximates the integral with trapezoids:

```
x = .000001:.00001:1;
y = log((((1 + x) · /(1 − x)) · ^ 2) · /x;  trapz(x, y)
```

which produces the result of 4.9327. This is "correct," too, as $\frac{1}{2}\pi^2 =$ 4.9348.

Both MATLAB and Electronics Workbench weren't even science fiction when I was an undergraduate in the late 1950s and early 1960s—if imagined in those days by some visionary, they would have been called *fantasy*. My students always look bewildered when I tell them this (or, when I'm really feeling nostalgic, when I mention *slide rules!*) Today's engineers, however, use such computer applications routinely, and so today's students need to see them as soon as possible in their studies. At the University of New Hampshire, for example, electrical engineering students begin to use MATLAB in their freshman year, and to learn Electronics Workbench in their sophomore year. I don't believe UNH is unique in that, and the presence of both applications in this book simply reflects how much undergraduate electrical engineering studies have changed in less than a single professional lifetime.

---

[2] The reason for the not-quite zero start value in the x-vector statement is to avoid evaluating the integrand at either $x = 0$ or $x = 1$. The integrand is, in fact, perfectly well-behaved everywhere (as you can show by making a power series expansion of the log function), but MATLAB will give a division-by-zero warning at $x = 0$ because of the $\frac{1}{x}$ factor. At $x = 1$ you'll get a similar message, too, because of the $1 − x$ in the denominator of the log function.

The style of the writing remains, without apology, exactly as in the original edition. This book has been written by an author who, when he hears Ferde Grofé's "On the Trail," instantly thinks of cigarettes and not of the Grand Canyon (Suite) because Philip Morris used it as the theme music for its old radio ads. You have been warned!

(Students may read this, too.)

Modern-day electrical engineering academic programs are generally regarded on most college campuses as among the toughest majors of all, but being tough doesn't mean being the best. EE programs are tough because there is so much stuff coming at students, all the time—with no let-up for four years—that many of these students are overwhelmed. In addition to the usual image of trying to drink from a firehose, the process has been likened to getting on a Los Angeles freeway at age eighteen, with no speed limit or off ramps, and being told to drive or die until age twenty-two. A lot of good students simply run out of gas before they can finish this ordeal.

Most electrical engineering faculty are aware of the crisis in the teaching of their discipline; indeed, each month engineering education journals in *all* specialties carry columns and letters bemoaning the problem. The typical academic response is to shuffle the curriculum,[1] a lengthy and occasionally contentious process that leaves faculty exhausted and nobody happy. And then, four or five years later, by which time nearly everyone is *really* unhappy (again), the whole business is repeated. A more innovative response, I believe, is to teach more of electrical engineering using the "top-down," "just-in-time" approaches.

Top-down starts with a global overview of an entire task and evolves into ever more detail as the solution is approached. Just-in-time means all mathematical and physical theory is presented only just before its first use in an application of substance, i.e., not simply as a means of working that week's turn-the-crank (often unmotivated) homework problem set. The traditional electrical engineering educational experience is, however, just the reverse. It starts with a torrent of mind-numbing details (hundreds of mathematical methods, analog circuit laws, digital circuit theorems, etc., etc., one after the other), details which faculty expect students to be able to pull out of their heads on command (i.e., upon the appearance of a quiz sheet). Faculty, who swear by the fundamental conservation laws of physics in the lab, oddly seem to think they can violate them in the classroom when it comes to education. They are wrong, of course, as you simply cannot pour twenty gallons of facts into a one-gallon head without making a nineteen-gallon puddle on the floor.

And what is the reward for the agony? Sometime in the third year of this amazing process the blizzard of isolated facts finally starts to come together with a simple system analysis or two. In the fourth (and final) year, perhaps a simple design project will be tackled. It seems to me to really be precious little gain for such an ocean of sweat and tears. But faculty appear unmoved by the inefficiency (and the horror) of it all. As one little jingle that all professors (but perhaps not all students) will appreciate puts it:

> Cram it in, jam it in;
> The students' heads are hollow.
> Cram it in, jam it in;
> There's plenty more to follow.

It is strange that so many engineering professors teach this way. As M. E. Van Valkenburg, one of the grand old men of American engineering education once observed: "engineers seem to learn best by top-down methods."[2] But, except for isolated, individual teachers going their own way, I know of not a single American electrical engineering curriculum that is based on the top-down idea, and for a simple reason. As Professor Van Valkenburg said: "I know of no engineering textbooks that follow a top-down format."

This is truly a Catch-22, chicken-and-egg situation.

There are no top-down books because there are no curricula for them, and there are no curricula because there are no top-down books. Top-down books, in electrical engineering, are a huge departure from presently accepted formats, and many potential authors simply see no financial rewards in writing such books. Instead, we get yet *more* books on the same old stuff (e.g., elementary circuit theory), written in the same old way;

how many ways are there, I wonder, of explaining Kirchhoff's laws!? The author of a modern bestseller was embarrassed enough by this to write in his preface "the well-established practice of revising circuit analysis textbooks every three years may seem odd," and then he went on to blame students' declining abilities for the endless revisions and the reburying of bones from one graveyard to another. And I had always thought it had something to do with undermining the used-book market. How foolish of me!

An example of the sort of introductory engineering book I do think worth writing is John R. Pierce's *Almost All About Waves* (MIT Press 1974), in which he ends with "Commonly, physicists and engineers first encounter waves in various complicated physical contexts and finally find the simple features that all waves have in common. Here the reader has considered those simple, common features, and is prepared, I hope, to see them exemplified [in more advanced books]." Pierce's book then, despite Professor Van Valkenburg's assessment, was perhaps the first top-down electrical engineering book, but published before anybody had a name for it! I view my book as a similar radical departure from ordinary textbooks, an *experiment* continuing Pierce's pioneering effort, if you will. My hope is that in years to come faculty will declare the present state of affairs in the education of electrical engineers as having been quaint, if not downright bizarre.

This is an attempt at a combination top-down, just-in-time "first course" electrical engineering book, anchored to the specifics of a technical and mathematical history of the ordinary superheterodyne AM radio receiver which, since its invention 85 years ago, has been manufactured in the billions (as the new millenium begins there are 700 million AM radios *in use*, in America alone). I have written this book for the beginning second-year student in *any* major who has the appropriate math/physics background. As I explain in more detail in the prologue, this means freshman calculus and physics. To those who complain this is an unreasonable expectation in, say, a history major, I reply that history faculties really ought to do something about that. To allow their students to be anointed with the Bachelor of Arts, even while remaining ignorant of the great discoveries of eighteenth and nineteenth centuries' natural philosophers, is as great a sin as would be committed, for example, by electrical engineering faculties allowing *their* students to graduate without taking several college-level English and history courses. There are, of course, more advanced books available to those readers who want more electronic circuit details, but your students will not have to "unlearn" anything from this book. I offer here what might be called an "advanced primer," with no simplifications that will fail them in the future.

Your students will see here, for example, not only the Fourier transform but the *Hilbert* transform, too, a topic not found in any previous second-year book (see Appendix H). I have two reasons for doing this. First, and less importantly, the Hilbert transform occurs in a natural way in expressing the constraints causality places on the real and imaginary parts of the Fourier transform of a time signal. Second, and more importantly, the Hilbert transform occurs in a natural way in the theoretical development of single-sideband radio (which, in turn, is a natural development of radio theory after AM sidebands and bandwidth conservation are discussed—and this was not slow to happen; as I explain in Chapter 20, single-sideband radio was developed in the 1920s!). This book discusses the Hilbert transform in both applications. The issue of the convergence of Fourier series is given more than the usual quick nod and a wink. And when I do blow a little mathematical smoke in front of a mirror—as in the discussion of impulse functions—I have tried to be explicit about the handwaving. I acknowledge the nontrivialness of doing such things as reversing the order of integration in double (perhaps improper) integrals, or of differentiating under the integral sign, two processes my experience has shown me befuddle even bright electrical engineering seniors. Something in the freshman calculus courses electrical engineers take these days simply isn't taking, and I have tried to address that failure in this book.

To make the book as self-contained as possible, I've included brief reviews of complex exponentials (Appendix A), of linear and time invariant systems (Appendix B), of Kirchhoff's laws and related issues (Appendix C), and of resonance (Appendix E) for those students who need them. These appendices are written, however, in a historical manner that I think will make them interesting reading even for those who perhaps don't actually need a review, but who nonetheless may learn something new anyway (as in Appendix G, where even professors may see for the first time how to derive Dirichlet's discontinuous integral using only freshman calculus).

I have also done my best to emphasize what I consider the intellectual excitement and beauty of the history and mathematical theory of radio. I frankly admit that the spiritual influences of greatest impact on the writing of this book were Garrison Keillor's wonderfully funny novel of early radio, *WLT, A Radio Romance* (Viking 1991),[3] and Woody Allen's sentimental 1987 movie tribute to World War II radio, *Radio Days*. Born in

1940, I was too young to experience first-hand those particular days, but I was old enough in the late 1940s and early 1950s to catch the tail-end of radio drama's so-called Golden Age. I listened, in fascination, to more than my share of "Lights-Out," "The Lone Ranger," "Little Orphan Annie," "Yours Truly, Johnny Dollar" (the insurance investigator with the "action-packed expense account!"), "The Jack Benny Show," "Halls of Ivy," "Boston Blackie" ("Enemy to those who make him an enemy; friend to those who have no friends!"), "The Shadow," and, most wonderful of all, the science fiction thriller "Dimension-X" (later called "X-Minus One").

I shared that fascination with lots of others. After all, who could fail to feel the hair rise on the back of their neck at the sound of "Inner Sanctum"'s creaking door, or at the maniacal voice of Orson Wells as "The Shadow" ("Who knows what evil lurks in the hearts of men? The Shadow knows!") followed by a crazy, spine-chilling laugh that must have sent all of America's children, and a lot of their parents, too, under the bed blankets? Who could fail to feel pride for our men and women of the law at the burst of machine-gun fire and the wailing siren that opened each episode of "Gangbusters" (the origin of the phrase describing super-enthusiasm as "Coming on like gangbusters"!), or to laugh every single time when the closet full of junk fell out on the floor—for what seemed *forever*—on "Fibber McGee and Molly"? I remember it all like it was yesterday.

And as I said, I wasn't alone. As another writer has recalled his love of radio as a ten year old in Los Angeles,[4] listening to KHJ in 1945 (just two years before I started to listen to the same station),

> How can sitting in a movie theater or sitting on a couch before my television duplicate the wonderful times I had when I was tucked safely in bed with the lights out listening to a small radio present me with drama, fantasy, comedy and variety, all for free, and all of it dancing beautifully in my imagination, day by month by year? There has never been anything quite like it and, sadly, I must say there will never be anything like it again. That's what radio...and the nineteen forties meant to me.

Three Pedagogical Notes

When the writing of this book reached the point of introducing electronic circuitry (Chapter 8), I had to make a decision about the technology to discuss. Should it be vacuum tubes or transistors? Or both?

I quickly decided against both, if only to keep the book from grow-
ing like Topsy. The final decision was for vacuum tubes. My reasons
for this choice are both technical and historical. Vacuum tubes are
single-charge carrier devices (electrons), understandable in terms of
"intuitive," classical freshman physics. Transistors are two-charge carrier (electrons *and* holes) devices, understandable really only in terms
of quantum mechanics. Electrical engineering professors have, yes, invented lots of smoke-and-mirror ways of "explaining" holes in terms of
classical physics, but these ways are all, really, seductive frauds. They're
good for their ease in writing equations and in thinking about how circuits work, but even though they *look* elementary I don't like to teach
a first course in electrical engineering with their use. They are really
short cuts for *advanced* students who have had quantum mechanics.
From a historical point of view, of course, it was the vacuum tube that
made AM broadcast radio commercially possible, and to properly understand the work of Fleming and De Forest, the vacuum tube is the
only choice. And finally, students should be told that the small-signal
equivalent-circuit model for the junction field-effect transistor (JFET)—
also a single-charge carrier device—is *identical* to that for the vacuum-
tube triode! *Plus ça change, plus c'est la même chose.*

A second decision had to be made about the Laplace transform.
Studying this mathematical technique for solving linear, constant coefficient differential equations has traditionally been a "rite of passage"
for sophomore electrical engineering students, but I have decided not
to use Laplace in this book. I have the best reason possible for this important decision—it just isn't necessary! The Laplace transform is without equal for situations involving transient behavior in linear systems,
yes, but the mathematics of AM radio is essentially *steady-state* ac theory. For that the Fourier transform is sufficient. In this book we will never
encounter time signals that don't have a Fourier transform (we do, of
course, have to use impulses in the frequency domain for steps and undamped sinusoids). Unbounded signals that require the convergence
factor of the Laplace transform do not play a role in this book's telling
of the development of AM radio (but the unbounded signal $|t|$ *does*
have a nonimpulsive Fourier transform and I show the reader how to
derive it with freshman calculus). After completing this book, the student will be well prepared for more advanced studies in engineering
and physics that introduce the Laplace transform.

And finally, because this book is written for first-semester sophomores, there is no discussion that requires knowledge of probability
theory. This means, of course, no discussion of the impact of noise on
the operation of radio circuits.

In a memoir of yesteryear, one radio actor explained how the plot of an
episode of "Lights-Out" required the sound of a man being turned inside-

out. This astonishingly gross noise was achieved by one sound-effects technician slowly peeling a tight rubber glove off of his hand while a colleague, just as slowly, crushed a strawberry box to simulate breaking bones. Such creativity goes a long way to explaining the last paragraph in that same author's book:

> Trying to analyze the reasons for the broad, universal appeal of radio drama I found it expressed best by a little seven-year-old boy who, during a recent survey on preferences of children, was asked which he liked better, plays on the radio or plays on television.
>
> "On the radio," he said.
>
> "Why?" he was asked.
>
> He thought for a moment, then replied, "Because I can see the pictures better."[5]

This explains, too, the success of two early programs based on skin color (Amos 'n' Andy) and a ventriloquist (The Edgar Bergen and Charlie McCarthy Show)—two gimmicks you can't *see*.

As an enthusiastic supporter of early radio wrote more than twelve years before I was born, "nothing could be creepier than human voices stealing through space, preferably late on a stormy night with a story of the supernatural...particularly when you are listening alone."[6] Now those programs are gone forever. One of the characters in George Lucas's nostalgic tribute to radio, the 1995 film *Radioland Murders*, says, "Radio will never die. It would be like killing the imagination." I'm afraid, though, that the corpse of radio *as entertainment* has long been cold. As Garrison Keillor wrote in *WLT*, "Radio was a dream and now it's a jukebox. It's as if planes stopped flying and sat on the runway showing travelogues." Today's mostly insipid radio fare, with its rock music, banal and often viciously nasty call-in talk shows, and all-news stations endlessly repeating themselves, is a pale ghost of those wonderful, long-ago broadcasts. But the technical wonder of radio, itself, continues.

It was that technical wonder, in fact, that attracted so many youngsters to electrical engineering from the 1920s through the 1960s; from building primitive radios, to more advanced electronic kits available through the mail, right up to the early days of the personal computer when high schoolers could, before college, get hands-on experience at what electrical engineers do. I still recall the fun I had building a Heathkit oscilloscope in 1957, and then using it, when I was a junior in high school. But as Robert Lucky has noted, the development of the totally self-contained VLSI chip has destroyed the kit market. As he writes,

> I hear that freshman enrollment in electrical engineering has been dropping steadily since those halcyon days of [kit building]. I'm looking at my non-distinctive, keep-your-hands off [personal computer], and I'm wondering—do you think there is any connection?[7]

I certainly do! As another electrical engineer recently wrote of how the wonder of radio changed his life,

> When I was about eight years old, my uncle showed me how to build a radio out of wire and silver rocks [crystals]. I was astounded. My dad strung a long wire between two trees in the backyard and I sat in the back on the picnic table listening to the BBC. This is what shaped my life and my chosen profession. I was truly a lucky child. From the time I was eight, I knew what I would be when I grew up. I asked my dad, "What kind of guy do you have to be if you want to work on radios?"
> "An electrical engineer," Dad said.
> That's what I'm going to be.[8]

In his excellent biography of Richard Feynman, James Gleick catches the spirit of what radio meant in its early days to inquisitive young minds:

> Eventually the art went out of radio tinkering. Children forgot the pleasures of opening the cabinets and eviscerating their parent's old [radios]. Solid electronic blocks replaced the radio set's messy innards—so where once you could learn by tugging at soldered wires and staring into the orange glow of the vacuum tubes, eventually nothing remained but featureless ready-made chips, the old circuits compressed a thousandfold or more. The transistor, a microscopic quirk in a sliver of silicon, supplanted the reliably breakable tube, and so the world lost a well-used path into science.[9]

A couple of pages later, on the fascination of the "simple magic" of a radio set, Gleick observes:

> No wonder so many future physicists started as radio tinkers, and no wonder, before *physicist* became a commonplace word, so many grew up thinking they might become electrical engineers ...

Figure 2   In the early days of broadcast radio, many home correspondence schools used the romantic image of the new technology to attract students from the ranks of those who felt trapped in depression-era, dead-end jobs. One of the biggest schools was the National Radio Institute, which ran ads in the pulp fiction magazines most likely to be read by young men. This art was part of such an ad that appeared in the October 1937 issue of the science fiction pulp *Thrilling Wonder Stories*. Two other similar pieces of art from the same time period and magazine appear later in this book, at places where some encouragement will perhaps help motivate "sticking with it!"

Times have changed, though. In another essay Lucky wrote of the time he asked a college student why she was majoring in materials engineering, rather than electrical engineering. The student looked at him with incredulity and disdain and replied, "You can see and touch things here." Then, with a glance (and a shiver) at the nearby electrical engineering building, she added "Nothing is real over there." Lucky found he had to agree with that student's assessment; as he correctly describes the state of electrical engineering today, "Most of our stuff is made of nothing at all. It

Figure 3   High-tech youngsters from yesteryear! From "It's Great to be a Radio Maniac," *Collier's*, September 12, 1924.

is made of software, of math, of conceptual thought. [Electrical engineers now] live mostly in a virtual world."[10] So, in response to that, I have picked the AM radio receiver as the centerpiece for my experiment in writing an introductory top-down, just-in-time electrical engineering book because I believe it is the simplest, common household electronic device that seems mysterious to an intelligent person upon their first encounter with it.

Consider, for example, the case of Leopold Stokowski who, when he died in 1977, was declared (in his *New York Times* obituary notice) to have been "possibly the best known symphonic conductor of all time." In an essay written for *The Atlantic Monthly* ("New Vistas in Radio," January 1935), Stokowski gloomily asserted "The fundamental principles of radio are a mystery that we may never fully understand." And an amusing story from the early days of broadcast radio has a technologically challenged Supreme Court justice perplexed by radio. When Chief Justice

William Howard Taft was faced with the possibility of hearing arguments about the government regulation of radio, he reportedly said "I have always dodged this radio question. I have refused to grant writs and have told the other justices that I hope to avoid passing on this subject as long as possible."[11] When asked why he felt this way, he admitted that, for him, "... interpreting the law on this subject is something like trying to interpret the law of the occult. It seems like dealing with something supernatural. I want to put it off as long as possible in the hope that it becomes more understandable." And, indeed, during his tenure as Chief Justice from 1921 to 1930, the Supreme Court heard not a single case dealing with radio.

If radio doesn't seem mysterious to someone, then that person simply has no imagination! But nobody can doubt, as they spin the dial, that radio is very real. There is nothing "virtual" about it. In its own way, then, perhaps this book (if it falls into the right hands) can spark anew a little bit of the wonder that has been lost over the years. That, anyway, is my hope.

Notes

1. See, for example, Robert W. Lucky, "The Curriculum Dilemma," *IEEE Spectrum*, November 1989, p. 12. Lucky is an electrical engineer who, after making impressive technical contributions to the electronic transmission of information, moved into upper-management at the AT&T Bell Laboratories.

2. In his column "Curriculum Trends," *Newsletter* of the IEEE Education Society, Fall 1987.

3. The call letters WLT stand for "With Lettuce and Tomatoes," a joke based on the fictional station being operated out of a sandwich shop. And in real life, too, station call letters could have equally silly meanings. In Chicago, for example, the *Chicago Tribune* began operating WGN—a subtle(?) plug for the "World's Greatest Newspaper."

4. Ken Greenwald, *The Lost Adventures of Sherlock Holmes*, Barnes & Noble Books 1993. More warm memories of the "golden age" of radio can be found in the books by Ray E. Barfied, *Listening to Radio, 1920–1950*, Praeger 1996, and Gerald Nachman, *Raised On Radio*, Pantheon 1998. See also Brock Brower, "A Lament for Old-Time Radio," *Esquire*, April 1960, pp. 148–150.

5. Joseph Julian, *This Was Radio*, Viking 1975. Here's another neat sound-effect trick for you: how do you make thunder on radio? Put a couple of BBs in a balloon, blow it up, and shake the balloon!

6. Roy S. Durstine, "We're On the Air," *Scribner's Magazine*, May 1928.

7. Robert W. Lucky, "The Electronic Hobbyist," *IEEE Spectrum*, July 1990, p. 6. The positive relationship between hobby activity and intellectual stimulation in young people was recognized very early, e.g., see Howard Vincent O'Brien, "It's Great to Be a Radio Maniac," *Collier's*, September 13, 1924, pp. 15–16.

8. Joe Mastroianni, "The Future of Ham Radio," *QST*, October 1992, p. 70.

9. *Genius: The Life and Science of Richard Feynman*, Pantheon 1992, p. 17.

10. Robert W. Lucky, "What's Real Anymore?". *IEEE Spectrum*, November 1991, p. 6.

11. Quoted from R.H. Coase, "The Federal Communications Commission," *The Journal of Law & Economics* 2, October 1959, pp. 1–40.

RADIO SWEEPING COUNTRY—1,000,000 sets in use
—Front-page headline, *Variety* (March 10, 1922)

The air is full of wireless messages every hour of the day. In the evening, particularly, there are treats which no one ought to miss. Famous people will talk to you, sing for you, amuse you. YOU DON'T HAVE TO BUY A SINGLE TICKET—You don't have to reserve seats.
—Radio ad in *Scientific American* (July 1922)

One ought to be ashamed to make use of the wonders of science embodied in a radio set, the while appreciating them as little as a cow appreciates the botanic marvels in the plants she munches.
—Albert Einstein, in his remarks opening the Seventh German Radio Exhibition at Berlin (August 1930)

Radio is almost a miracle.

That's right—the little box by your bedside, or on the refrigerator in the kitchen, or in the study next to the sofa, or behind the fancy buttons on the dash of your car, is an invention of near supernatural powers. Now, quickly, before every physicist and electrical engineer reading these words dismisses this book as the work of an academic mystic, I wish to point out those two all-important qualifiers *almost* and *near*. The wonder of AM

Calvin and Hobbes   by Bill Watterson

Figure 4   Calvin and Hobbes, copyright 1989 Watterson. Reprinted with permission of Universal Press Syndicate. All rights reserved.

(amplitude modulation) radio, in fact, can be understood through physics and mathematics, not sorcery or theology—but that is a fact that many of my students are not so sure about. That's why I have written this book. I want to take the mystery, what some of my students (like Calvin's dad in Figure 4) even think the spookiness, out of radio.

Well, you perhaps say, it's a little late for me to be worrying about *that*—there are already plenty of books available on radio theory. That's right, there are, but for my purposes here they have two characteristics that limit them. First, they are generally physically big books of several hundred pages, written for advanced students in the third or fourth year of an electrical engineering major; such books are specifically published as textbooks.[1] Second, they are essentially 100% theory, with very little historical development or, even more likely, simply none at all. In those books, radio springs forth total and complete like Adam from the clay.

This book is different on both counts. First, as you can tell at a glance, it is relatively short. I've made conciseness a specific goal not because I'm lazy (it is far easier to use too many words than to search for those that are sufficient), but because I want you to see that reading this book will not be the Thirteenth Labor of Hercules. Not so immediately obvious is that you don't have to be a third- or fourth-year electrical engineering major to read this book. Indeed, you can be a second-year student majoring in anything (chemistry, biology and, yes, even *history*), just so long as you've had freshman courses in calculus and electrical physics. Anything else you need to know, I'll teach you here.

This book is also different from others in its presentation of the history of radio. There are good, modern radio history books available,[2] and I believe all electrical engineering and physics students (and their professors) would benefit from reading them. But as good as those books are, they are not technical books. They are books by historians treating the social history of the broadcast industry, and the intellectual history of the technical and scientific inventions that make radio possible. To take just one example, the term *superheterodyne* is mentioned in those books, but only to indicate it is a crucial concept in modern radio and to detail the vicious patent fights that raged over its implementation. This is fascinating and historically important material to read (and you'll find some of it here, too), but by itself it isn't radio theory and the authors of those books didn't intend their books to be understood as even beginning to present mathematical theory. This book, however, in addition to discussing the history of the superheterodyne concept, gives it a precise mathematical formulation and shows you how it is actually achieved in real circuitry. It is the mathematics in this book that further distinguishes it from yet a different sort of book—the sort that presents radio theory in a quasi-technical yet mathematics-free way for the hobbyist.[3]

I have become convinced, after nearly 30 years of college teaching, that most electrical engineering students spend at least three of their four undergraduate years secretly wondering just what electrical engineering is about. Their initial course work, immense in detail but devoid of almost any sense of global direction, tells them little about where it is all heading. My goal in this book is to develop, quickly, in the second year of college, a start-to-finish answer-by-example of the sort of *system* an electrical engineer deals with, and how it is different from what a mechanical engineer (or, for that matter, an electrician) is normally concerned about.

I have a quick test to see if you're ready for the next 300 or so pages: Have you studied calculus to the point of understanding the physical significance of the derivative of a function and the area interpretation of an integral? Can you write Kirchhoff's equations for electrical circuits (and solve them for "simple" situations)? Do you know what electrons are? (They are *not*, as the Pulitzer Prize winning editor at the *Miami Herald* David Barry once wrote, "very small objects that carpet manufacturers weave into carpets so they will attract dirt"!) If you can answer yes to these questions, then you are ready for this book.

This book takes the view that electrical engineers and physicists think with mathematics, and a quick flip through the following pages will show just how strongly I hold that belief. The pure historical approach is the prose approach, and while I encourage you to read what modern historians of technology have written, prose alone is simply not enough. There are some, however, who actually believe mathematics somehow detracts from the inherent beauty of nature. Such people might well argue that radio is more wonderful *without* mathematics, much as famed essayist Charles Lamb declared at the so-called Immortal Dinner. There he toasted a portrait containing the image of Isaac Newton with words describing Newton as "a fellow who believed nothing unless it was as clear as the three sides of a triangle, and who had destroyed all the poetry of the rainbow by reducing it to the prismatic colors."

The Immortal Dinner was a party given on December 28, 1817, at the home of the English painter Benjamin Haydon. In attendance were such luminaries as the poets Wordsworth and Keats. Lamb was described by Haydon as having been "delightfully merry" just before he made his toast, which I interpret to mean he was thoroughly drunk. Certainly, if sober, an intelligent man like Lamb wouldn't have made such a silly statement.

I don't agree with Lamb; he may have been a great writer but he evidently understood very little about mathematics and its relationship to physical reality. My sympathies lie instead with the "master mathematician" in H. G. Wells' powerful story "The Star," in which life on earth appears doomed by a cataclysmic collision with an enormous comet (published in 1897, 101 years before the 1998 movie *Deep Impact*.) The mathematician has just calculated the fatal orbit, and gazes up at the on-rushing mass: "You may kill me, but I can hold you—and all the universe for that matter—in the grip of this little brain. I would not change. Even now."

Even today, many professors probably don't realize that the science-and-math curriculum of a modern undergraduate electrical engineering program is a relatively new development. Before the Second World War, electrical engineering education was heavily dominated by nuts-and-bolts technology (e.g., power transmission, and ac machinery), and tradition (e.g., surveying and drafting). Back in the 1920s and 1930s, many faculty members were convinced that electrical engineers didn't need to know Maxwell's equations in vector form unless they were going to be PhDs. As a specific example of what I mean, let me quote from the 1938 book *Fundamentals of Radio*, by Frederick E. Terman, then a professor of electrical engineering at Stanford (later Dean of Engineering, and even later Provost). The chapter on antennas opens with this astonishing statement: "An understanding of the mechanism by which energy is radiated from a circuit, and the derivation of equations for expressing this radiation quantitatively, involve conceptions that are unfamiliar to engineers." No electrical engineering textbook on radio theory (including this one) that made such an assertion could be published today, but in 1938 Terman knew his audience. It was only in graduate school, in those days, that you could perhaps find an electrical engineer who knew how to solve Maxwell's equations for the electromagnetic fields inside a waveguide, or how to calculate the probability density function of a sum of random variables.

Today, that sort of thing is required junior (or even sophomore) year material. But before the war it wasn't, and when the war came, with its dramatic need for technical people able to apply basic science principles and high-level mathematics to new problems that didn't have "cookbook" solutions, most electrical engineers simply came up short. It was found, for example, that physicists were far better equipped to handle the technical challenges of not only the atomic bomb, but of microwave radar and the radio proximity fuse, too. Physicists, of course, have delighted for decades in telling this story. (Electrical engineers can salvage some pride, however, in knowing that the Director of the Office of Scientific Research and Development, with oversight of *all* war research, was the MIT electrical engineer Vannevar Bush, who reported directly to President Roosevelt.)

That painful, embarrassing lesson wasn't lost on electrical engineering faculties, and after the war great educational changes were made. One of the personal side benefits of writing this book, however, is the opportunity it gives me to tell students that radio was developed almost entirely by electrical engineers [and even one electrical engineering *student*— Edwin H. Armstrong (1890–1954) who, in 1912 while an undergraduate at Columbia University, invented the regenerative feedback amplifier and oscillator]. These were men who received their formal training in electrical engineering and who called themselves electrical engineers, not physicists.

One of the early "modern" pioneers of radio was Armstrong's friend Louis A. Hazeltine (1886–1964), who invented the neutrodyne radio receiver in 1922. Hazeltine was a professor of electrical engineering at Stevens Institute of Technology in Hoboken, New Jersey, until his "retirement" in 1925 (years later Stevens reappointed him as a professor of mathematical physics). In an interview article that appeared in the October 1927 issue of *Scientific American*, the central point was that Hazeltine did "all his creative work with a notebook, a fountain pen and a slide rule, thus avoiding trial and error methods." So innovative and striking did the magazine find this (contrasting greatly with the by then near-mythical Edisonian method of "try everything until you trip over the answer") that the interview's head-line boldly declared "A College Professor Solves a Mathematical Problem and Becomes a Wealthy Inventor." When asked what was the secret of his success, Professor Hazeltine responded, "the first requisite is a thorough knowledge of fundamental principles."

> Armstrong was also the inventor of the superheterodyne radio receiver, later became a full professor of electrical engineering at Columbia and, if anyone deserves the title, was the "Father of Modern Radio." And yet, even though he could read an equation as well as most electrical engineers, he always retained a cautious skepticism about too much reliance on mathematics. For Armstrong, physical experiment was the bottom line. In his later years, after he'd demonstrated frequency modulated (FM) radio was useful even though some mathematically inclined engineers had declared it wasn't, Armstrong set the record straight in a paper that must have surprised a few people: "Mathematical Theory vs. Physical Concept," *FM and Television*, August 1944.

The late Richard Feynman (who as a boy had a reputation for being a formidable "fixer of busted radios,"[4] and who shared the 1965 Nobel Prize in physics), was agreeing with Hazeltine when he said, "It is impossible to explain honestly the beauties of the laws of nature [to anyone who does not have a] deep understanding of mathematics," (The Character of Physical Law, MIT Press 1965). But keep in mind my promise that your mathematics, here, doesn't have to be all *that* deep; just freshman calculus. Now and then I do mention such advanced mathematical ideas as *contour integration* and *vector calculus* but these are never actually used in the book and are included strictly for intellectual and historical completeness.

Still, Lamb's ill-advised praise for technical ignorance dies hard. The previously mentioned *Miami Herald* humorist Dave Barry twice described how radio works; once as "by means of long invisible pieces of electricity (called "static") shooting through the air until they strike your speaker and break into individual units of sound ("notes") small enough to fit inside your ear!" And again when he wrote "When you turn on the radio, you take it for granted that music will come out; but do you ever stop to think that this miracle would not be possible without the work of scientists? That's right: There are tiny scientists inside that radio, playing instruments." Barry was of course merely trying to be funny, but I suspect not just a few of his readers either took him at his word, or believe in some other equally bizarre "weird science explanation."

This book will not correct all the weird misconceptions about radio held by the "average man on the street," but I hope it will help engineering and science students who also don't yet quite have it all together. All too often I've had students come to me and say things like, "Professor, I've just had a course in electromagnetic field theory, and learned how to solve Maxwell's equations inside a waveguide made of a perfect conductor and filled with an isotropic plasma. I know how Maxwell 'discovered' radio waves in his mathematics. There are just two things left I'd really like to know: what is actually happening when a radio antenna radiates energy, and how does a receiver tune-in that energy?"

The plea in those words reminds me of an anonymous bit of doggerel I came across years ago, while thumbing through the now defunct British humor magazine, *Punch*. Entitled "A Wireless Problem," it goes like this:

Music, when soft voices die,
Vibrates in the memory,
But where on earth does music go
When I switch off 2LO?[5]

"2LO" refers to the call letters of the first London radio station (operating at 100 W from the roof of a department store), which went on the air in August 1922 at a frequency of 842 kHz. Just to be sure this notation is clear, "kHz" stands for kilohertz (and "MHz" denotes megahertz), where hertz is the basic unit of frequency. Just to show what an old fogey I am, let me state here, loudly, that the old frequency unit of cycle per second was perfectly fine! The most logically named radio station I know of, Radio 1212 ("Radio Twelve-Twelve"), used its very frequency as its identification. Radio 1212 was an American clandestine instrument of psychological warfare against the German population and regular army troops in the Second World War. It operated on a frequency of 1212 kHz, or 1.212 MHz. The relationship between frequency (f) and wavelength (λ) is $\lambda f = c$ (the speed of light, 3×10^8 m/sec.) The wavelength of Radio 1212, then, was about 250 meters.

I know what such puzzles feel like, too. I went to a good undergraduate school, with fine professors, who showed me how to cram my head full of all sorts of neat technical details; but when I received my first engineering degree I still didn't *know*, at the intuitive, gut level, what the devil was *really* going on inside a kitchen radio. That, like my thickening girth and thinning hair, came with time.

The ordinary kitchen radio is today so common we take it for granted and find it hard to appreciate what enormous impact it had (and continues to have) on society, and on individuals, too. It is perhaps the single most important electronic invention of all, surpassing even the computer in its societal impact (the telephone doesn't depend on electronics for its operation, and television is the natural extension of radio—one name for it in the 1920s was "radiovision"; indeed, television's video and audio signals *are* AM and FM radio, respectively). Even if we drop the "electronic" qualifier, only the automobile can compete with radio in terms of its effect on changing the very structure of society.

One of history's greatest intellects took time off from his physics to comment on this. The last quote that opens this prologue is from Einstein's opening address to the 1930 German Radio Exhibition. In that same address he also stated, "The radio broadcast has a unique function to fill in bringing nations together ... Until our day people learned to know each other only through the distorting mirror of their own daily press. Radio shows them each other in the liveliest form ... "[6]

Because we take radio for granted, many simply don't appreciate how young radio is. Just think—there was no scheduled radio until the Westinghouse-owned station KDKA-East Pittsburgh broadcast (with just 100 watts) the Harding–Cox presidential election returns on November 2, 1920! And it wasn't until nearly two years later that commercial radio (i.e., broadcasts paid for by sponsors running ads) appeared on WEAF-New York. There are many people still alive today who were teenagers (even college students perhaps like you) before the very first regular radio programs for entertainment were broadcast. The *very first*, when movies were still silent and long before music videos, laser disks, and computer games. Some of these people have never forgotten how radio affected them. One of them, R. V. Jones, recalled it this way:

> There has never been anything comparable in any other period of history to the impact of radio on the ordinary individual in the 1920s. It was the product of some of the most imaginative developments that have ever occurred in physics, and it was as near magic as anyone could conceive, in that with a few mainly home-made components simply connected together one could conjure speech and music out of the air.[7]

While November 2, 1920 is the traditional date of the "start" of broadcast radio, the real history is actually a bit more complex. 8MK-Detroit (later WWJ) had been on the air regularly two months before KDKA, and two AT&T experimental stations (2XJ-Deal Beach, NJ and

2XB-New York) had been transmitting to all who cared to listen since early 1920. And years before, in 1916, 2ZK-New Rochelle, NY was regularly broadcasting music. And years before that, in 1912, KQW-San Jose, CA could be heard regularly in earphones. About that same time Alfred Goldsmith, a professor of electrical engineering at the City College of New York who later was chief consulting engineer to RCA, operated the broadcast station 2XN. What set KDKA apart from all those earlier efforts (besides being the first station to receive a U.S. government license), however, was its owner's intent to provide a freely available commodity (radio transmissions) that would induce the purchase of a product (radio receivers) made by that same owner (Westinghouse). Later, this striking concept would be replaced by an even bolder one. As the price of radio sets dropped (and so too their profit margins), the sale of receivers as a direct producer of corporate wealth became inconsequential, i.e., kitchen and bedroom radios today aren't worth fixing and have literally become throwaway items. What has become profitable to sell is the radio time, itself, i.e., advertising (in 1922 the rate was just ten dollars a minute). Or so is the case in America—in England, where radio is a state-controlled monopoly, broadcasting costs are covered by listener-paid fees (an approach considered, but rejected, in the early days of American radio—see R.H. Coase, *British Broadcasting*, Longmans, Green and Co. 1950). For a discussion of the early concerns over whether American radio should be public or private, see Mary S. Mander, "The Public Debate about Broadcasting in the Twenties: An Interpretive History," *Journal of Broadcasting* 28, Spring 1984, pp. 167–185.

One of America's most famous radio sportscasters, Walter ("Red") Barber, who started his career at the University of Florida's 5,000-W station WRUF while a student in 1930, put it much the same way (*The Broadcasters*, Dial 1970):

Kids today flip on their transistor radios without thinking … and take it all for granted. People who weren't around in the twenties when radio exploded can't know what it meant, this milestone for mankind. Suddenly, with radio, there was instant human communication. No longer were our homes isolated and lonely and silent. The world came into our homes for the first time. Music came pouring in. Laughter came in. News came in. The world shrank, with radio.

And John Archibald Wheeler, Feynman's graduate advisor more than half a century ago (and today an emeritus professor of physics at Princeton University), a few years ago recalled his fascination with radio at age thirteen (in 1924):

Living in the steel city of Youngstown, Ohio, I delivered *The Youngs-town Vindicator* to fifty homes after school. A special weekly section in the paper reported the exciting developments in the new field of radio, including wiring diagrams for making one's own receiver. And my paper-delivery dollars made it possible for me to buy a crystal, an earphone, and the necessary wire. The primitive receiver that I duly assembled picked up the messages from KDKA ... what joy![8]

Other listeners like Wheeler just couldn't get enough of radio (radio was such a rage in the 1920s it even inspired a movie—the 1923 *Radio-Mania*, in which the hero tunes-in Mars!) The following is typical of the letters that poured into early broadcasting stations:

I am located in the Temegami Forest Reserve, seven miles from the end of steel in northern Ontario. I have no idea how far I am from [you], but anyway you come in here swell.... Last week I took the set back into the bush about twenty miles to a new camp.... Just as I thought—in you came, and the miners' wives tore the head-phones apart trying to listen in at once. I stepped outside the shack for a while, while they were listening to you inside. It was a cold, clear, bright night, stars and moon hanging like jewels from the sky; five feet of snow; forty-two below zero; not a sound but the trees snapping in the frost; and yet ... the air was *full of sweet music.* I remember the time when to be out here was to be out of the world—isolation complete, not a soul to hear or see for months on end; six months of snow and ice, fighting back a frozen death with an ax and stove wood, in a seemingly never-ending battle. But the long nights are long no longer—you are right here ... and you come in so plain that the dog used to bark at you.... He does not bark any more—he knows you.[9]

In its earliest days, radio spoke to the masses who couldn't read, both the millions of new immigrants and the simply uneducated. But radio had enormous power over all (see Figure 5), even the educated, multi-generational American. Any who doubt this need only read the front-page headlines of almost any newspaper in the land for the morning of October 31, 1938. That was the "morning after" of Orson Welles's CBS radio dramatization, on his Mercury Theatre, of H. G. Wells's 1898 novella *The War of the Worlds.* As listeners tuned in to the previous evening's Halloween eve, coast-to-coast broadcast, they heard the horrifying news: Martians had invaded the earth, their first rockets landing in the little town

Figure 5   "When Uncle Sam Wants to Talk to All the People." From the May 1922 issue of *Radio Broadcast*.

of Grovers Mill near Princeton, New Jersey! Hundreds were already dead! Panic and terror literally swept the nation. Radio had spoken, and people believed.[10] The wild public response so stunned the government that the Federal Communications Commission (FCC) announced it would hold hearings on whether the "public trust" had been violated. So serious was the threat of censorship taken that CBS apologized, which was enough to mollify the FCC.

The continuing importance of radio, even in the modern age of the ubiquitous television set, was specifically acknowledged in an editorial in *The Boston Globe*. Published on August 21, 1991, two days after a powerful hurricane had blown through New England (and right over *my* home in Dover, New Hampshire) at the same time the second Russian revolution was blowing the Soviet Union away, "The Power of Radio" declared,

> Thousands of New Englanders, darkened by the power blackouts, got much of their news about the Gorbachev ouster and Hurricane Bob from battery-operated radios. It was a reminder of the immediacy and power of this medium.... Television pictures are attention grabbing, but the true communications revolution occurred not when the first TV news was broadcast but a generation earlier, when radio discovered its voice.

The word *magic* has appeared more than once in this prologue, but that is exactly how many sophisticated people first thought of radio. *And still do.* Here's how one modern-day print journalist recently wrote of his youthful days as a newspaper reporter in the Mississippi Delta:

> I don't believe in magic, but I do know that sitting in my car in the middle of Mississippi and listening to a signal that traveled more than a thousand miles, over nearly a dozen states, and came down into my car through a metal pole antenna and two paper-cone speakers, was as near to a magical experience as ever I'm likely to have.[11]

Calvin's dad thought electric lights and vacuum cleaners to be magic, and radio would surely be *super*magic to him. This is really just another form of the famous "Clarke's Third Law" (after science fiction writer Arthur C. Clarke): "Any sufficiently advanced technology is indistinguishable from magic." A superheterodyne radio receiver would have been magic to the greatest of the Victorian scientists, including James Clerk Maxwell himself, the man who first wrote down the equations that give life to radio. In the Middle Ages such a gadget would have gotten its owner burned at the stake—what else, after all, could a "talking box" be but the work of the Devil? I hope that when you finish this book, however, you'll take Calvin's mom's advice, forget magic, and agree with me when I say: radio is *better* than magic.

Notes

1. Since I first wrote those words there has been a remarkable new book published that I must mention. After reading the first edition of this book, Professor David Rutledge at the California Institute of Technology sent me a copy of *his* nontraditional sophomore radio course notes, called *The Electronics of Radio* (those notes are now being published as a book of the same title by Cambridge University Press). Rutledge writes in the preface to his book: "Today's electrical-engineering students have usually not built stereos or tinkered with cars, and this means that they do not know the smoke and smell of construction, or the excitement of electronic circuits coming to life. Many universities encourage this trend, with exercises where students switch components in and out of a circuit, never even heating up the soldering iron." His book bucks that backward trend, and takes the pragmatic view that "it is valuable for students to learn to put knobs on without destroying the screw heads." Buy Professor Rutledge's book, put it next to this one, and you'll have a *set* of *two* radical sophomore electrical engineering books.

2. I have in mind Susan Douglas's *Inventing American Broadcasting, 1899–1922,* Johns Hopkins 1987; the two volumes by Hugh Aitken, *Syntony and Spark: The Origins of Radio,* Wiley 1976, and its sequel *The Continuous Wave: Technology and American Radio, 1900–1932,* Princeton 1985; Michael B. Schiffer, *The Portable Radio in American Life,* University of Arizona Press 1991; and George H. Douglas' *The Early Days of Radio Broadcasting,* McFarland 1987. Also recommended are the biographical treatments by Tom Lewis, *Empire of the Air: The Men Who Made Radio,* Harper, Collins 1991, and Michele Hilmes' two books *Hollywood and Broadcasting: From Radio to Cable,* University of Illinois Press 1990, and *Radio Voices: American Broadcasting, 1922–1952,* University of Minnesota Press 1997.

3. Two excellent examples of such books (which I highly recommend) are David Rutland, *Behind the Front Panel: The Design & Development of 1920's Radios,* Wren 1994, and Joseph J. Carr, *Old Time Radios! Restoration and Repair,* TAB 1991.

4. See Feynman's funny recounting of how some common sense could go a long way in fixing a radio in the 1920s and 1930s, in his autobiographical essay "He Fixes Radios by Thinking!" (in *Surely You're Joking, Mr. Feynman!,* W. W. Norton 1985).

5. *Punch,* May 25, 1927, p. 573.

6. Quoted from an article on the front page of *The New York Times,* August 23, 1930.

7. R. V. Jones, In his exciting memoir *Most Secret War,* Hamish Hamilton 1978. Reginald Victor Jones (1911–1997) was a key player in British Scientific Intelligence during the Second World War. It was Jones who, in Winston Churchill's words, "broke the bloody [radio] beam" used by the Germans as an electronic bombing aid during the Battle of Britain.

8. John Archibald Wheeler, *A Journey into Gravity and Spacetime,* Scientific American Library 1990. Norman Rockwell's May 20, 1922 cover on *The Saturday Evening Post* also captured the wonder of early radio, showing an elderly couple (who appear to have been born about 1850, when Maxwell was still a teenager) listening to a radio in amazement even as the future Professor Wheeler delivered his papers.

9. Bruce Barton, "This Magic Called Radio," *The American Magazine,* June 1922.

10. An interesting treatment of this amazing event in radio history (along with a complete text of the radio play) is in Howard Koch's *The Panic Broadcast,* Little, Brown 1970. Koch, not Orson Welles, was the actual writer of the play, entitled "Invasion from Mars." See also Robert J. Brown, *Manipulating the Ether: The Power of Broadcast Radio in Thirties America,* McFarland 1998.

11. For this insightful essay on the psychology of radio listening, see Richard Rubin, "It's Radi-O! The Medium That Can Turn Anywhere into Somewhere," *The Atlantic Monthly,* January 1998. For an extended, book-length treatment of the same subject, the definitive work is Susan Douglas, *Listening In: Radio and the American Imagination from Amos 'N' Andy and Edward R. Morrow to Wolfman Jack and Howard Stern,* Times Books 1999.

Problem

1. The sun radiates energy over a very broad range of frequencies. If one measures the sun's energy output as a function of wavelength, the maximum occurs at the wavelength of green light, i.e., at $\lambda = 550 \times 10^{-9}$ meters (550 nanometers); this is determined strictly by the sun's surface temperature of 5700°C. This is often claimed to be indirect evidence in support of the theory of evolution, as the human eye has maximum sensitivity at *just* that wavelength. What is the frequency of green light, in MHz? Are you surprised by the answer, now that you know the eyes of your "significant other" are more than just limpid pools of beauty, but are also efficient detectors of incredibly high frequency (*far* higher than those of radio) electromagnetic radiation?

Mostly History and a Little Math

Solution to an Old Problem

Speech is one of the central characteristics that distinguishes humans from all the other creatures on Earth. There are other means of communication, of course, as anyone who has shared living space with a cat knows, but even the closest human–cat relationship always ends with the puzzle of "I wonder what that darn cat is *thinking*?" You cannot simply ask the cat; it simply does not know how to answer. You *can* ask another human.

Along with this ability to communicate by speech, it seems to be the case that most humans have a powerful desire to actually do so—and with as many fellow humans as possible. Before Alexander Graham Bell's invention of the telephone in 1876, "real-time" speech communication was limited to how loudly you could shout (and how well the other fellow could hear). Long-distance communication required the written word or the telegram. Long-distance communication is expensive, however, and so it is no surprise that governments, not individuals, sponsored the development of such communication systems. And it should also be no surprise that it was the point-to-point transmission of *private* information that was the original interest of governments. Such obvious and primitive long-distance communication systems based on beacon fires and smoke signals (weather permitting) and thundering jungle drums clearly fail on that score. It was, in fact, the *telegraph* that proved to be the first successful answer to the desire of governments to quickly send relatively secure signals over long distances.

The history of the telegraph (a word that literally means "writing at a distance") can be traced at least as far back as 1753, when its principles were described in a letter to the *Scots' Magazine* in Edinburgh. Authored by someone now lost to history,[1] who signaled his name only as C. M., the letter described a set of twenty-six wires, one for each letter of the alphabet, which would be individually electrically charged at the sending end so as to attract an electrified ball at the receiving end. This "electrostatic telegraph" does not appear to have gone anywhere. More successful was Claude Chappe (1763–1805), the French inventor of the semaphore telegraph.

Chappe's system, a form of optical telegraphy (and so the first modern operational telegraph was wireless!), used a wooden beam pivoted at its center. The motions of the beam could be used to send coded information, and in 1793 the French government authorized Chappe to construct a line of fifteen semaphores connecting Paris to Lille, 145 miles to the north on the Belgian border. Operators at each semaphore station observed the incoming message through a telescope. Chappe's system, which was never used for either commercial or private purposes, continued to be expanded until 1823. When it was finally shut down in 1852, it incorporated 556 stations distributed over a total path length of more than 2500 miles.

Because of the Napoleonic wars with France at the end of the eighteenth century, the British government soon felt the need to respond with its own system for sending, quickly, military messages. In 1796 it purchased the rights to the *shutter telegraph* of Lord George Murray (1761–1803). Murray's system used a frame, atop a high tower, of six octagonal wooden shutters, mounted vertically in three rows of two shutters each. Each shutter rotated around its horizontal axis and could be independently opened or shut, with the aid of ropes, by an operator on the ground. There are $2^6 = 64$ possible combinations of open/shut, and so Murray's optical telegraph, which linked London to the major port cities of Dover, Portsmouth, Plymouth, and Yarmouth on the English southern coast facing France across the English Channel, was the first binary-coded transmission link.[2]

One curious testament to those original optical telegraphs is the name they often bestowed on their tower sites—Telegraph Hill or Signal Hill—names that most modern residents who don't know the history probably find as mystifying as they do charming.

In America the first telegraph system was the 40-mile *electric* line built by congressional appropriation in 1844, between Baltimore, Maryland and Washington, D.C. The brainchild of Samuel Morse (1791–1872), its first message "What Hath God Wrought" was sent in the now famous "Morse code" of dots-and-dashes, a code still in use to this day. Morse's claim to fame is now known to be weak; the real scientific and engineering brains behind the American electric telegraph were those of the American physicist Joseph Henry (1799–1878) after whom the unit of inductance is named and who educated Morse on the nature of electricity, and a young technician employed by Morse, Alfred Vail (1807–1859). It was Vail, in fact, *not* Morse, who created the telegraph code that should really be called "Vail code." Morse was the administrative and political force behind the electric telegraph, just as Pope Julius was a similar force behind the ceiling of the Sistine Chapel in the Vatican. In that case, however, credit for the work is rightfully given to Michelangelo *who did the work*, not to the Pope.

The electric telegraph seemed a miracle of the mid-nineteenth century, and even a professional visionary like Jules Verne (today we would call him a "futurologist") couldn't imagine anything beyond it. In his 1863 work *Paris in the Twentieth Century* (rejected for publication as being unbelievable in its predictions, and not rediscovered until 1989), Verne imagined that in the world of 1963 world wide communication would *still* be by electric telegraph. And for him, telegraphy's greatest virtue was its *secure*, point-to-point communication. "Broadcasting," a term radio borrowed from its use in agriculture to describe the wide-flung spreading of seed, would have seemed to Verne to be pointless.

So, by the 1850s rapid, long-distance, *wired* electrical communication was a reality (in 1861 the Western Union Telegraph Company completed a 2000 mile line connecting St. Joseph, Missouri to Sacramento, California), and by 1866 the transatlantic undersea telegraph cable had linked America and Europe. But even in the case of the electric telegraph, which required the use of intermediaries (telegraph operators and delivery boys), communication still wasn't "real time." With Bell's telephone, however, that changed. By 1884, with the completion of one of the earliest long-distance circuits, a husband in Boston could actually talk, instantly, with his wife in New York. Contrary to what most people today believe, however, the early telephone was not limited to simple point-to-point, one-on-one communications.

Today we are use to such concepts as conference calls, broadcast advertising, and subscription stereo music (e.g., the innocuous background noise on telephones when you're put on hold, are riding in elevators, and waiting in doctors' offices), but we all too quickly assume these applications of broadcasting are inherent to modern radio. That is simply not true. All these concepts quickly achieved reality with the use of the telephone, all as means to satisfy an apparently quite basic human desire: to instantaneously communicate with large numbers of other people for either entertainment or profit (or both).

A cartoon published in an 1849 issue of *Punch* indicated that even at that early date the *telegraph* was being used to transmit music over long distances by unnamed experimenters in America. The text accompanying the illustration stated "It appears that songs and pieces of music are now sent from Boston to New York by Electric Telegraph.... It must be delightful for a party in Boston to be enabled to call upon a gentleman in New York for a song." Unfortunately, no specifics were given about who was doing this. Certainly by 1874, however, Elisha Gray had conducted tests of such telegraphic electroharmonic broadcasting.

Later, in 1878, live opera was transmitted over wire lines to groups of listeners in Switzerland,[3] and in 1881 the French engineer Clement Ader (1840–1925) wired the Paris Opera House with matched carbon microphone transmitters and magneto-telephone receivers to allow stereo listening from more than a mile distant. In 1893 wired broadcasting was a booming commercial business in Budapest, Hungary, with the operation of the station Telefon-Hirmondo ("Telephonic Newsseller"). With over 6,000 customers, each paying a fee of nearly eight dollars a year, and with an advertising charge of over two dollars a minute, this was a big operation. The station employed over 200 people and was "on the air" with regular news and music programming twelve hours a day. A similar operation appeared in London, in 1895, under the control of the Electrophone Company. By 1907 it had 600 individual subscribers, as well as 30 theatres and churches as corporate customers, all linked through 250 miles of wired connections. Such regular wired broadcasting, suggestive of today's cable television, didn't appear in America until after the turn of the century. Some individual, special events, however, were covered in the U.S. by wired broadcasts in the last decade of the nineteenth century; for example, the Chicago Telephone Company broadcast the congressional and local election returns of November 1894 by wire, reaching an audience that may have exceeded 15,000.

Still, while representing a tremendous engineering achievement, wired broadcasting was simply too cumbersome, inconvenient, and expensive in hardware for commercial expansion beyond local distances. A way to reach out on a really wide scale, without having to literally run wires to everyone, was what was needed. As a first step toward this goal, at least two imaginative American tinkerers looked to space, itself, as the means to carry information. These two men, Mahlon Loomis (1826–1886), a Washington, D.C. dentist, and Nathan Stubblefield (1859–1928), a Kentucky melon farmer, both constructed wireless communication systems that used inductive effects.[4] Such a system uses the energy in a *non*-propagating field (the so-called induction or near field); while wireless in a restricted sense, such a system is really not radio. The story of Loomis' system, which used a kite to loft an antenna over the Blue Ridge Mountains of Virginia in 1866, is shrouded in some mystery. He did, apparently, achieve a crude form of wireless telegraphy.[5] The story of Stubblefield is better documented, and there is little doubt that he actually did achieve wireless voice transmission as early as 1892.[6] Neither man, however, had built a radio, which utilizes high-frequency electromagnetic energy *radiating* through space.

That historic achievement was the success of the Italian Guglielmo Marconi (1874–1937), who used a pulsating electric spark to generate radio waves and who used these waves to transmit telegraphic Morse code signals. Marconi's system used a true radiation effect, and he could transmit over a distance of miles; in theory, the transmission distance is unlimited. One human could, at last, in principle speak to every other human on earth at the same time. Marconi's work, for which he shared the 1909 Nobel Prize in physics with the German Karl Ferdinand Braun (1850–1918), was the direct result of the fundamental experiments of the tragically fated German Heinrich Hertz (1857–1894), which were performed in a search for the waves predicted by the theory of the equally grim-fated Scotsman James Clerk Maxwell (1831–1879).[7] This theoretical work, among the most brilliant physics in the history of science, is described in the next chapter.

The claim sometimes made that the Russian Alexander S. Popov (1859–1905) invented radio is important to mention here. Popov's work is perhaps less appreciated than is Marconi's because of the excessive zeal of the old Soviet state in appearing to claim *everything* was invented by Russians. There is, however, evidence that Popov did in fact do much independent work that closely paralleled Marconi's, and that in fact Popov was the first to fully appreciate the value of (and to use) an antenna. Oddly enough, Popov's use of large antennas was based on the then commonly held but *incorrect* view that wireless communication had to be line-of-sight. When asked in 1901 about Marconi's claim to have signaled across the Atlantic, for example, he replied that he had his doubts—Popov thought such a feat would require fantastically tall antennas to satisfy the (false) line-of-sight requirement (see Problem 2.3 for more on this). This, and other observations, are in an essay by the Russian historian K.A. Loffe, "Popov: Russia's Marconi?" *Electronics World and Wireless World*, July 1992. And finally, let me make the personal observation that Marconi's Nobel Prize is one of those very occasional ones that looks increasingly less deserved with time. Certainly the fundamental theoretical physics that underlies radio is due to Maxwell, while it was Hertz who did the basic engineering of spark gap transmitters and receivers. And it was the English scientist Oliver Lodge (1851–1940) who has priority in developing the fundamental ideas of frequency selective circuits to implement tuning among multiple signals. By 1909 both Maxwell and Hertz were long dead, and the prize is never given posthumously. But Lodge was alive—so why did *Marconi* get the Nobel Prize? The answer may lie buried beneath nearly a century's weight of paper in the archives of the Nobel committee, but there can be no question that politics and self-promotion played very big roles. His co-winner, K.F. Braun (1850–1918), who was honored for his fundamental *alterations* in Marconi's spark-gap transmitter (I discuss what he did in Chapter 4), was a far more obvious choice as he was also the inventor in 1897 of the cathode-ray oscilloscope—called originally the *Braun tube*—an instrument that no electrical engineer today could possibly imagine not existing. Marconi was certainly a successful and energetic businessman, but that ain't physics! For more on this, see my book *Oliver Heaviside, Sage in Solitude*, IEEE Press 1988, pp. 263, 278, and 281.[8]

By starting my history of radio where I have, I am of course ignoring the speculations by some that the first radio actually dates back to the Old Testament and the Ark of the Covenant! The Ark, built by Moses according to detailed instructions from God (*Exodus* 25) to hold the stone tablets of the Ten Commandments, is described in various ancient Jewish legends as being surrounded by sparks and so was perhaps electrical in nature. Further, when Uzzah touched the Ark (*II Samuel* 6:6–7) he immediately died (electrocuted?). In *Exodus* 25:22 the Lord tells Moses he will speak to him from the Ark, and the view of the Ark as a radio was cleverly woven into the first Indiana Jones movie (the 1981 *Raiders of the Lost Ark*): the central villain tells Indy "It's a transmitter. It's a radio for speaking to God!" Well, I love a mystery as much as the next person, but in this book I'll leave such stuff to Steven Speilberg, George Lucas, and Erich von Däniken.

I cannot resist concluding this chapter with a comment on an amusing summary of radio history, written by two lawyers(!).[9] Their book is actually quite interesting, but at one point they say, in a footnote (p. 47), "Once Marconi invented the wireless telegraph [then] creating radio and television were comparatively simple engineering tasks." That is simply not so. What it actually required to get from Marconi's brute force spark-gap radio (which he did *not* invent) to Armstrong's beautiful superheterodyne radio were conceptions of *pure genius*. When you finish this book, I think you'll agree.

One other person whose name is occasionally put forth as the inventor of radio is Nikola Tesla (1856–1943), who was born in Croatia and later became a naturalized American citizen. Tesla's was an intuitive, erratic personality, and his rightful fame among electrical engineers is for the discovery of the rotating magnetic field principle behind the synchronous ac induction motor. He was also a force in the early development of multi-phase ac power distribution. The unit of magnetic flux density is named after him. Others, however, not satisfied with Tesla's true achievements, find it necessary to claim he did all sorts of other things as well (which, curiously, not even the full scientific might of the Pentagon can duplicate, such as Tesla's famous 1934 'radio death ray' that he said could destroy 10,000 planes 250 miles away and annihilate, in an instant, an army of 1,000,000). It seems more likely that Tesla, unable to repeat his early triumphs, looked for other ways to get back into the limelight he so coveted; he began to make astonishing claims to wealthy potential patrons who, knowing next to nothing of science, could be easily dazzled. One of these claims was that he had "invented radio." Tesla was, without question, very skillful at generating large, noisy sparks with the aid of step-up transformers tuned to resonance (the famous *Tesla coil*) and he seems to have really believed that, since Marconi used sparks in *his* wireless work, then he too must be a wireless pioneer. There is, however, not a shred of credible evidence that Tesla did anything more than just *talk* about radio (in 1901, for example, he claimed that two years before he had received radio signals from Mars[10]), and nothing in the historical record supports his grandiose claims. It is clear, in fact, from what he did write, that Tesla actually had only the slightest (if that) understanding of electromagnetic radio physics; he claimed, for example, that "his" electric waves were both immune to the inverse-square law and that they traveled faster than light. Tesla does appear to have sincerely believed his own outrageous statements; he lived in a delusional world of self-aggrandizement that became increasingly cut off from reality. His only human joy seems to have been feeding the pigeons of New York City, where he died in a hotel room a lonely, bitter man. Modern biographers of Tesla (none of whom have any technical training) continue to muddy the historical record, however, and so let me be quite clear: *Tesla did not invent radio*, although his flowery talk about it no doubt inspired many youngsters at the start of the twentieth century to become interested in "the new wireless."

Notes

1. On p. 41 of Walter Kellogg Towers' book *Masters of Space* (Harper & Brothers 1917) there appears the following passage: "The identity of C.M. has never been established, but he was probably Charles Morrison, a Scotch surgeon with a reputation for electrical experimentation, who later emigrated to Virginia." Towers incorrectly gives the date of C. M.'s article as 1755, but a facsimile of it (displaying the date 1753) appears in the scholarly essay by G. R. M. Garratt, "The Early History of Telegraphy," *Philips Technical Review* 26, 1965, pp. 268–284.

2. Optical telegraphy (or *aerial* telegraphy, as it occasionally is called) has a history predating Chappe by millennia, e.g., soldiers in the Peloponnesian wars sent prearranged messages by torch light. All such early systems, however, were limited to *anticipated* messages, while Chappe's system was the first to allow messages about unexpected situations to be transmitted. More on this, including an information theoretic analysis of optical telegraphy, can be found in Alexander J. Field, "French Optical Telegraphy, 1793–1855: Hardware, Software, Administration," *Technology and Culture* 35, April 1994, pp. 315–347.

3. See Elliott Sivowitch's two articles "Musical Broadcasting in the 19th Century," *Audio*, June 1967, pp. 19–23, and "A Technological Survey of Broadcasting's 'Pre-History,'" 1876–1920, *Journal of Broadcasting* 15, Winter 1970–71, pp. 1–20.

4. Such induction communication was well known in the early 1890s; see, for example, the letter by E.A. Grissinger in *Electrical World* 24, November 10, 1894, p. 500.

5. See Otis B. Young, "The Real Beginning of Radio: the neglected story of Mahlon Loomis," *Saturday Review*, March 7, 1964, pp. 48–50. Loomis received perhaps the first "wireless" patent (U.S. Patent No. 129,971 on July 30, 1872 for "A New and Improved Mode of Telegraphing").

6. See Thomas W. Hoffer, "Nathan B. Stubblefield and His Wireless Telephone," *Journal of Broadcasting* 15, Summer 1971, pp. 317–329. More on Loomis is in this article, as well. Another now forgotten pioneer of early "wireless" inductive communication is Amos Emerson Dolbear (1837–1910), a professor of physics and astronomy at Tufts College. Unlike Stubblefield, Dolbear received patents ("Electric Communication") for his invention (British Patent No. 1367 received March 21, 1882, and U.S. Patent No. 350,299 received October 5, 1886).

7. Hertz and Maxwell both died young, in agony; Hertz of blood poisoning after enduring several operations for terrible jaw, teeth, and head pains, and Maxwell after long months of suffering from abdominal cancer.

8. For more on Marconi and Lodge, and who did what, when, see Sungook Hong, "Marconi and the Maxwellians: The Origins of Wireless Telegraphy Revisited," *Technology and Culture* 35, October 1994, pp. 717–749. For more on Marconi and Popov, see Ralph Barrett, "Popov Versus Marconi: the Centenary of Radio," *Proceedings of the Royal Institution of Great Britain* 68, 1997, pp. 317–326.

9. Thomas G. Krattenmaker and Lucas A. Powe, Jr., *Regulating Broadcast Programming*, MIT and AEI Presses 1994.

10. This claim was not taken seriously by many (who argued convincingly that if Tesla was receiving anything it was certainly of terrestrial origin), but it did not pass without some lasting literary impact. H.G. Wells noticed it, and mentioned Tesla's supposed Martian contact in his novel *The*

First Men in the Moon published that same year (1901). In Wells' story one of the characters sends wireless telegraphy Morse code from the Moon back to Earth. Ironically, Tesla's own words (in his patent application of 1897) show that he was really not thinking of true radio at all, but rather of a *conducting* system (see *IEEE Transactions on Antennas and Propagation* 38, October 1990, pp. 1723–1726). The most recent "explanation" for what Tesla claims to have heard is that he was detecting electromagnetic radiation caused by the magnetic field of Jupiter. This is a speculation that at best seems (to me) pretty far-fetched for an 1899 radio receiver, an objection that Tesla advocates have anticipated. Their response is not unexpected—Tesla's 1899 radio was a superadvanced design that nobody else on Earth had even dreamed of!

Problem

1.1. *Tesla's radio death ray.* As was typical of Tesla, he provided no details for his death ray other than, in some unspecified way, it would use 50,000,000 volts. From that, and Tesla's professional background, it seems reasonable to assume the ray was electromagnetic in nature. In this problem you are to estimate the energy required to achieve Tesla's claim that the ray could "instantly annihilate an army of 1,000,000." Let's assume that the "rapid-kill' mechanism is that of over-heating the blood of each soldier. This sounds awful, I know, but that's exactly the "death ray" requirement put forth by the British Air Ministry in the 1930s for killing an approaching enemy bomber pilot; elevating the temperature of eight pints of *water* (a human body contains eight pints of blood and so everybody knew what was really meant). To give Tesla the benefit of the doubt at every step, assume his death ray doesn't really have to be instantaneous but instead can take ten seconds to kill. A soldier will certainly be dead at a blood temperature of 100° C (boiling!), but let's say Tesla's death ray only has to get it to 60° C. Also assume all energy generated, focused, and transmitted is totally absorbed by the soldiers' blood with no losses. You may find the following numerical data useful in your calculations (where water and blood are considered to be interchangeable):

- normal human blood temperature = 38° C.
- 1 liter ≈ 2 pints.
- 1 liter of blood weighs about 1 kilogram.
- 1 Calorie of energy is required to raise the temperature of 1 kilogram of blood by 1° C.
- 1 Calorie = 4200 joules.
- 1 watt = 1 joule/second.

(a) What is the total death ray energy (in joules) required to kill an army of 1,000,000?

(b) What is the power level required during the 10 second "kill time" to deposit the energy of part a in the army?

(c) A good-sized nuclear power plant running at full throttle can generate something like 5000 megawatts. How many such nuclear power plants would be required to operate Tesla's death ray?

(d) How would you characterize Tesla's claim, and why do you think his biographers continue to assert that there has been a vast conspiracy, for decades, to "cover-up" Tesla's "secret papers" on his death ray?

2

Preradio History of Radio Waves

As of the middle of the nineteenth century, scientific knowledge of electricity and magnetism was mostly a vast collection of experimental observations. There had been few previous theoretical analyses of electricity, with Germany's Wilhelm Weber's (1804–1891) incorrect extension in the 1840s of Coulomb's inverse-square force law to include velocity and acceleration dependency indicative of the state of the art. The greatest electrical experimentalist of the day was Michael Faraday (1791–1867) after whom the unit of capacitance is named, a man of intuitive genius who invented the idea of the field; but he was also a man totally unequipped for the enormous task of translating a tangle of experimental data into a coherent mathematical theory.

Given a point electric charge, Faraday thought of the space around the charge as permeated with a spherically symmetric field of radial electric lines of force (pointing away from a positive charge and pointing towards a negative charge). A positive (negative) charge in the field of another charge experiences a force in the direction (opposite to the direction) of the field. Thus, the familiar "like charges repel, unlike charges attract." Faraday also visualized a field of magnetic lines of force around a magnet, beginning on the North Pole and terminating on the South Pole (as beautifully displayed by iron fillings on a piece of

> paper placed over the magnet). These lines of force were mechanically interpreted by Faraday who thought of them as stress in the ether, a mysterious (now debunked) substance once thought to fill all space.

As Faraday stood perplexed among a multitude of apparently unconnected facts, two new players appeared on the scene, each armed with the mathematical skills Faraday lacked. These two Scotsmen, William Thomson (1824–1907) and his younger friend James Clerk Maxwell, had both made it their goal to find the unifying theoretical structure beneath the myriad of individual facts. Thomson, who was the technical genius behind the first proper mathematical analysis of the Atlantic undersea cables and who later became the famous Lord Kelvin, eventually fell by the wayside in this quest after some early, limited successes. Maxwell, however, was successful beyond what must have been his own secret hopes.

In a series of fascinating letters[1] written in the middle 1850s to Thomson, Maxwell outlined his ideas on where to start on the path that would lead to a mathematical theory of the ocean of loosely connected experimental facts that Maxwell called a "whole mass of confusion." In particular, Maxwell felt the key was Faraday's intuitive idea on inductive effects; what Faraday called the *electrotonic state*. Maxwell's first step toward an electromagnetic theory started, therefore, with a paper in 1856 on that vague, ill-formed concept. He wrote this paper, "On Faraday's Lines of Force," when he was just 24. In his paper Maxwell borrowed Thomson's 1847 idea of calculating a vector from another vector using the vector curl operation. This initial paper was followed by "On Physical Lines of Force," published in four parts during 1861 and 1862. The electrotonic state was further clarified in terms of a mechanical model of the ether, and yet more mathematical machinery was introduced; in particular, the famous integral theorem named after Cambridge mathematician George Stokes (a friend of both Maxwell and Thomson) that was first published on an exam (taken by Maxwell) given by Stokes in 1854. And finally, in 1865, came "The Dynamical Theory of the Electromagnetic Field."

The field concept of Faraday, expressed mathematically by Maxwell, replaced the older action-at-a-distance idea. Instead of thinking as did Weber of an electric charge reaching somehow instantaneously across space to directly exert an inverse-square Coulomb force on another charge, Faraday and Maxwell imagined that a charge somehow modifies space. This modified space then extends outward with finite speed and locally interacts with all remotely located charges. Modified space is said to contain an electric field. Charge A does not directly interact with charge B, but rather charge A interacts with the *field* of charge B *as that field exists at the location of charge A*, and vice versa. This may seem like mere a word game, and at this point it is, but Maxwell later showed Faraday's fields actually have physical reality.

Faraday's field concept was one of the great intellectual breakthroughs in human thought, and all modern physical theories are field theories. A similar process has been applied, for example, to Newton's direct action-at-a-distance formulation of gravity. The modern theory of gravity is Einstein's *field* theory of curved four-dimensional spacetime, and although requiring more advanced mathematics than does Maxwell's vector field theory (tensors, of which vectors are a special case), Maxwell's theory inspired Einstein.[2] As Einstein himself wrote, "Since Maxwell's time, Physical Reality has been thought of as represented by continuous fields, governed by partial differential equations." Einstein then went on to declare the field concept to be "the most profound and the most fruitful that physics has experienced since the time of Newton."

Maxwell's third paper, with the unnecessary mechanical model of the ether in the second paper deleted, presents his field theory in its essentially final abstract form. What had started in Faraday's wonderfully imaginative mind as the electrotonic state had become Maxwell's electromagnetic momentum. Today we call it the vector potential, a term first used by Maxwell; the curl of the vector potential is the magnetic field vector. The third paper (as does the second), however, presents the mathematical theory in a manner that a modern physicist or electrical engineer, used to vectors and tensors, wouldn't recognize. Maxwell presented his theory as twenty(!) equations, expressed in a hodge-podge mix of component and quaternionic (a sort of vector) notation.

The mathematics might look strange, but the physics is all there and the main conclusion was simply astounding: electromagnetic effects travel through space at the speed of light. Indeed, light itself is a propagating electromagnetic field. And, astonishingly, Maxwell could actually calculate the speed of light from the laboratory measurement of two electrical constants! As Maxwell himself wrote of those measurements, "The only use of light in the experiment was to see the instruments." Maxwell had achieved the second great unification in physics by showing the science of light and optics is merely a branch of electromagnetism. (The first unification is usually attributed to Newton's extension of gravity from a mere earthly phenomenon to one operational throughout the entire universe, i.e., the claim that the mechanics of earth and of the heavens are one in the same.) It was in this paper, too, that Maxwell stated that the energy of electromagnetic phenomena resides not just in electrified bodies, but also in the space surrounding such bodies.

What Maxwell expressed mathematically in his famous set of partial differential equations is (1) electric lines of force are created either by electric charge or by time-varying magnetic fields, and (2) magnetic lines of force are created either by currents (moving electric charge) or by time-varying electric fields. The last half of (2) is uniquely Maxwell, as it represents his famous displacement current. That a time-varying electric field in space could produce a magnetic field, just like a conduction current in a wire, was an audacious statement by Maxwell because at that time *there simply was not the slightest experimental evidence for it.* Today, electrical engineering and physics professors derive the displacement current term by showing that, without it, the rest of the equations are inconsistent with the conservation of electric charge.[3] Maxwell did not reason this way, however, and his hazy physical arguments[4] are now merely of historical interest; but no matter, his intuitive genius guided him to the correct result anyway.

Maxwell's equations are *differential* equations for the electric (*E*) and magnetic (*H*) field vectors because these fields, at every point in space, for every instant of time, can be related to the fields at nearby points in space and time. They are *partial* differential equations because there are multiple independent variables, i.e., time, and at least one space variable.

Any inconsistency with charge conservation is undetectable in a closed circuit (and only closed circuits had been studied; after all, what sense could an open electrical circuit make?) Without a closed path for current to flow along, how could anything happen? But with the displacement current, an open circuit does make sense, and it is the displacement current that gives life to radio, television, and radar signals, light, and X-rays, all of which are electromagnetic energy, at various frequencies, propagating through space.

A detailed mathematical analysis of Maxwell's equations is required to fully appreciate how the fields propagate, but even the prose descriptions given above in (1) and (2) provide insight. If one imagines that somehow a time-varying (oscillating) *E* field has been generated in space (and in the next chapter I'll tell you how that is done), then (2) says an oscillating *H* field will then be created. But then this new *H* field will in turn, because of (1), give rise to a new oscillating E field, which then generates an oscillating H field, etc. This endless spawning of new fields is not perfectly confined to the same region of space; instead the fields are continually spreading ever outward. Indeed, the fields spread at the speed of light. An analysis of Maxwell's equations also tells us that the oscillating E- and H-field vectors are mutually perpendicular, and that both are in turn perpendicular to their common direction of propagation. Electromagnetic waves are therefore *transverse* waves, as opposed to longitudinal waves (such as sound or pressure waves in matter) in which oscillations occur along the direction of propagation. The intimate coupling—or "mutual embrace," as Maxwell often romantically put it[5]—of the E- and H-field vectors is why we talk not of the electric and magnetic fields separately, but rather of a unified electromagnetic field.

In 1873 Maxwell brought all his ideas together in book form, in his famous *A Treatise on Electricity and Magnetism*. It was clearly his intention to pursue the implications of his equations (e.g., in his *Treatise* Maxwell presented the theoretical prediction of radiation pressure, i.e., electromagnetic radiation carries momentum and therefore light pushes!) but he had run out of time. In 1879, only 48 years old, he was dead of the same type of cancer that had killed his mother at the same age. When he died Maxwell's theory of electricity and magnetism was only one of several, however, including Weber's (which Maxwell had called "a mathematical speculation which I do not believe"). It was only with Heinrich Hertz's sensational experimental discovery in 1887 of electromagnetic radiation at microwave frequencies (between 50 and 500 MHz), as predicted by Maxwell, that there could at last be no doubt that Maxwell's theory was the correct one. Hertz's experiments verified that electromagnetic radiation possessed all the usual optical properties of light, e.g., that it could be reflected and refracted, that it traveled in straight lines, and that it could be polarized. The only difference between his radiation and visible light was the frequency.

The prediction by Maxwell's theory that light is an *electromagnetic* wave phenomenon traveling through air (or even through the vacuum of space, unlike sound waves) quickly provoked the obvious question—how can such waves at *any* frequency be generated? To produce a wave at visible frequencies is, of course, easy; just start a fire, or pass an electrical current through a wire (such as the filament in a light bulb) so as to make it hot enough to glow. Such visible electromagnetic waves are very short, however, because they are of extraordinarily high frequency. Green light, in the middle of the visible spectrum, for example, has a frequency of

600 *million* MHz! Even higher frequencies could be produced in the late nineteenth century by allowing an accelerated beam of electrons to strike matter, such as is done in an X-ray tube (the first medical X-ray, of a human hand, was made in 1896). How to produce the much *lower* frequencies of the radio spectrum was a much more puzzling question, however, one that attracted the attention of such well-known experimenters as Oliver Lodge in England and, of course, Hertz in Germany.

The key idea for the first (and eventually successful) approach to generating radio frequency (rf) waves came from the Irish physicist George Francis FitzGerald (1851–1901). In what may perhaps be the shortest important scientific paper ever published, FitzGerald suggested (in 1883) charging a capacitor (e.g., with a static electricity generator) to a high voltage, and then letting it discharge through an inductive circuit. The resulting circuit current is oscillatory (as will be shown in Chapter 4) and, by controlling the circuit parameter values, the frequency of oscillation can be controlled.[6] In the next chapter you will see how Hertz used FitzGerald's suggestion to experimentally demonstrate that Maxwell's theory is correct.

There were those who hadn't had to wait for Hertz, however; these true believers were members of a small group that has become known as the "Maxwellians." They included, for example, the Englishman John Poynting (1852–1914) who, in 1883, discovered how Maxwell's theory predicts that a propagating electromagnetic field transports energy through space. This result, in particular, is what makes radio possible. What Poynting found was that if there are E and H fields at a point in space, then there is a flow of energy at that point.[7] The rate at which energy flows is power and the Poynting vector P gives the power density in units of power per unit area (in MKS units the Poynting vector has dimensions of joules/second/square meter). Specifically, $P = E \times H$ where the indicated operation is the vector cross product. As you'll see in the next chapter, radio antennas broadcast energy into space by creating E and H fields that result in a P vector that always points away from the antenna, i.e., there is a unidirectional flow of energy from the antenna into space.

Recall from your math courses that the magnitude of P is the product of the magnitudes of E and H, and of the sine of the angle between E and H. Since E and H are perpendicular for elecromagnetic waves, this factor is unity. The direction of P is given by the so-called right-hand rule; rotate E into H, with the sense of rotation that of a right-hand-threaded screw, and then P points in the direction the screw advances. For example, looking downward at E and H in a plane, if E points upward (in the plane) and H points to the right, then the rotation is clockwise and P points into the plane (away from you). (Don't you think the Poynting vector has the *perfect* name?) There is also a P vector that alternately points away and then toward the antenna. This P vector is important only near the antenna, as its amplitude decays very rapidly with distance (an exact analysis of the Maxwell equations for an elementary antenna shows that this P vector decays as the inverse fourth and fifth powers of distance from the antenna). It represents an alternating energy flow into and out of the antenna, the so-called near- or induction-field (recall the systems of Stubblefield and Loomis from the previous chapter). This energy always remains coupled to the antenna. The unidirectional P vector, on the other hand, decays more slowly with distance and represents the far or radiation field that makes radio possible (see Problem 2.2). It represents energy that has decoupled from the antenna, never to return. It is propagating energy.

The technical stage was now clearly set, as we can see from the vantage point of the present, for radio. But it wasn't at all clear *at the time*, as the following anecdotes illustrate. Even before Hertz began planning his great experiments in Germany, in America the social commentator Edward Bellamy was penning his short story "The Blindman's World," published in the November 1886 issue of the *Atlantic Monthly*. It tells of how a "professor of astronomy and higher mathematics" suddenly and unexplainably finds himself on Mars, where he is greeted by friendly creatures. They are glad to see him, in fact, and tell him "For ages we have been waiting for you to improve your telescopes so as to approximate ours, after which communication between the planets would be easily established." Telescopes, not radio, is what Bellamy thought to be the coming technology of long-distance communication (messages written in sand?)!

Soon after he went on to write the utopian novel *Looking Backward*. This is the fantastic story of a young man who goes to sleep one night in the Boston of 1887, and wakes up in the year 2000. He finds that all of humankind's social problems have been solved, and that some remarkable technical advances have occurred over the last 113 years, as well. In Chapter 11, for example, the hero is shown a twenty-four-hour music distribution system based on the telephone, a system remarkably similar to the ones described in our Chapter 1. In this particular prediction, Bellamy was five years ahead of his time, but even his active imagination boggled at the idea of anything better. His hero clearly thought he had seen the last word: "It appears to me ... that if we [the technology of 1887] could have devised an arrangement for providing everybody with music in their homes, perfect in quality, unlimited in quantity, suited to every mood, and beginning and ceasing at will, we should have considered the limit of human felicity already attained, and ceased to strive for further improvements." What would Bellamy have thought of radio?—alas, like Maxwell he died young, at age 48 in 1898, and just missed what even he would have thought of as being simply magic.

Looking Backward was published in 1888, and in its first year sold a quite respectable 10,000 copies. The next year, however, its sales soared to over 300,000 copies, a number not often reached even with today's inexpensive paperbacks. Besides just selling, Bellamy's book also had enormous impact worldwide. The great Tolstoy himself, for example, arranged for the Russian translation, and when several famous men of letters (including the editor of the *Atlantic Monthly*) looked back in 1935 at the books of the past half-century having the greatest influence on the intellectual thought of the world, *Looking Backward* was declared second only to Marx's *Das Kapital*. It is still in print. Not everybody could accept Bellamy's timetable for the arrival of his happy future, however. In a newspaper review of the book, for example, the *Boston Transcript* thought the world of 2000 should have more realistically been set 75 *centuries* into the future! In 1890 Alvarado Fuller published the remarkably awful novel *A.D. 2000* in which the hero, like Bellamy's, arrives in the future after a century-long snooze in a vault. In this story we find something called the "sympathetic telegraph system," a gadget "to be" invented in 1892. It is a true radio in everything but name because "there is no wire, metal, or tangible connection of any kind between the [sending and receiving] instruments" and it works over vast distances (and so it isn't an induction-field device). So, how *did* Fuller explain his gadget? He didn't—after one character exclaims "How could mortal man have discovered such a secret of nature!" we are told that the inventor never revealed the science behind it and, alas, that "now" in the year 2000, "The secret is lost!"

Since the pioneering work of the Maxwellians, Maxwell's equations have been studied more than a century further, and the equations have proven to be one of the most successful theories in the history of science. For example, when Einstein found that Newtonian dynamics had to be modified to be compatible with the special theory of relativity, he also found that Maxwell's equations were already relativistically correct. This is so because magnetic effects *are* relativistic effects produced by moving charges (see Problem 3.2), and so Maxwell had automatically built relativity (before Einstein's birth) into his equations. And Maxwell's original belief in the

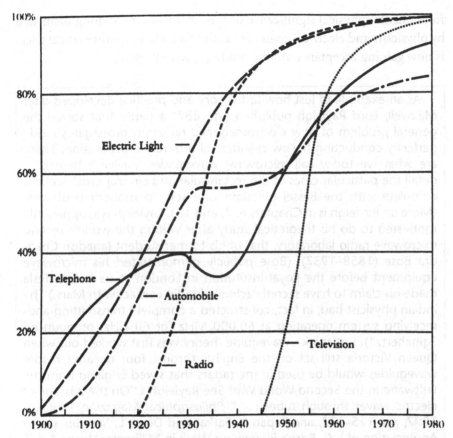

Figure 2.1   In these graphs we see how five common domestic technologies permeated twentieth-century American society. The dips in the curves for the telephone and the automobile are due to the Great Depression, but even those catastrophic times couldn't stop the inexorable growth of the electrical and electronic gadgets. Perhaps most astonishing is the market saturation achieved by radio; as recently as 1980 radio was in more households than even the electric light! Radio (and television) could penetrate the marketplace quickly because of its low infrastructure requirements. The electric light, telephone, and automobile in contrast, all required enormous distribution networks (wires, cables, and highways) before they could spread. The infrastructure issue is also why *cable* television spread much less quickly than did broadcast television. The same factor may well be the case for inhibiting the spread of high-speed Internet video, as cable modems will require extensive infrastructure upgrades on the Net. Source: U.S. Bureau of the Census, taken from Claude S. Fischer, *America Calling: A Social History of the Telephone*, University of California Press 1992.

fundamental physical significance of the vector potential, long dismissed by physicists and electrical engineers as simply a clever mathematical trick, is now gaining acceptance among modern theoreticians.

As an example of just how fast theory and practice developed after Maxwell, Lord Rayleigh published (in 1887) a paper that solved the general problem of how electromagnetic radiation propagates inside perfectly conducting hollow cylinders of arbitrary cross-section. These are what we today call microwave waveguides. Rayleigh treated in detail the particular cases of the rectangular and circular cross-section, complete with the Bessel functions one sees in modern textbooks. (More on Rayleigh is in Chapters 6, 7, and 13). Rayleigh was apparently motivated to do his theoretical study after visiting the world's second microwave radio laboratory, that of his former student Jagadish Chandra Bose (1858–1937). (Bose *publicly demonstrated* his microwave equipment before the Royal Institution in London *years* before Tesla made his claim to have secretly achieved radio contact with Mars.) The Indian physicist had, in fact, constructed a complete transmitting-and-receiving system operating at 60,000 MHz (or 60 GHz, pronounced "gigahertz"). Just think—waveguide theory was first worked out when Queen Victoria still sat on the English throne, four decades before waveguides would be used in the radars that saved England from the Luftwaffe in the Second World War! See Rayleigh's "On the passage of electric waves through tubes ...," *Philosophical Magazine*, February 1897, pp. 125–132, and Tapan K. Sarkar and Dipak L. Sengupta, "An Appreciation of J. C. Bose's Pioneering Work in Millimeter Waves," *IEEE Antennas and Propagation Magazine* 39, October 1997, pp. 55–63.

Surely Richard Feynman was correct when he declared, in his famous *Lectures on Physics*, "ten thousand years from now—there can be little doubt that the most significant event of the nineteenth century will be judged as Maxwell's discovery of the laws of electrodynamics. The American Civil War will pale into provincial insignificance in comparison." James Clerk Maxwell had wrought better than he knew.

Notes

1. See "The Origins of Clerk Maxwell's Electric Ideas as Described in Familiar Letters to W. Thomson," *Proceedings of the Cambridge Philosophical Society* 32 (Part 5), 1936, pp. 695–750.

2. See Einstein's essay "Maxwell's Influence on Theoretical Physics," in *James Clerk Maxwell: A Commemoration Volume*, Macmillan 1931.

3. For the details of this see my book *Oliver Heaviside: Sage in Solitude*, IEEE Press 1988, pp. 90–91.

4. For how Maxwell did reason, see Bruce J. Hunt, *The Maxwellians*, Cornell University Press 1991.

5. M. Norton Wise, "The Mutual Embrace of Electricity and Magnetism," Science, March 30, 1979, pp. 1310–1318.

6. "On the possibility of originating wave disturbances in the ether by means of electric forces," in *The Scientific Writings of the Late George Francis FitzGerald*, Longmans, Green & Co. 1902, p. 92. The possibility of an oscillatory nature to a capacitive discharge actually predates FitzGerald by more than a century, as the early English electrical experimenter Benjamin Wilson (1721–1788) had observed such discharges in 1750. It was FitzGerald's contribution to suggest the use of such discharges to generate Maxwell's electromagnetic waves.

7. This comes from a direct mathematical manipulation of Maxwell's equations; see my previously cited book (Note 3) on Heaviside, pp. 129–132.

8. J.A. Ratcliffe, "Scientists' Reactions to Marconi's Transatlantic Experiment," *Proceedings of the IEE* (British) 121, September 1974, pp. 1033–1038. This paper presents very strong technical analyses in support of the conclusion that Marconi's equipment simply *could not have done what he claimed!* Did Marconi deceive himself into thinking he heard the "S" signals—or did he deceive *everybody?* You decide. For more on this issue, see the last shaded box in Chapter 5. I should tell you, however, that by October 1907 Marconi's wireless company did succeed in establishing a daily transatlantic wireless telegraph service (with an average speed of 5 words per minute); "Marconi Dispatches" from London appeared in each Sunday edition of the *New York Times*.

Problems

2.1. There is an E field in the space between the terminals of a battery. If you hold a bar magnet near the battery, there is also an H field in that space. Thus, there is a P vector and so, according to Poynting, there is an energy flow! Where does that energy come from (and where does it go)? Don't forget—energy is conserved at every point, at every instant. Hint: see Feynman's *Lectures on Physics*, vol. 2, Chapter 27, Addison-Wesley 1964.

2.2. The unidirectional, radiation (or far-field) P vector that represents propagating energy decays as the inverse square power of distance from the antenna (a point that Tesla, as I mentioned in the previous chapter, failed to understand). As mentioned in the text, however, the induction or near-field P vector decays faster, at least as fast as the inverse fourth power. Show how these decay rates for the far- and near-field P vectors explain why (as far as classical, nonquantum physics is concerned) the near field cannot be detected at "sufficiently great" distances no matter what the state of technology may some day be, while the far field can, in principle, be detected at any distance. Hint: consider the energy that can be intercepted by an antenna subtending a fixed solid angle, and how that energy depends on distance and field decay rate.

2.3. As mentioned in Chapter 1, radio frequency energy was at first
thought to travel strictly along straight lines, and that it certainly
couldn't travel *around* the earth. It seemed as though terrestrial radio
would be restricted in range by the earth's horizon. The laboratory-
scale experiments of Hertz and Bose seemed to confirm this. It was
thus a great surprise when Marconi apparently was able to transmit a
brief Morse-code signal—three dots for the single letter "S"—on December 12, 1901 from Poldhu in Cornwall, England (six miles north
of Lizard Point, a place described by Marconi as "hard and bleak")
to Signal Hill at St. John's, Newfoundland. Many scientists simply
did not believe the claim because the separation distance of about
2200 miles was so great that the signal *must* have "curved around"
the planet.[8] The presence of the ionosphere, a layer of ionized atmosphere about 60 miles or more above the surface of the earth which
can act as a "radiation mirror" and so reflect radio signals around the
curvature of the planet, was not yet known. Indeed, it was Marconi's
claim that *started* speculations about the possible existence of such
a "mirror." In fact, however, the phenomenon of a signal traveling
beyond the horizon had been observed even earlier, *two years* before,
during British naval exercises in 1899 involving Marconi equipment.
An observer noticed that two ships, 60 miles apart, had successfully
communicated. What made this surprising was that, with both ships
using 150-foot-high antennas, the horizon was fifteen miles from each
ship and so the maximum range "should have been only 30 miles." As
that observer noted, the radio signal "must have passed through or
over a mass of sea water about 500 feet high and 30 miles thick."

(a) Taking the earth to be a sphere of radius R, show that the maximum line-of-sight separation between the tops of two anten-
nas of heights h_1 and h_2 is $\sqrt{2Rh_1} + \sqrt{2Rh_2}$. Make the reasonable assumption that h_1 and h_2 are both far smaller than R. For
$h_1 = h_2 = 150$ feet, and $R = 3960$ miles, is this formula consistent with the first part of the British naval observer's comment?

(b) For the special case of $h_1 = h_2 = h$, call the distance from each
antenna to the horizon d. Then, let *each* antenna be moved *away*
from that horizon by an additional distance of d, i.e., the two
antennas are now separated by distance $4d$. A straight line path
from one antenna to the other will now obviously have to pass
through the earth. Show that the point along this path that is fur-

thest from the surface is at a depth of approximately $3h$. Is this result consistent with the second part of the British naval observer's comment?

(c) Assuming $h_1 = h_2 = h$, what is the minimum value for h (in *miles*!) for line-of-sight signaling (no penetration of the earth) between Poldhu and St. John's? Also, calculate the maximum depth beneath the surface of the earth of the straight communication path ($h = 0$) between Poldhu and St. John's. Do your numbers explain the argument, often used to ridicule Marconi's plans to transmit radio signals across the Atlantic, that there was "a hump of water more than a hundred miles high" blocking the way, and that the only way to do it would be to use antennas two hundred miles tall?

Antennas as Launchers and Interceptors of Electromagnetic Waves

Modern atomic theory portrays matter as made of various kinds of particles, two of which are the negatively charged electron and the positively charged proton. Matter, in bulk, is normally electrically neutral because there are usually an equal number of protons and electrons present. Protons are found inside the nuclei of atoms, tightly locked into place, while electrons are found in orbits around the nucleus. A simple image of an atom is that of a miniature solar system, with the nucleus as the central sun and the electrons as orbiting planets. It is extremely difficult to "get at" the protons buried inside the nucleus, but the electrons are quite easy to manipulate. The energies required to influence a proton bound to a nucleus are quite literally of the magnitude required to split or fission an atom, while the exterior electrons can be influenced by very much smaller energies. All the reactions we see and experience countless times each day, from the striking of a match to the digestion of the food we eat, are *chemical* reactions that involve only electrons (and even then only the electrons furthest away from the nucleus, those electrons most weakly bound to the atom in the outermost, so-called *valence* orbits).

The solar-system model of the atom is usually called the *Bohr atom*, after the Danish physicist Niels Bohr (1885–1962), who received the 1922 Nobel Prize in physics for his work with the model. With the development of quantum wave mechanics in the 1920s, however, Bohr's image of the atom was discarded as a legitimate representation of reality. Still, as long as one realizes it shouldn't be taken too literally, the Bohr atom can be a helpful image.

The engineering science of controlling the electrons in matter and space, through the use of applied electric and magnetic fields, is called *electronics*. One of the most elementary examples of such control is the manipulation of the electrons in a wire, i.e., in a radio antenna. If electrons in a wire are made to move back and forth along the wire, then those electrons are accelerated; and Maxwell's theory predicts accelerated charges will radiate energy. That is, the antenna will launch electromagnetic waves into space, which will then travel away from the antenna at the speed of light.

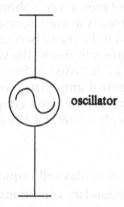

Figure 3.1   A center-driven dipole antenna, so called because the upper and lower halves of the antenna form a *dipole* (two charges of equal magnitude but opposite sign). The flat plates at the ends of the dipole provide what is called *capacitive loading*; by increasing capacitance, increased energy can be stored in the antenna. As a historical note, Hertz used both spheres and plates, but they served the same function.

We can accelerate some of the electrons in a wire (the ones most weakly bound to the atoms of the wire) by connecting the wire, at its center, to a generator of alternating voltage. This will create an oscillating electric field in the wire, which will drive the electrons back and forth along the wire. The wire, in total, is always electrically neutral, with zero net charge, but the generator essentially changes the *distribution* of the charge in the wire. The two halves or poles of the wire alternatively change polarity from plus to minus, as negatively charged electrons surge back and forth along the wire. There are a vast number of ways to build an antenna, but the one shown in Figure 3.1 is perhaps the simplest. It is called a center-driven dipole. It was just such an antenna that Hertz used with his transmitter in his pioneering experiments[1] that verified Maxwell's prediction of the existence of electromagnetic waves. In fact, a *short*, straight antenna is often called, by physicists and electrical engineers, a *Hertzian dipole*.

> We measure distance, in radio, in units of wavelength. If λ and f denote wavelength and frequency, respectively, then $\lambda f = c$ (the speed of light). So, since $c = 3 \times 10^8$ meters/sec in space, the wavelength of Bose's 60,000 MHz microwave system was 5 millimeters. Hertz's waves, by contrast, were at least a hundred times longer. Marconi's 1901 trans-Atlantic receiver was tuned to a wavelength of 366 meters, which is in the modern AM frequency band; at 820 kHz. Later, in his commercial operations, Marconi used wavelengths ten times *longer* and so, of course, frequencies ten times *lower*. "Short" means a fraction of a wavelength, and "long" means at least several wavelengths. Any real antenna can be considered to be many Hertzian dipoles in series (each with a slightly different current to model the varying current along the length of a real antenna, as discussed at the end of this section). The total effect of the real antenna is simply the sum of the individual effects of the individual, short dipoles. See, for example, John D. Kraus, *Electromagnetics* (4th edition), McGraw-Hill 1992 (in particular, pp. 741–764).

The mathematical solution of Maxwell's equations in the space around the dipole, matched to the boundary conditions on the surface of the antenna, is outside the scope of this book. Still, it is possible to form a simple image of how such an antenna works. We start with the discovery in 1820, by the Danish experimentalist Hans Christian Oersted (1777–1851), that electricity can generate magnetism. In particular, Oersted found that the H field generated by a current flowing in a straight length of wire is of the form of closed circular loops around the wire, as shown in Figure 3.2. The discovery of the inverse of Oersted's effect, that magnetism can generate electricity, took another eleven years; it was in 1831 that Faraday made his momentous discovery of electromagnetic induction (a *time-varying* magnetic field can produce an electrical current in a conductor).

Now, what is the *electric* field of a dipole? To understand the structure of the electric field in the space around the antenna, we can use the simple

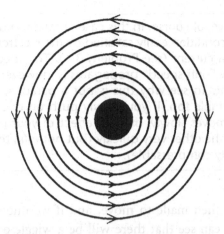

Figure 3.2 Oersted discovered the circular nature of the magnetic field lines around a straight wire carrying a dc current by the simple means of holding a compass near the wire. The solid circle at the center is a cross-section of the conductor, with the current flowing out of the paper toward you.

image of the electric field due to a single electron (which, as far as anyone knows today, is a *point* charge with no detectable spatial extent). As discussed in the previous chapter, the electric field of a stationary (or slowly moving) point charge is radially symmetric, as shown in Figure 3.3. Faraday thought of the field lines as real physical entities, as *continuous* lines of stress or force in space.

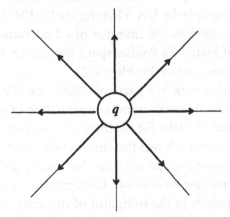

Figure 3.3 The electric field of a stationary positive charge *q*. The field is drawn as radially symmetric in two dimensions, but only because of the limitations of the planar nature of the page. The field is, of course, radially symmetric in all three space dimensions.

> *Slowly* is relative, of course. In this case, it means much less than the speed of light, a condition easily satisfied by the electrons in a wire that form the antenna current (see Problem 3.1). The speed at which the *effect* of this slow "drift speed" propagates is, however, comparable to the speed of light. To see why this is not a paradox, simply imagine a yardstick on a table top. If you slowly push on one end, each part of the yardstick moves slowly, but the *effect* of your push propagates from one end to the other very quickly (but not *infinitely* fast since the yardstick is slightly compressable).

If the charge is then made to move, and if we impose continuity on the lines, then we can see that there will be a wiggle or kink sent along each field line, much as would happen if you snapped one end of a rope that is fixed or tied down at the other end. The idea of a field line kink as the basis for electromagnetic radiation is due to Sir J. J. Thomson (1856–1940), who won the 1906 Nobel Prize in physics for his 1897 discovery of the electron. Thomson presented the kink concept as an explanation for the origin of the X-rays produced by the sudden deceleration of an electron beam upon hitting a metal plate inside a vacuum tube. He publicly discussed this idea during his May 1903 Silliman lectures at Yale University, which were soon after published as the book *Electricity and Matter*, Charles Scribner's Sons 1904. See, in particular, his Chapter III ("Effects Due to Acceleration of the Faraday Tubes")—in the terminology of those times, *tubes* meant the electric field lines. The kink idea eventually showed up in the engineering literature, as in the third edition (1916) of *The Principles of Electric Wave Telegraphy* by J. A. Fleming (1849–1945), a man who will appear later in this book as the inventor of a first vacuum tube; he also was the designer of Marconi's Poldhu spark transmitter used for the 1901 transatlantic transmission (see Problem 4.2).

In Figure 3.4 such a kink in a particular field line of a moved charge is shown. Now, try to hold in your mind a combination of Figure 3.2 (the *H*-field lines) and Figure 3.3 (the *E*-field lines), for the case where the dipole antenna current is upward. Since the current is upward, then *effectively* a positive charge q is moving upward (but the *actual* moving charges, *negative electrons*, are moving *downward*). Concentrating your attention on a point in space broadside to the midpoint of the antenna, you should see that we have a magnetic field vector H pointing *into* the paper and an electric field vector E with a component pointing *downward* (at the kink) in the plane of the paper (parallel to the antenna). Thus, the Poynting vec-

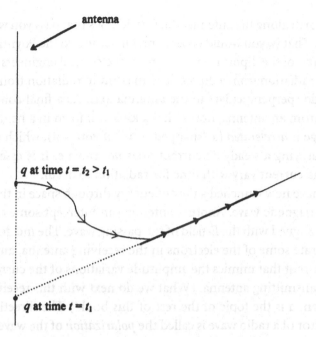

Figure 3.4   Suddenly moving an electric charge causes a kink in a *continuous* electric field line to propagate outward from the charge.

tor $P = E \times H$ gives P pointing *outward*, away from the antenna (in the plane of the paper). When the current in the antenna reverses direction, then the direction of H will also reverse. But so, too, will the direction of the E-field kink, and so the Poynting vector will continue to point outward, away from the antenna. Besides this component of the Poynting vector that always points outward (the *radiation* component), there are also components of the Poynting vector that alternately point inward and outward (the *induction* component), and yet others that *encircle* the antenna. These two *non*radiation components (which are difficult, if not impossible to "see" with our simple kink imagery) can be studied only by a detailed mathematical analysis of Maxwell's equations.

The intensity of the radiation from an antenna depends on where we detect it. As we move further away from the antenna the intensity will decrease, which is probably obvious. Slightly more subtle is the observation that even if we stay the same distance from the antenna, but move off of dead-center broadside (the geometry of Figure 3.4), the intensity will also decrease. As a limiting example of this, consider what you would observe if you were in line with the antenna, i.e., sitting at some distance from one end and on axis with the dipole. Then, as electric charges are accelerated back and forth along the antenna, the E-field lines that pass you will *always be straight*. That is, you would observe no kinks, and so the Poynting vector along the axis of the dipole is zero. There is, as electrical engineers put it, no "end-fire" radiation from a dipole. The maximum radiation from a dipole is broadside (perpendicular) to the antenna axis. As a final comment on radiation from an antenna, notice that a kink will form in a field line only if the charge is *accelerated* (a "snapped rope," if you will), which means an antenna carrying a steady dc current does *not* radiate. It is essential that the antenna current vary with time for radiation to occur.

So, we have now launched a flow of energy through space in the form of an electromagnetic wave. Another antenna can intercept some of this energy if it is aligned with the E-field of the passing wave. The incident E-field will accelerate some of the electrons in the receiving antenna, and thereby create a current that mimics the amplitude variations of the current in the original transmitting antenna. (What we do next with this received signal in the antenna is the topic of the rest of this book!) The direction of the E-field vector of a radio wave is called the *polarization* of the wave. Because the maximum radiation from a dipole is in the broadside direction, commercial AM radio stations always construct their antennas as vertical structures. This sends the station's radiated energy mostly out horizontally over the earth, where all the listeners are. And because the E-field, broadside, is parallel to the antenna, commercial AM radio signals are vertically polarized. For maximum signal interception, then, a receiver's antenna should be vertical, too, and a peek inside an AM radio receiver will reveal either a vertically mounted planar coil of wire or an elongated coil with its axis horizontal—both geometries make it difficult to orient the receiver *without* some part of the antenna having a vertical component.

The frequency at which an antenna efficiently radiates can be related directly to its physical length with the aid of Figure 3.5. There we have a vertical antenna of height h, with a signal source at the base. The inductance and capacitance of the antenna are *distributed* parameters, rather than lumped circuit elements; if we think of the macroscopic antenna as a series connection of Herztian dipoles each of differential length, then the distributed inductance is the inductance of each dipole, and the distributed capacitance is the capacitance of each dipole to ground (shown in dashed lines). Now, as we follow the current up the antenna, it continuously decreases as it "shunts off" through the distributed capacitance to ground (where it then returns to the signal source). This results, at every instant of time, in maximum current at the base and zero current at the

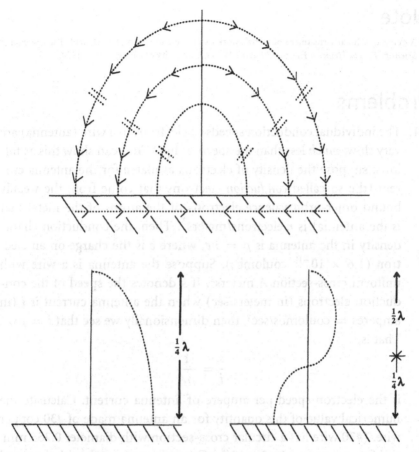

Figure 3.5 The current distribution along a grounded monopole antenna, with distributed capacitance to ground, such that the current at the top is always zero. On the left is the case $h = \frac{1}{4}\lambda$, and on the right is the case for $h = \frac{3}{4}\lambda$.

top (if it wasn't zero, where would it go?!) If λ is the wavelength of this current variation (which is, in fact, nearly sinusoidal), then these two *electrical boundary* conditions are satisfied if $h = \frac{1}{4}\lambda$, and the antenna is often called a *grounded quarter-wave monopole* (*not* a dipole, as one of the dipole "arms'" has been replaced by the earth). For AM broadcast radio, this physical condition results in a tall antenna (at 1 MHz, $\lambda = 300$ meters and a $\frac{1}{4}\lambda$ antenna is over 245 ft high). This isn't the only possibility, of course, only the one that gives the *minimum* h for a given λ. Can you see that the next possibility (with maximum current at the base and zero at the top) is $h = \frac{3}{4}\lambda$?

Note

1. A detailed, scholarly treatment of Hertz's work can be found in Jed Z. Buchwald, *The Creation of Scientific Effects: Heinrich Hertz and Electric Waves*, University of Chicago Press 1994.

Problems

3.1. The individual conduction speeds of electrons in a wire (antenna) are very slow, *much* less than the speed of light. You can show this as follows: suppose the density of electrons available for the antenna current (the so-called *conduction* electrons that come from the weakly bound outer-orbit valence electrons of the atoms of the metal that *is* the antenna) is n electrons/meters^3. Then, the conduction charge density in the antenna is $\rho = ne$, where e is the charge on an electron (1.6×10^{-19} coulombs). Suppose the antenna is a wire with uniform cross-section A meters^2. If s denotes the speed of the conduction electrons (in meters/sec) when the antenna current is i (in amperes = coulombs/sec), then dimensionally we see that $i = \rho s A$. That is,

$$\frac{s}{i} = \frac{1}{\rho A}$$

is the electron speed per ampere of antenna current. Calculate the numerical value of this quantity for an antenna made of #30 copper wire. (#30 wire has a circular cross-section with diameter 0.255 mm, and for copper $n = 8.43 \times 10^{28}$.) Compare your result with the speed of light ($c = 3 \times 10^8$ meters/sec, in a vacuum) for $i = 0.1$ ampere and $i = 1000$ amperes. (Don't worry about the "practical" problem of running a thousand amperes through #30 wire!) Can you see why the conduction electrons are said to "drift" in a wire even for extremely large currents?

3.2. In the previous chapter the statement was made that magnetic effects are *relativistic* effects produced by moving charges. But how, you may be wondering, can that be (if you did the last problem correctly you now know that the conduction electrons are, even for *very* large currents, hardly moving at all!)? As you may recall from freshman physics, special relativity generally introduces a correction factor of the form $\sqrt{1 - (\frac{v}{c})^2}$ to the answers given by Newtonian physics. For $v \ll c$, this correction factor is essentially equal to $1 - \frac{1}{2}(\frac{v}{c})^2$. For $v = 30$ meters/sec., to pick a value that represents an *enormous* current with hard-to-overlook magnetic effects, this correction factor is 0.999999999999995. This seems awfully close to one, and so how can such a "correction" factor explain electromagnetic cranes so powerful that, when energized, they can pick up hunks of iron weighing *tons*?

3.3. The concept of energy flowing through apparently empty space is really quite an astonishing claim. We can write equations all day long showing mathematically how it goes, but if you try to form a physical image of such a flow I think you will fail (at least *I* fail!) Even with a totally mechanical system, the flow of energy is sufficiently abstract that it bothered a genius like Hertz. In 1891 he presented the simple system shown in Figure 3.6: a paddlewheel turning (via a belt) a distant electrical generator connected to a light bulb. It seems clear that energy is transported *from* the paddlewheel *to* the light bulb—after all, unless the paddlewheel is turned by a flow of water the light bulb does not glow, but as soon as we let water flow over the paddlewheel the light bulb shines brightly. Somehow, energy goes from the paddlewheel to the light bulb. But *how*? Notice the tensed part of the belt

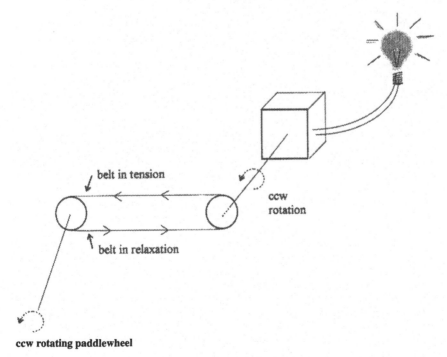

Figure 3.6   Hertz's puzzle of the flow of energy in a mechanical system.

is moving from the light bulb to the paddlewheel, the direction *opposite* to the supposed flow of energy! What do *you* think is going on? For a good discussion of Hertz's puzzle, see Jed Z. Buchwald's *From Maxwell to Microphysics*, University of Chicago Press 1985, pp. 41–43.

Early Radio

To see what G. F. FitzGerald had in mind with his 1883 suggestion to generate Maxwellian waves by letting a charged capacitor discharge through an inductive circuit, consider Figure 4.1. The circuit shown there has a charged capacitor in series with an inductor and a resistor. The total instantaneous energy in the circuit is the energy stored in the magnetic field of the inductor and in the electric field of the capacitor. The resistor, of course, stores no energy—a resistor only *dissipates* energy by getting hot. Thus, if we use E to denote the total stored energy at any time t, then with $i(t)$ as the circuit current,

$$E = \frac{1}{2}Li^2 + \frac{1}{2}Cv_c^2,$$

where v_c is the voltage drop across the capacitor. Differentiating E gives

$$\frac{dE}{dt} = Li\frac{di}{dt} + Cv_c\frac{dv_c}{dt} = Li\frac{di}{dt} + Cv_c\frac{i}{C}$$

or,

$$\frac{dE}{dt} = iL\frac{di}{dt} + iv_c = i\left[L\frac{di}{dt} + v_c\right] = i[v_L + v_c],$$

where v_L is the voltage drop across the inductor. But, from Kirchhoff's voltage law, $v_c + v_L + v_R = 0$ (where v_R is the voltage drop across the resistor) and so we can write

$$\frac{dE}{dt} = i[-v_R] = i[-iR] = -i^2R \leq 0$$

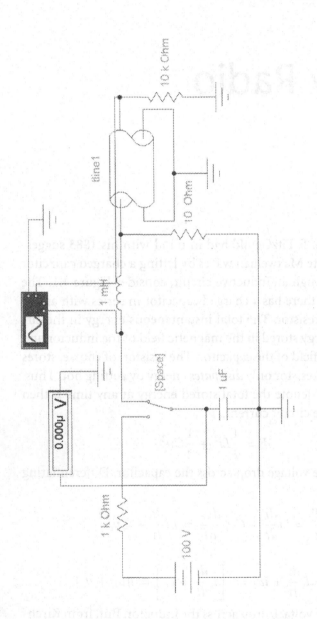

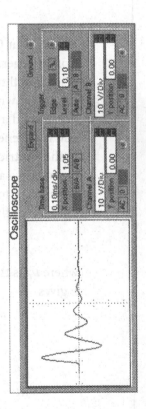

Figure 4.1 This circuit shows the decaying oscillations produced by a charged capacitor discharging through a series inductor and resistor. Initially, the switch is in the left position, which allows the battery to charge the capacitor (the voltmeter lets you watch the charging of the capacitor up to 100 volts). Next, the switch is flipped to the right (by hitting the "space" key). The resulting decaying oscillations of the discharge current are displayed on channel A of the oscilloscope via the voltage drop across the 10-ohm resistor to ground. Notice that the 'scope's horizontal sweep is externally triggered by the switch being flipped to the right, i.e., the sweep starts when the external signal passes through 0.1 vertical divisions (which is 1 volt) in a positive-going direction. That would mean the 'scope would "miss" the very first part of the discharge current, but this can be avoided by running that signal through a perfect transmission delay line (which has been adjusted to have a delay of 0.1 millisec). Notice that to avoid reflection echos, the line must be terminated in its characteristic impedance (which is 10,000 ohms for this demonstration, sufficiently large to avoid loading the 10-ohm resistor).

as $i^2 \geq 0$ and $R \geq 0$ for *the case where* R *is a passive circuit element.* For the case of $R = 0$ the total circuit energy is constant, but if $R > 0$ then the total circuit energy will steadily decrease with time. That is, the presence of positive circuit resistance is the physical mechanism by which this circuit *loses energy* (the "i^2R" term is called the ohmic power loss in the resistor and it appears as heat).

The case of $R < 0$ might seem to be a physically nonsensical situation, perhaps of interest to mathematicians but not to "practical" electrical engineers. In fact, it is the $R < 0$ case that is of *primary* interest to radio engineers, and it was first experimentally observed in the nineteenth century in the behavior of electric arcs. Later, with the invention of the triode vacuum tube and the transistor, it is again the $R < 0$ case that allows the electronic generation of constant amplitude radio frequency oscillations. We'll take up these issues in the next several chapters.

In the business world what counts, in the end, is the *patent*. And the fact is that Marconi, not Fitzgerald or Lodge or Tesla or anybody else, received the world's first radio patent; British Patent No. 12039, "Improvements in Transmitting Electrical Impulses and Signals, and in Apparatus Thereof," applied for on June 2, 1896 and received July 2, 1897. In his application he boldly declared "I believe that I am the first to discover and use any practical means for effective telegraphic transmission and intelligible reception of signals produced by artificially-formed Hertz oscillations." When you get to Figure 5.1, you'll see this was not true; Marconi's patent was based on no new principles. Every bit of it used ideas and devices developed by others who had not had Marconi's business sense to claim legal rights. This patent, however, did not involve a tuned, resonant system even though the fundamental experiments on such a "syntonic" (as resonance was then called) system had been demonstrated two years earlier, by Oliver Lodge, before the Royal Institution just five months after the death of Hertz. It was Lodge's great mistake not to patent the resonance circuitry before publishing (which, under British law, blocks patentability), but he did later apply for a U.S. patent in 1897 (No. 674,846, received May 21, 1901). Again, however, Marconi was first, receiving on April 26, 1900 British Patent No. 7777 (now famous as the "four sevens" patent) for a *tuned* radio system. In 1943 the U.S. Supreme Court ruled that Marconi had of course been anticipated by Lodge and others, but of course at that late date it hardly mattered *what* the Court decided.[1] Certainly not to either Lodge or Marconi, who were by then both dead.

I have been able to say something quite specific about the behavior of this circuit without working out the details of $i(t)$, but now let's look a bit more closely at what is really happening. If we write Kirchhoff's voltage law out in detail, then (if V_0 is the initial voltage drop across the C at time $t = 0$)

$$-V_0 + \frac{1}{C} \int_0^t i(u)\, du + L\frac{di}{dt} + iR = 0$$

or, after differentiation (see Appendix F),

$$\frac{d^2 i}{dt^2} + \frac{R}{L}\frac{di}{dt} + \frac{1}{LC}i = 0.$$

We can solve this second-order differential equation with the trick of *assuming* $i(t)$ is of the form Ie^{st}, where I and s are some (perhaps complex) constants. The justification for this trick is that it works! In fact, since this trick works in so many other commonly encountered differential equations, it is more usually called a *method*. (See Appendix C for more about this, if necessary.) Then, substituting back into the equation and cancelling the Ie^{st} factor which appears in each term, we arrive at

$$s = \frac{1}{2}\left(-\frac{R}{L} \pm \sqrt{\left(\frac{R}{L}\right)^2 - \frac{4}{LC}} \right).$$

Notice that if $R = 0$, then there are two pure imaginary values for s,

$$s = \pm j\sqrt{\frac{1}{LC}} = \pm j\omega_o, \quad \omega_o = \sqrt{\frac{1}{LC}}$$

where of course $j = \sqrt{-1}$. Thus, the most general solution for $i(t)$, for $R = 0$, is

$$i(t) = I_1 e^{j\omega_o t} + I_2 e^{-j\omega_o t}.$$

Since the circuit current is zero just before we close the switch in Figure 4.1, then the circuit current is zero just after we close the switch (because the current through the inductor cannot change instantaneously). Thus, $i(0+) = 0 = I_1 + I_2$, and so $I_1 = -I_2$.

Dropping subscripts, then, we can write (with the aid of Euler's identity)

$$i(t) = I\left(e^{j\omega_o t} - e^{-j\omega_o t}\right) = 2jI\sin(\omega_o t).$$

We can determine the value of I by again using the observation that $i(0+) = 0$. This means there is no voltage drop across R at $t = 0+$, and so at that instant the initial capacitor voltage must appear across the L. This may seem a silly statement because we are assuming $R = 0$ and so *of course* there is no voltage drop across R. However, in just a few more sentences we will have $R > 0$ and it will be important then to understand that the voltage drop across R will *still* be zero at $t = 0+$ [because, as argued, $i(0+) = 0$]. Thus,

$$L\frac{di}{dt}\bigg|_{t=0+} = V_o$$

or,

$$\frac{di}{dt}\bigg|_{t=0+} = \frac{V_o}{L}.$$

Since

$$\frac{di}{dt} = 2jI\omega_o \cos(\omega_o t),$$

then substitution for $t = 0+$ gives

$$j2\omega_o I = \frac{V_o}{L}$$

or,

$$I = \frac{V_o}{j2\omega_o L}.$$

And so, finally, we have for the special case of $R = 0$,

$$i(t) = \frac{V_o}{\omega_o L} \sin(\omega_o t), t \geq 0,$$

where

$$\omega_o = \frac{1}{\sqrt{LC}}.$$

A subtle, easy-to-miss, but extremely important aspect of the analysis just done for $i(t)$ is that the initial conditions $i(0)$ and $i'(0)$ play crucial roles in determining the steady-state (i.e., $t \gg 0$) behavior of $i(t)$. This is because of the linearity of the mathematics. But nothing in real electrical circuits is truly linear, and if currents and voltages get large enough then nonlinearities become too important to ignore. Such concerns are generally quite difficult to handle as they involve nonlinear differential equations, which is the historical reason for the emphasis on *linear* circuits in sophomore and junior level courses. With the aid of modern computers, however, we can easily squeeze a lot of information out of such differential equations. In Chapter 7, for example, you'll see that the early negative resistance arc and vacuum tube radio transmitters quickly led electrical engineers to the famous nonlinear *Van der Pol equation,* and with the aid of MATLAB you will also see that its solutions are nonsinusoidal oscillations with a steady-state amplitude that depends only on the circuit component values and is *independent* of initial conditions.

Thus, the circuit current oscillates with frequency $f_o = \omega_o / 2\pi$ Hz (remember, ω_o itself has units of radians/sec). In Figure 4.2 the oscillating magnetic field around the L is shown linked, via a second inductance immersed in this field, to an antenna. The oscillating magnetic field is said to *couple* the two inductors because the field, created by an oscillating current in one inductor, induces a new oscillating current in the other inductor via Faraday's law of electromagnetic induction, mentioned in the previous chapter. (There is much more on magnetic coupling in Appendix C.) This, then, would seem to be a way to achieve either the grounded quarter-wave monopole (as shown) or the center-driven dipole antennas discussed in the previous chapter. It *almost* is, but alas, it also has a crucial fault.

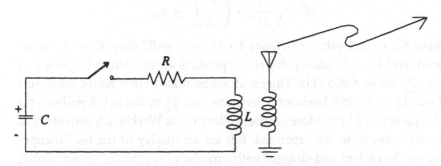

Figure 4.2   Magnetically coupling an oscillating circuit to an antenna.

If the circuit in Figure 4.2 would, in fact, work, then by definition it would lose energy (in the form of a radiated signal), and so the lossless energy condition $R = 0$ is *a priori* violated! In fact, the *better* the antenna radiates energy the *more* the $R = 0$ assumption is violated. The desired result of radiated energy is not an ohmic, heat-loss mechanism of a physically present resistance, but the net effect is *as if* a positive R is in the circuit. So, we *must* use the condition $R > 0$ for a realistic derivation of $i(t)$ in the circuit of Figure 4.1. Backing up, then, to just before we set $R = 0$, we return to the expression for s,

$$s = \frac{1}{2}\left[-\frac{R}{L} \pm \sqrt{\left(\frac{R}{L}\right)^2 - \frac{4}{LC}}\right].$$

From this we see that if $(\frac{R}{L})^2 > \frac{4}{LC}$, then both values of s are purely real (and negative), and so there will be no oscillations in $i(t)$. So, we assume instead that the condition $(\frac{R}{L})^2 < \frac{4}{LC}$ is satisfied. Then we can write

$$s = \frac{1}{2}\left(-\frac{R}{L} \pm j\sqrt{\frac{4}{LC} - \left(\frac{R}{L}\right)^2}\right) = \sigma + j\omega'$$

where $\sigma < 0$ and $\omega' > 0$. Then, by repeating the arguments of all the previous steps done for the $R = 0$ case, you can show that (assuming, don't forget, that $R < 2\sqrt{\frac{L}{C}}$)

$$i(t) = \frac{V_o}{\omega' L}e^{-\frac{R}{2L}t}\sin(\omega't), \; t \geq 0,$$

where

$$\omega' = \sqrt{\frac{1}{LC} - \left(\frac{R}{2L}\right)^2} \leq \omega_o.$$

Using the circuit values of Figure 4.1 ($L = 1$ millihenry, $C = 1$ micro-farad, and $R = 10$ ohms) this result predicts an oscillation frequency of $f = \frac{\omega'}{2\pi}$ Hz $= 4.905$ kHz. The oscilloscope trace in that figure shows just about $2\frac{1}{2}$ cycles in 5 horizontal divisions, i.e., $2\frac{1}{2}$ cycles in 0.5 millisec, for a frequency of 5 kHz. More precisely, Electronics Workbench provides two timing cursors (in the expanded, full-screen display of the oscilloscope) that can be clicked-and-dragged with a mouse to any part of the waveform. When I did this, marking consecutive peaks of the trace, I found the period to be 205 microsec. That corresponds to a frequency of 4.878 kHz, an error of just 27 Hz (about half of 1%).

The presence of an $R > 0$, therefore, has *two* effects. First, $i(t)$ is exponentially decaying (or *amplitude damped*) and, second, the oscillation frequency is lower than that for the $R = 0$ case (this effect is often called *frequency damping*). If we magnetically couple the oscillations of $i(t)$ to an antenna (see Figure 4.2 again), then R corresponds to the combined effects of the actual circuit resistance and the effective *additional* resistance due to the energy lost by radiation. If we can arrange to periodically inject new energy into the circuit, then we have a means for radiating a sequence of exponentially damped electromagnetic waves. And in the early days of what is called *spark radio* this is, in fact, precisely what was done.

An idealized spark radio transmitter circuit is shown in Figure 4.3. Depressing the telegraph key produces a sudden spark and discharge current in the primary of a so-called oscillation transformer (see Appendix C for a review of transformers). This transformer couples the energy of the spark transient into the antenna, which is a series resonant LC circuit (see Appendix D for a review of resonance). The two horizontal arms at the bottom of Figure 4.3 terminate in plates that form an "unfolded" C (the arms

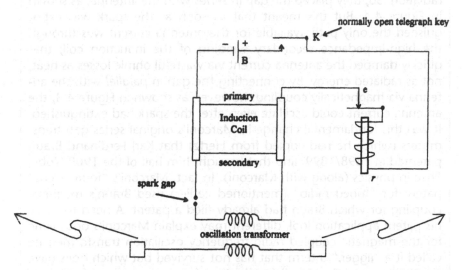

Figure 4.3   A telegraph-keyed spark-gap radio transmitter for sending Morse code signals. The low-voltage source B, in series with depressed telegraph key K, provides dc current to the primary winding of the induction coil through the winding of an electromagnetic relay r in series with a normally closed contact e. This causes r to *break* the circuit, thus interrupting the current in the primary and so inducing a much larger voltage pulse in the secondary winding which appears as a spark (i.e., as a sudden dumping of energy into the oscillation transformer). After the primary current is interrupted, then r will close and thus allow e to again complete the primary circuit. This entire process then repeats, *for as long as the key is closed*, at a rate determined by the *mechanical* parameters of the relay, e.g., the physical spacing of contacts, the spring constants, and the inertia of each moving part. It is important to realize that the frequency of this mechanical vibration is *not* the radiated signal frequency. The function of the mechanical system is simply to rapidly and periodically inject energy into the antenna circuit, which then oscillates at *its* resonant frequency. By depressing K for long or short intervals of time, the transmitter radiates signals in Morse code (dashes and dots). It is amusing to note that the primary circuit, alone, is the basis for the common *door buzzer* (the noise of contact e opening and closing is the "buzz"). As a final note, the early radio pioneers usually referred to the induction coil as a *Ruhmkorff coil*, invented in 1855 by the German Heinrich Ruhmkorff (1803–1877).

themselves also serve as the dipole antenna). The intrinsic inductance of the antenna wires and the inductance of the oscillation transformer's secondary form the L. The effect of the transient spark on the antenna circuit is much like that of hitting a bell with a hammer—it vibrates (oscillates) at its resonant frequency.

Many of the early radio pioneers, Marconi included, failed to understand that the spark gap was simply the means by which to quickly inject energy into a resonant LC circuit. They believed that the continual presence of the spark, itself, was essential to generation of antenna radiation. So, they placed the gap in *series* with the antenna, as shown in Figure 4.4. But this meant that as soon as the spark was extinguished the only path available for the antenna current was through the high-impedance secondary winding of the induction coil; that quickly damped the antenna current via wasteful ohmic losses as heat, not as radiated energy. By connecting the gap in *parallel* with the antenna via magnetically coupling, however, as shown in Figure 4.3, the antenna current could oscillate even after the spark had extinguished. It was this fundamental change in Marconi's original series gap transmitters (which he had copied from Hertz) that Karl Ferdinand Braun patented in 1898/1899, and that brought him half of the 1909 Nobel Prize in physics (along with Marconi). In fact, Marconi's "four sevens" patent for "tuned radio" (mentioned earlier) used Braun's magnetic coupling for which Braun had already filed a patent. A need to make his patent application look different may explain Marconi's odd name for the magnetic coupled radio frequency oscillation transformer; he called it a "jigger," a term that has not survived but which does have a picturesque origin. Visualizing the transformer as a means for "pouring" energy from part of a circuit to another, Marconi adopted the bartender's name for measuring the volume of $1\frac{1}{2}$ ounces of alcohol! Braun's modification was a great safety advance because the transformer action of the magnetic coupling isolated the exposed antenna from direct connection to the very high—and very dangerous—spark-gap voltages. It was with a Braun-type transmitter that Marconi claimed to have made the historic 1901 trans-Atlantic transmission from Poldhu to St. John's.

While amplitude-damped spark-gap radio waves are suitable for transmitting keyed on–off Morse coded-signals, such waves are *not* suitable for voice (or music) transmission. The reason for this can be understood once it is realized that for information to be transmitted it is necessary for something to *change*. If nothing changes, then the receiver of the signals can predict with perfect accuracy what the future nature of the signals will be; exactly what they *were*! It is unnecessary, actually, to transmit anything if

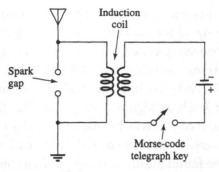

Figure 4.4   This sketch, based on a slide used by Marconi in a lecture to the Royal Society of Arts titled "Syntonic Wireless Telegraphy" in May 1901, clearly shows the spark gap in series with the antenna. Notice also, that this transmitter as drawn *could not send Morse code!* Unlike the Braun-type spark transmitter of Figure 4.3, which radiates as long as the key is depressed, the above transmitter sends only a brief pulse with the closing of the key, and a second brief pulse when the key is opened (those are the only times at which the current is changing in the primary winding of the induction coil). *Holding* the key down for varying time intervals, to try to send dots and dashes, only results in a *steady* dc current in the coil primary which gives a *zero* induced voltage across the spark gap. What is missing is the self-interrupting "buzzer" circuit in the primary. The astonishing conclusion is, then, that Marconi seems either to have been rather careless in preparing his slide or, *perhaps*, not to have really understood[2] how a spark transmitter actually works!

nothing changes. With a radio wave, there are only a few parameters available for transmitting change—most obvious is the amplitude (frequency is another, but that is another book). Before you are through with *this* book, you will see how we can insert low-frequency voice signals onto another higher frequency signal (called the *carrier*) by varying the amplitude of the carrier (a process called *amplitude modulation*) to mimic the amplitude variations of the voice signal. But an amplitude-damped spark-gap signal is inherently amplitude modulated even before we attempt to insert voice signal amplitude variations, and so the two amplitude modulations become inseparably mixed. There is no way (or at least no *easy* way) a receiver could distinguish between the inherent spark amplitude variations, and the voice signal variations.

Acceptable voice transmission by amplitude modulation requires a source of constant amplitude oscillations, which spark gaps are simply unable to generate. It is true that in December 1900 the Canadian-born American electrical engineer Reginald Fessenden (1866–1932) succeeded in sending voice with a spark transmitter, but the quality of the received signal was far too poor to be acceptable for commercial uses. It was that

failure that caused Fessenden to appreciate the importance of using a
constant amplitude signal for the carrier, and he will appear in the next
section with his astonishing concept for *mechanically* generating such a
carrier. Spark gap transmitters continued to be used as late as the beginning of the 1920s, and were implemented as truly impressive gadgets
that reached power levels as high as 300,000 W. But the future of AM
voice transmission and commercial broadcast radio lay elsewhere. In 1923
spark transmitters were legally banned, and by the end of the 1920s spark
transmitters could be found mostly in scrap piles, although even as late as
the Second World War they were kept as emergency backups on ships at
sea. Still, while now gone in form, spark has left a lasting mark; it is not
uncommon in war novels and war movies, and even in real-life military
units, to learn that the radio operator is nicknamed "Sparks" or "Sparky."

> I don't often ask students to read a science fiction story as part of
> their electrical engineering studies, but the 1952 story "Sail On! Sail
> On!" by Philip José Farmer is the exception. In the parallel world of Friar
> Sparks, the radio operator on *his* world's version of the *Santa Maria*(!),
> we find that technology developed in a way quite different from our
> world. The laws of nature in that parallel world are different, too; we
> get an early hint of this when we learn the experiments at the Leaning
> Tower of Pisa, by Angelo Angelei (*not* Galileo Galilei), proved that dif-
> ferent weights fall at different speeds. It is *most* interesting, as well, to
> read Friar Spark's explanation of how spark radio works (in *his* world).
> You can find this wonderful tale reprinted in volume 14 of the series
> *Isaac Asimov Presents the Great SF Stories*, DAW Books 1986.

Before giving up on spark radio Fessenden attempted to overcome the
inherent limitations of the damped waves produced by sparks with the elementary approach of simply increasing the sparking rate to as high as
20,000 sparks per second (*how* he did this is mentioned in Chapter 7).
There *is* a certain logic to this, as increasing the spark rate periodically
"shocks" the resonate antenna circuit with fewer intervening cycles of decaying amplitude oscillations. But a simple calculation shows that even
that prodigious sparking rate is far too low. If the antenna circuit resonates
at 500 kHz, for example, there would still be twenty-five radio frequency
cycles between consecutive sparks and, as will be discussed in Chapter 11,
the amplitude of the twenty-fifth cycle could easily be vanishingly small
compared to the amplitude of the first cycle. In an elaboration to his lecture titled "The Work of Hertz" given to the Royal Institution on June 1,
1894, Oliver Lodge put it this way: "even if sparks were made to succeed

one another at the rate of 1000 per second, the effect of each would have died out long before the next one came. It would be something like plucking a wooden spring, which, after making three or four vibrations should come to rest in about two seconds; and repeating the operation of plucking regularly once every two days."

Despite its inability to support acceptable voice transmissions, spark telegraphy was excellent for Morse Code signaling and it didn't take long for the military to appreciate the potential of spark radio in warfare. Indeed, plans to use the 'new wireless' were made in South Africa before the start of the bloody Anglo-Boer War of 1899–1902. Using Marconi's spark-gap equipment contracted through the London office of Siemens Bros. & Co. and manufactured in Germany, the fortification of Pretoria was to be enhanced by radio-telegraphic communications that would be immune to being either cut or wire-tapped as could land cables. The equipment *was* shipped, but was never used by the Boers as it was seized by custom authorities upon its arrival in Durban. The British cannibalized some of it for their own use (the British War Office had also bought Marconi equipment, directly from Marconi's company), and after the war the rest was sold at auction in Cape Town. For more on this fascinating, little-known episode of early radio history, see Duncan C. Baker and Brian A. Austin, "Wireless Telegraphy Circa 1898–99: The Untold South Africa Story," *IEEE Antennas and Propagation Magazine* 37, December 1995, pp. 48–58. The first military conflict to actually use wireless was the Russian–Japanese War of 1904–5.

Notes

1. A very interesting discussion of just what the court decided is in the essay by A. David Wunsch, "Misreading the Supreme Court: A Puzzling Chapter in the History of Radio," *Antenna* (Department of History, University of Colorado at Denver) 11, November 1998, pp. 8–9. In particular, Wunsch writes "The Supreme Court never determined that Tesla invented radio."

2. I have taken this illustration from one that appears in R. W. Simons, "Guglielmo Marconi and Early Systems of Wireless Communication," *GEC Review* 11 (No. 1), 1996, pp. 37–55. The author, a longtime employee of the Marconi Company, makes no mention of the severe technical deficiencies in Marconi's transmitter circuit.

Problems

4.1. Here's a pretty little problem that has become a classic in electrical engineering. Suppose two capacitors of equal value are charged to V_1 and V_2 volts, as shown in Figure 4.5, where $V_1 > V_2$. That is, there is more electric charge on the left C than on the right C. If the switch is then closed, the charges in the system will redistribute so as to make the two capacitor voltages equal. Since electric charge is conserved (this is a fundamental conservation law, on equal footing with conservation of energy), this can be used to calculate the final (equal) capacitor voltages.

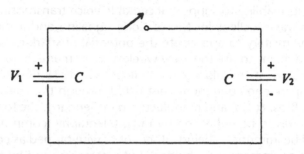

Figure 4.5 A classic puzzler!

(a) Calculate the initial and final stored energies in the system, W_i and W_f, respectively. Use your results to show $W_i > W_f$, i.e., that there is more energy in the system before the switch is closed than there is after the switch is closed.

(b) Since we believe in the conservation of energy as well as that of electric charge, however, there seems to be a puzzle here. Where is the missing energy? In older textbooks it was sometimes claimed the missing energy was lost by radiation, i.e., it was claimed in effect that suddenly switching two charged capacitors together was a way to build a radio transmitter. Show this is incorrect by analyzing a more realistic circuit that contains some (any) series resistance, R. Calculate the total heat energy dissipated (W_d) by R over the time interval zero to infinity (your answer should be *independent* of R!) Finally, show that $W_f + W_d = W_i$, which means there actually is no "missing" energy because every real circuit has *some* resistance. Alas, building a radio transmitter isn't *that* easy! Some final comments: After the appearance of the first edition of *The Science of Radio* I received more correspondence about this problem than I did on any other part of the book. Indeed, this problem periodically resurfaces in the electrical engineering and physics literature, to start anew the debate on what is "really" going on. What is going on, of course, is that the original problem is what mathematicians call *ill defined*. For example, as discussed in Appendix C, the voltage drop across a capacitor cannot change instantly. And so, at the instant the switch closes, the circuit of Figure 4.5 demands that the voltage drop across both capacitors be V_1 *and* V_2 (which makes no physical sense). Including a nonzero R "separates" the two capacitors and eliminates that problem. If you want to read more advanced discussions of this circuit, then look up "Ideal Capacitor Circuits and Energy Conservation" by K. Mita and M. Boufaida in *American Journal of Physics*, August 1999, pp. 737–739, "The Two-Capacitor Problem Reconsidered" by R. P. Mayer, J. R. Jeffries, and G. F. Paulik in *IEEE Transactions on Education*, August 1993, pp. 307–309, and "Two-Capacitor Problem: A More Realistic View" by R. A. Powell in *American Journal of Physics*, May 1979, pp. 460–462.

4.2. In July 1900 the Board of Directors of the Marconi Wireless Telegraph and Signal Company approved Marconi's plan to attempt to send radio signals across the Atlantic. This project, called the "Big Thing" by Marconi, faced some daunting theoretical challenges. One was the sheer radio frequency power required. At that time Marconi's most powerful induction coil spark transmitter produced between 100 and 200 watts of radiated power, and had a range of about 200 miles.

(a) Using the inverse-square law (as did Marconi's own technical experts), estimate the power required to send a signal over 2000 miles of ocean from Poldhu to St. John's.

(b) Marconi's original conception of the transatlantic transmitter was to use a capacitor of 0.02 μfd charged to a voltage sufficient to create a 2-inch spark. If it takes 30,000 to 40,000 volts to form a 1-centimeter spark in air, then estimate the energy stored in the capacitor just before the spark jumped its gap. (Note: you should get a pretty big voltage, and capacitors that function at such voltages are not "off-the-shelf" devices. For the "Big Thing" transmitter, its designer J. A. Fleming used 12 sheets of glass, 16 × 16 inches, separated by zinc plates, all submerged in linseed oil in a wooden box. The resulting capacitance of 0.033 μfd was 65% larger than Marconi's original value so, for the same energy, the capacitor voltage is reduced.)

(c) Using the results of parts (a) and (b), and assuming each spark completely discharges the capacitor, estimate the number of sparks per second required. What do you think a transmitter creating 2-inch sparks at this rate would sound like? What do you think it would smell like? (An electrical discharge through air creates ozone—the odor commonly experienced immediately after an electrical storm—a gas poisonous in large concentrations.) Would there be any advantages to operating the arc gap in a vacuum chamber? Any disadvantages?

5

Receiving Spark Transmitter Signals

A simple receiver circuit capable of detecting the damped sinusoidal Morse code signals from a spark-gap transmitter is shown in Figure 5.1. (I won't actually tell you how this receiver works until a bit later in this chapter.) The only component in that circuit that needs special explanation is the diode, a device that conducts electricity in one direction only. A nearly ideal diode is either a low-resistance path when conducting ("on") or a high-resistance path when not conducting ("off"). The ratio of the "off" resistance to the "on" resistance for a real diode is typically 10,000 or more. The simplest possible "model" for an ideal diode is that when "on" it is a short circuit, and that when "off" it is an open circuit. The arrow in the diode symbol points in the direction of conduction, while the vertical line is a metaphor for "hitting a wall" in the reverse direction.

The on and off states of the diode are determined only by the voltages at its terminals: when the arrow end is more positive than the other end (forward bias) the diode is on, otherwise (reverse bias) it is off. Today we understand the physics of such devices quite well, either in the form of a vacuum tube or as a solid-state junction. In the early days of radio, however, it was only known that certain crystals (e.g., lead sulphide or *galena*, and silicon carbide or *carborundum*) had this peculiar property (e.g., Bose used a galena detector in his 60 GHz microwave system), but not *why* they possessed it.

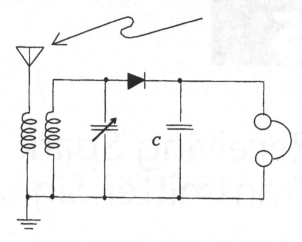

Figure 5.1   A simple crystal radio receiver circuit.

Indeed, even before the use of crystals in radio circuits, an even more mysterious gadget had been used to detect the Morse code signals of spark transmitters. First reported in the French literature in 1890, it was called a "radio conductor" by its inventor Edouard Branly (1844–1940). It commonly took the form of a glass tube filled with iron and nickel filings, with a connection terminal at each end. When there was a voltage drop across a coherer's terminals (either dc or ac) that exceeded a threshold value (on the order of at least a few tenths of a volt) the ordinarily loosely packed filings offering an initially high resistance path suddenly clumped together to produce an equally sudden drop in resistance.[1] The passage of a radio wave over an antenna connected to the terminals of a coherer was sufficient to cause clumping. The change in resistance could be dramatic, e.g., 80,000 ohms to 10 ohms. It was Oliver Lodge who, impressed by the clumping action, introduced the name "coherer" in 1894. Much to Branly's displeasure, it was Lodge's terminology that was adopted by radio engineers and not his "radio conductor."

It is quite important to understand that Branly's coherer was not a "one-way only" diode. Whatever signal initiated clumping, ac or dc, passed through the coherer which was connected in series with the antenna (and perhaps a battery, too, with a voltage not quite sufficient to induce clumping by itself, so that even a very tiny additional ac antenna signal would be enough to cross the clumping threshold). The sudden drop in coherer resistance could then be used to activate a parallel-connected relay circuit that recorded the clumping event, i.e., recorded the reception of a signal in the antenna. When the relay closed, its contacts could complete yet another separately powered circuit that activated a bell or an automatic ink pen-and-paper recorder (as shown in Figure 5.2).

A big disadvantage of the coherer was the need to apply a mechanical shock to "uncoher" the device, to reset it, so to speak, so that it could receive the next incoming antenna signal. The coherer was fundamentally a bistable, self-latching device and, while useful for receiving *slow*-speed Morse code, it was useless as a detector of information stored as a smoothly varying amplitude, e.g., a human voice. The search was quickly on, there-

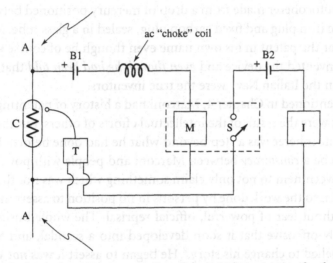

Figure 5.2   In September 1894, at a meeting of the British Association for the Advancement of Science, Oliver Lodge demonstrated spark radiotelegraphy nearly two years before Marconi applied for his first patent. Lodge's receiver circuit, which used a mechanically restored Branly coherer, works as follows. When there is no signal in the antenna (A) the coherer (C) presents a high resistance and so battery B1 cannot force enough current through electromagnet M to close the switch contact S. When a signal is present in A, however, the coherer switches to a low resistance and the current from B1 does energize M and so closes S. That allows battery B2 to force current through both I (an ink pen that makes a mark on a paper strip moving slowly and continuously via a clock mechanism) *and* through the coherer tapper T that hits C and so mechanically "decohers" C. C thus switches back to a high resistance, the electromagnet de-energizes and S opens. If the signal is still present in A then the above sequence repeats and generates another ink mark on the paper. This process continues until the signal in A ceases. A long duration dash will thus leave more marks than will a short duration dot. The mechanical inertia of this electromechanical feedback receiver obviously imposes an upper limit on the speed with which signals can be recorded. Notice, carefully, the presence of the so-called ac choke coil to keep the ac antenna signal from bypassing the coherer through the dc electromagnet path (although the magnet coil often served that purpose by itself).

fore, for something better; at the least for a coherer that would automatically and rapidly decoher (what became called, somewhat oddly, an "autocoherer").

Indeed, it was with an autocoherer that Marconi at first claimed to have heard the transatlantic signals in December 1901, using a design that Marconi himself held the patent on (British Patent No. 18,105 "Improvements in Coherers or Detectors for Electrical Waves," applied for on September 10, 1901). It was revealed soon after the claimed reception of the "S" signals, however, that Marconi was *not* the inventor. In the summer of 1901 Marconi had been given, by a friend in the Royal Italian Navy, a newly developed autocoherer made from a drop of mercury positioned between an adjustable iron plug and fixed carbon plug, sealed in a glass tube. Marconi applied for the patent in his own name even though he of course knew he had not invented the device and *even though he had been told* that experimenters in the Italian Navy were the true inventors.

As I mentioned in Chapter 4, Marconi had a history of patenting inventions that were the result of the intellectual efforts of others, but the patent for the autocoherer was a step beyond what he had done before. This appeared to be a *conspiracy* between Marconi and people with power in the Italian government to not only claim something which was not theirs, but to lay claim to the work done by persons in no position to assert their legal rights without fear of powerful, official reprisal. The whole business was sufficiently offensive that it soon developed into a scandal, and Marconi felt compelled to change his story.[2] He began to assert it was *not* with the autocoherer that he had heard the "S" signals at St. John's but, rather, he had heard the "S" signals with an ordinary coherer (he also changed his patent application, declaring now that it was really for an invention that had been *communicated* to him).

There was, in fact, some technical rationale for Marconi's change of story. The St. John's receiver was intended to be a *tuned* receiver, and for such a receiver the coherer would be placed in *parallel* with a resonant circuit connected to the antenna (see Figure 5.3) because it is the voltage drop across the unclumped coherer *at a specific, desired frequency* that was to actuate the coherer. This, in turn, means the unclumped coherer should have a high resistance so as not to interfere with the tuning in the first place.

The mercury autocoherer, however, had a relatively low resistance in its "decohered" state and that was Marconi's stated reason for later discounting its use in the reception of the "S" signals. According to Marconi's own assistant at St. John's, however, what actually took place was quite differ-

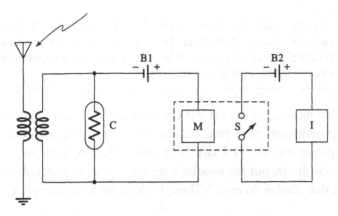

Figure 5.3   Marconi's tuned circuit, originally intended to be the St. John's receiver, looked quite similar to Lodge's 1894 circuit. The exception is the absence of the electromechanical feedback coherer-tapper, no longer necessary because the coherer was intended to be the "Italian Navy" autocoherer and not a Branly coherer.

ent from Marconi's changed tale. The attempt to use a tuned receiver was defeated by the antenna, itself, which was a wire floated aloft by a kite. The winds caused the kite to rise and dip so violently that the resulting variations in capacitive coupling to ground were large enough to ruin any possibility of tuning. So, the mercury autocoher was simply placed in *series* with the untuned antenna along with a series-connected telephone head-set with which to hear the signals with the highly sensitive human ear, as opposed to using a less sensitive electrically actuated ink pen.

Such an automatically recording receiver *was* used shortly after the original trans-Atlantic experiment. Stung by critics who didn't believe he had really heard what he claimed to have heard, Marconi placed an ordinary coherer in parallel with a tuned antenna circuit on the ship *SS Philadelphia* in February 1902 and departed England. Because the received signals were *recorded* by ink-and-paper, there is no doubt of their authenticity. But they could only be received at a distance from Poldhu of 700 miles (daytime) and 1560 miles (nighttime). The word *deceptive* is not a pleasant one, but the fact is that later the same year Marconi is known to have misled no less than Czar Nicholas of Russia. In the summer of 1902 the King of Italy put his new warship, the cruiser *Carlo Alberto*, at Marconi's disposal for continuing at-sea radio experiments. When Nicholas came aboard for a visit, Marconi gave him a radio message addressed specifically to the czar, one Marconi said had come straight from Poldhu. It had, however, been sent by a *much* less distant transmitter, one secretly tucked away in another part of the *Carlo Alberto*. The view among most radio experts, today, is that

Marconi simply could *not* have heard the "S" signals at St. John's. It doesn't help his claim either that the signals had been *prearranged*, both as to content and transmission time. No one today would consider such an experimental arrangement proper.

The evolving state of the art left coherers behind soon after the trans-Atlantic experiment, and true diodes came into use as detectors of radio-frequency signals.[3] I'll talk later about Fleming's discovery of the vacuum-tube diode, but for now let's ignore the physical details and simply accept that diodes do exist.[4] Then, looking back at Figure 5.1, the rest is easy.

An arriving signal produces a damped, oscillatory current in the antenna, which is magnetically coupled through a transformer to a tuned (via the variable capacitor) resonant circuit. Maximum voltage develops across the resonant circuit when it is tuned to match the frequency of the arriving signal (see Appendix E). Only the positive half-cycles of the oscillations in this tuned circuit are passed by the diode. Thus, a sequence of closely spaced positive pulses (the result of the diode's *rectification*—often called *detection*—of the damped oscillations in the resonant circuit) is applied to the capacitor/headphone combination. The pulses occur at a rate determined by the resonant frequency of the tuned antenna circuit, and this frequency is typically "high," in the many tens to hundreds of kilohertz. These damped oscillations are periodically generated in the antenna at the *sparking rate* of the transmitter, which is typically in the hundreds of hertz range (say 500 to 1000 sparks per second).

Each short burst of high-frequency pulses rapidly dumps charge into the C, which then much more slowly discharges through the headphones (which has resistance R). This discharge continues until the next high-frequency burst of diode current pulses recharges the C. The C can't discharge back through the antenna circuit because that direction is blocked by the diode. The headphone signal, then, is a low-frequency sequence of exponential capacitor discharges which appears in the headphones as an audio tone at the spark rate frequency. Different spark transmitters, with different sparking rates, would therefore produce different frequency tones in the headphones. Even slight shifts in frequency can be detected by the human ear, and an experienced operator could learn to identify the origin of a specific signal (the specific transmitter being used, and even occasionally the specific human operating the telegraph key) simply from the audio tone frequency in the headphones.

For the parallel/diode headphone/capacitor combination (called an *envelope detector*—see Figure 5.4) to operate properly, the capacitor should be nearly discharged before the next high-frequency current pulse burst arrives (if the capacitor doesn't have sufficient time to so discharge, then the capacitor voltage won't vary enough to produce an audible tone). That is, we don't want to pick too large a value for C. On the other hand, we don't want to pick C too small, either, or the capacitor voltage will simply "follow" the shape of the individual high-frequency pulses and again fail to produce an audible tone). Notice carefully that there are different reasons for this same end result. Too large a C produces a tone in the audio

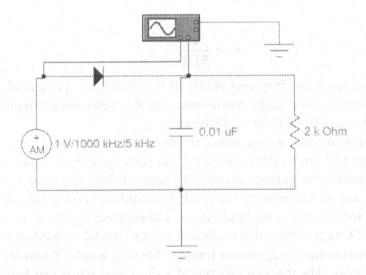

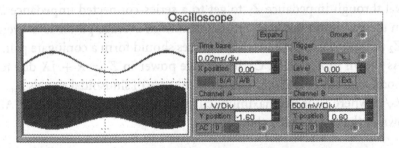

Figure 5.4 The input (lower trace) to this envelope detector circuit is a 1000 kHz (= f_c) sine wave, amplitude modulated by a 5 kHz (= f_m) sine wave, i.e., it is of the form $[1 + m^* \sin(w_m^* t)]\sin(w_c^* t)$. The upper trace shows the signal presented to the resistor ("earphones" of Figure 5.1) and the motivation for the name "envelope detector" should now be clear. The value of m is called the modulation index, and in this example $m = 0.2$ (see Figure 6.2 for more on the modulation index).

frequency range, but of low amplitude, i.e., the tone is inaudible because it is weak. Too small a C produces a strong signal *at the high-frequency pulse rate*, i.e., the tone is now inaudible because its frequency is too high to hear.

A common engineering design rule is to pick the time interval between the high frequency bursts to be $5RC$ (you should verify for yourself that 1 ohm-farad = 1 sec). This ensures the C is nearly discharged between pulse bursts, but keeps C large enough that the capacitor voltage does not follow the individual pulses. Thus, if f denotes the spark rate frequency, we have

$$\frac{1}{f} = 5RC$$

or

$$C = \frac{1}{5fR}.$$

For a typical spark rate frequency of 500 Hz (i.e., 500 sparks per second at the transmitter), and a headphone resistance of $R = 8000$ ohms, this gives the quite reasonable value for C of 0.05 μF.

There is no internal energy source in the circuit of Figure 5.1, and so all of the energy (from each pulse burst) that ends up in the headphones must originate in the antenna circuit. This generally isn't much energy to start with, and so it is important to transfer as much as possible (on average) from the antenna to the headphones. The engineering answer to the question of how to achieve this maximum energy transfer is based on the so-called maximum average power transfer theorem, which I'll now state and then prove. The theorem says that if a sinusoidal signal that has to travel through impedance Z_1 to get to a series-connected impedance Z_2, then the condition that ensures the maximum average power is delivered to Z_2 is $Z_2 = Z_1^*$, i.e., the two impedances should form a conjugate pair.

As shown in Appendix C, the average power in $Z = R + jX$ due to a periodic signal is $P = I_{rms}^2 R$, where for the case of sinusoidal signals $I_{rms} = I_M/\sqrt{2}$ (where I_M is the maximum current in Z). That is, $P = \frac{1}{2}I_M^2 R$. Also shown in that appendix is that

$$I_m = \frac{V_m}{\sqrt{R^2 + X^2}}$$

and so

$$P = \frac{1}{2}\frac{V_m^2 R}{R^2 + X^2},$$

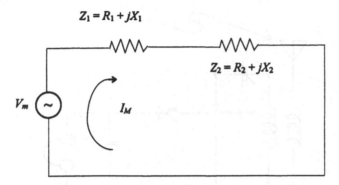

Figure 5.5   Theoretical model for impedance-matching calculation.

where V_m is the maximum voltage drop across Z. So, we have the situation shown in Figure 5.5, where Z_1 is identified with the antenna circuit (V_m represents the maximum antenna voltage), and Z_2 is identified with the diode/capacitor/headphone circuitry. We thus have

$$I_M = \frac{V_m}{\sqrt{(R_1 + R_2)^2 + (X_1 + X_2)^2}}$$

and so the average power in Z_2 is

$$P = \frac{1}{2} \frac{V_m^2 R_2}{(R_1 + R_2)^2 + (X_1 + X_2)^2}.$$

It is not uncommon in undergraduate electrical circuits books to see P maximized by laboriously calculating $\partial P/\partial R_2$ and $\partial P/\partial X_2$, and then setting both equal to zero. This is unnecessary, and P can be maximized in a much more direct way. Simply observe that, for *any* given values of R_1 and R_2, we clearly achieve the largest P by choosing $X_2 = -X_1$. With that choice, we then have

$$P = \frac{1}{2} \frac{V_m^2 R_2}{(R_1 + R_2)^2}.$$

This choice is possible because passive Xs come in both signs. We cannot use the same reasoning to argue that for given values of X_1 and X_2 we achieve the largest P by choosing $R_2 = -R_1$, because passive Rs do *not* come in both signs! Finally, we select R_2 to maximize P (set $dP/dR_2 = 0$), which results in $R_2 = R_1$. Thus, $Z_2 = R_1 - jX_1 = Z_1^*$, and this proves the maximum average power transfer theorem.

This optimal adjustment of the impedance levels in the receiver (called "impedance matching") results in half of the total available energy in the

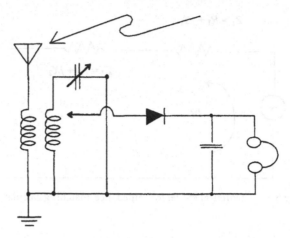

Figure 5.6   A sophisticated, impedance-matched crystal radio receiver circuit.

antenna circuit being sent to the output stage (and so under the best of conditions, with perfect matching, a crystal radio set was only 50% efficient). Figure 5.6 shows a slightly modified Figure 5.1 (now with *two* adjusting controls) that allows the receiver to be tuned in frequency, and also to attempt to impedance-match the antenna circuitry with the diode circuitry. One simply fiddles with the two knobs until the maximum average power transfer theorem is *experimentally* satisfied as nearly as possible (as determined by establishing the loudest signal possible in the headphones). Problem 5.3 discusses the mathematics of maximizing the average transferred power when there are various constraints on how much we can adjust Z_2.

Notes

1. The coherer was viewed from the start as "enigmatic," and many attempted to explain its operation. One such researcher was William Henry Eccles (1875–1966), best known to electrical engineers today as co-inventor of the famous Eccles–Jordan electronic multivibrator and triggered bi-stable circuits used in early high-speed counting instruments, based on the vacuum-tube triode (discussed in the next chapter). Eccles was a former assistant of Marconi who built the oscillation transformer "jiggers" described in the last chapter. The mystery of the coherer was considered sufficiently deep that Eccles' studies earned him a 1901 Doctor of Science degree from the University of London, but the physics underlying the Branly coherer has, to this day, not been satisfactorily explained. The mystery of the coherer even found its way into the *fiction* of the day; in his 1902 story "Wireless," Rudyard Kipling has one of his characters ask another about electricity, the stuff that makes radio "go":

    "But what *is* it?" I asked.

In reply, he is told

> "Ah, if you knew *that* you'd know something nobody knows. It's just It—what we call Electricity, but the magic—the manifestations—the Hertzian waves—are all revealed by *this*. The coherer, we call it."
>
> He picked up a glass tube not much thicker than a thermometer, in which, almost touching, were two tiny silver plugs and between them an infinitesimal pinch of metallic dust. "That's all," he said, proudly, as though himself responsible for the wonder. "That is the thing that will reveal to us the Powers—whatever the Powers may be—at work—through space—a long distance away."

It is thought that minute welding between the particles of "metallic dust" is part of the explanation, but no solid theoretical analysis exists, to my knowledge. For more general commentary on this still mysterious gadget, see Thomas H. Lee, *The Design of CMOS Radio-Frequency Integrated Circuits*, Cambridge University Press 1998, pp. 3–4. For a nice discussion of the historical antecedents of Branly's coherer, see Leonid Kryzhanovsky, "The Coherer: Preparing the Way for Wireless," *Elecronics World + Wireless World*, March 1992, pp. 212–214. For even more technical details (including how to *make* your own coherer—be the first on your block!), see George Pickworth, "Coherer-Based Radio," *Electronics World + Wireless World*, July 1994, pp. 563–567.

2. For a scholarly account of this episode from early radio history, as well as an account for the physical behavior (rectification) of the mercury autocoherer, see V. J. Phillips, "The 'Italian Navy Coherer' Affair: A Turn-of-the-Century Scandal," *Proceedings of the IEEE* 86, January 1998, pp. 248–258. Even more on the scandal can be found in the companion paper by Probin K. Bondyapadhyay, "Sir J. C. Bose's Diode Detector Received Marconi's First Transatlantic Wireless Signal of December 1901 (The 'Italian Navy Coherer' Scandal Revisited)," *Proceedings of the IEEE* 86, January 1998, pp. 259–285.

3. For technical details in a readable account, see Desmond Thackeray, "When Tubes Beat Crystals: Early Radio Detectors," *IEEE Spectrum*, March 1983. The definitive early history of detectors is by Vivian J. Phillips, *Early Radio Wave Detectors*, Peter Peregrinus 1980.

4. There were a number of other non-coherer, non-diode gadgets used for radio wave detection. For an excellent discussion of one of them, Reginald Fessenden's *barretter*, that shows how extremely clever were the radio pioneers, see George Pickworth's two-part article "Detection Before the Diode" in *Electronics World & Wireless World*, December 1994, pp. 1003–1006 and January 1995, pp. 28–30. The name for this gadget, a *liquid* electrolytic device based on the movement of ions through an acidic solution, derived from the French word for *exchanger*. That is, the device "exchanged" ac for dc!

Problems

5.1. Explain what a listener would hear in the headphones if the diode in Figure 5.1 is replaced with a wire, i.e., if the tuned resonant circuit is directly connected to the parallel capacitor/headphones. What would she hear if the diode is reversed in direction?

5.2. In Figure 5.7 the values of V, R_1, and R_2 are fixed, but R can be varied.

   (a) Show that maximum power is developed in R when R is equal to the parallel equivalent of R_1 and R_2 (i.e., $R = R_2 \parallel R_2$). Note: Electrical engineering students will recognize this result as an immediate consequence of *Thévenin's theorem* (discussed in

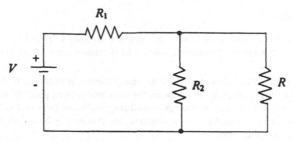

Figure 5.7   When is the power in R maximum?

Appendix D) and the maximum average power transfer theorem. You, however, should NOT use those theorems to answer this question. Indeed, your analysis here, based only on Ohm's and Kirchhoff's laws, will serve as a confirmation of those theorems (for this particular circuit).

(b) Suppose R is adjusted to the value of part (a). Let P_R denote the (maximized) power in R, and let P_B denote the battery power. Derive an expression for $E = \frac{P_R}{P_B}$, and use it to conclude that $E < \frac{1}{2}$ for $R_2 < \infty$, and that $E = \frac{1}{2}$ only if $R_2 = \infty$. Partial answer: if $R_1 = 30$ ohms and if $R_2 = 150$ ohms, then your expression should give $E = \frac{5}{14}$, i.e., if R is adjusted to $30 \parallel 150 = 25$ ohms, then maximum power will be developed in R but that power will be only slightly more than one-third of the power delivered to the circuit by the battery. The lesson here is that under conditions of maximum power transfer, the *maximum* possible efficiency is 50% but, depending on the details of the circuit, can be much less.

5.3. As shown in the text, if we can adjust R_2 and X_2 at will then $Z_2 = Z_1^* = R_1 - jX_1$ is the choice that gives the maximum average power in Z_2. Suppose now, however, that there are constraints on how we may adjust Z_2. What then? There are many different possible situations one can imagine, but here are three in particular. To understand them, first recall the situation of the unconstrained case. We have a *given* Z_1, which is a fixed vector in the complex plane, i.e., Z_1 has a fixed magnitude $|Z_1|$ and a fixed angle, α. Then, we pick $Z_2 = Z_1^* = |Z_1| \angle - \alpha$. Now,

(a) Suppose that $|Z_2|$ can be varied at will but the angle of Z_2 (call it θ) cannot be varied at all. Show that the condition that maximizes the average power in Z_2 is $|Z_2| = |Z_1|$. Hint: $Z_1 = |Z_1|\angle\alpha = |Z_1|\cos(\alpha) + j|Z_1|\sin(\alpha)$ and $Z_2 = |Z_2|\angle\theta = |Z_2|\cos(\theta) + j|Z_2|\sin(\theta)$, and note that $|Z_2|$ is the only variable.

(b) Suppose that the angle θ of Z_2 can be varied at will but $|Z_2|$ cannot be varied at all. Show that the condition that maximizes the average power in Z_2 is

$$\theta = -\sin^{-1}\left\{\frac{2|Z_1||Z_2|}{|Z_1|^2 + |Z_2|^2}\sin(\alpha)\right\}.$$

Notice that in the special case of $|Z_1| = |Z_2|$ this reduces to $\theta = -\alpha$, which is what we'd expect.

(c) Suppose that neither of R_2 and X_2 can be varied over all possible values (i.e., $0 < R_2 < \infty$ and $-\infty < X_2 < \infty$), but rather each is confined to a finite interval. Show that the procedure that maximizes the average power in Z_2 is to first select X_2 as close as possible to $-X_1$, and then to select R_2 to be as close as possible to $\sqrt{R_1^2 + (X_1 + X_2)^2}$. Hint: Consider separately the two cases $R_1 < R_{2\,\mathrm{min}}$ and $R_1 > R_{2\,\mathrm{max}}$, where $R_{2\,\mathrm{min}} \le R_2 < R_{2\,\mathrm{max}}$.

CHAPTER
6

Mathematics of AM Sidebands

You can understand the theoretical reason behind Fessenden's failed attempt to send acceptable quality voice by spark-gap transmission (mentioned at the end of Chapter 4) once it is understood that *amplitude modulation* (AM) spreads energy over an interval of frequencies. If you start with a constant amplitude sine wave of a given frequency, then you have what is called a *pure tone*. But as soon as you begin to vary the amplitude of that pure tone you create *additional* tones (sidetones), both higher and lower in frequency than the original frequency. This fact immediately establishes that amplitude modulation is not the result of a linear, time-invariant system (see Appendix B). In such systems, the output frequencies are limited to the input frequencies (with amplitudes perhaps modified as a function of frequency, as discussed in Appendix C).

In AM radio, however, there are frequencies present at the output of the transmitter that are *not* present at the input. These new frequencies are created by either a nonlinearity, or a time-varying process, inside the transmitter. We can establish this fact very quickly (and I will before the end of this chapter), but it is astonishing to read the older technical literature and see how even some of the "big names" in early radio simply didn't believe it even *after* being shown why they were wrong! Still interesting reading today, for example, is the essay by Sir John Ambrose Fleming, "The 'Wave Band' Theory of Wireless Transmission," *Nature*, January 18, 1930, pp. 92–93. Fleming (a Fellow of the Royal Society who had studied at Cambridge under the direction of the great Maxwell, himself, the designer of Marconi's Poldhu transmitter, and the inventor of the vacuum tube diode to be discussed in Chapter 8) called the elementary mathematical analysis in this chapter "a kind of mathematical fiction [that] does not correspond to any reality in nature." For Fleming to write this in 1930 was astonishing, because the observation of the sidetone phenomenon had occurred long before, in 1875. That year, the American physicist Alfred Mayer (1836–1897) *heard* them in an acoustical experiment in which he mechanically interrupted the sound of a vibrating tuning fork. Later, in 1894, when Lord Rayleigh published the second edition of his enormously influential book *Theory of Sound*, he took notice of Mayer's experiment and explained it mathematically. In light of this it isn't surprising that Fleming's essay prompted a flood of replies, all declaring him to be dead wrong.

But he remained unconvinced (see *Nature*, February 8, pp. 198–199; February 22, pp. 271–273; March 1, pp. 306–307). What *really* makes Fleming's essay hard to understand however is that, at the time he wrote it, the first commercial single-sideband (SSB) AM radio link between New York City and London had been in operation for three years! According to Fleming's position, such a radio (discussed in great detail in Chapter 20) simply couldn't work, and the fact that it did work apparently never caused him to reconsider his position. Old ideas die hard.

As mentioned in the last chapter, the problem with spark gaps is that, even before you attempt to modulate their transmissions with voice amplitude variations, their signals are *created* with AM inherently present because the antenna circuit oscillations are exponentially damped. The conclusion then (if you accept the premise of the first paragraph), is that spark-gap radio transmitters are *sloppy*, literally spilling energy into frequencies that can be considerably different from the nominal resonant frequency of the antenna circuit. Later, in Chapter 11, I'll work through the precise mathematical details of the sloppy radiation behavior so characteristic of spark-gap transmitters, which was a flaw sufficiently awful as to be a fatal one for their future.

Suppose, then, contrary to the state of technology even as the twentieth century began, that we have a stable source of constant amplitude, radio-frequency oscillations (at the frequency $\omega_c = 2\pi f_c$ radians/sec). The frequency f_c (in hertz) is called the *carrier frequency*. For commercial AM broadcast radio, f_c is in the interval 540 kHz to 1600 kHz. I'll write the carrier signal as simply

$$c(t) = \cos(\omega_c t),$$

which makes the assumption that the carrier has unit peak amplitude. That is, whatever the actual peak amplitude of the carrier signal may be, it will serve as our amplitude reference.

Next, suppose we have a very simple message signal we wish to transmit —the sound of a person whistling at the frequency $\omega_m = 2\pi f_m$. From experience, we know f_m is at most on the order of a few kilohertz, and so $f_m \ll f_c$. Such signals as a person whistling (or talking), with frequency content varying from dc (0 Hz) to some maximum upper frequency (in the low kilohertz range for human speech) are called *baseband* signals. If we denote the amplitude of this "whistle signal" by A_m, then the message signal is

$$m(t) = A_m \cos(\omega_m t).$$

In fact, because $c(t)$ and $m(t)$ are certainly not synchronized, we should actually toss in an arbitrary phase angle, ϕ, to allow for the (almost certain) possibility that $m(t)$ and $c(t)$ are not simultaneously at their peaks at $t = 0$. That is, let's more generally write

$$c(t) = \cos(\omega_c t + \phi) \quad \text{and} \quad m(t) = A_m \cos(\omega_m t).$$

Now, as shown in Figure 6.1, imagine that we apply the carrier signal to a multiplier, along with the message signal (added to a dc shift of unity) applied to the other multiplier input. Then writing the multiplier's output signal as $r(t)$, i.e., the signal to the radiated, we have

$$r(t) = [1 + m(t)]c(t) = [1 + A_m \cos(\omega_m t)] \cos(\omega_c t + \phi)$$
$$= \alpha(t) \cos(\omega_c t + \phi)$$
$$\alpha(t) = 1 + A_m \cos(\omega_m t).$$

At this point I should point out that in real AM transmitters the multiplication of the dc shifted message signal $(1 + m(t))$, with the carrier signal, is *not* actually done with a multiplier. Building good multiplier circuits that work at radio frequencies is not easy. There are very clever circuits (called *balanced modulators*) that produce the same effect, however, and they will be discussed in Section Three, after I have established the necessary mathematics required for you to understand how those ingenious electronic circuits work.

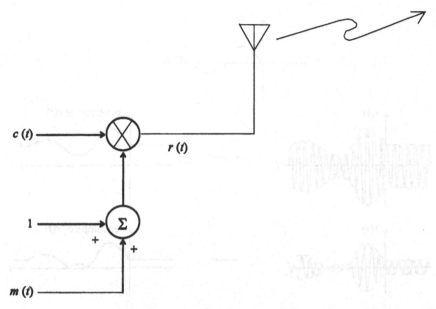

Figure 6.1   Generating an AM wave.

We can think of $r(t)$ as a high frequency signal (at the carrier frequency) with a time-varying amplitude $a(t)$ (at the modulation frequency), *if* $A_m \leq 1$. This condition ensures that the amplitude of $\cos(\omega_c t + \phi)$ is always nonnegative. The upper part of Figure 6.2 shows $r(t)$ for the case of $A_m \leq 1$. The so-called *envelope* of $r(t)$, shown in dashed lines and to the right, varies between $1 + A_m$ and $1 - A_m$, and this variation takes place at the modulation frequency ω_m. We can now see why the carrier is called the *carrier*—the information-bearing signal literally rides piggyback on (or is *carried* by) the high-frequency signal at frequency ω_c.

But what about the perfectly possible case of $A_m > 1$? Then it appears that the signal applied to the antenna would have a *negative* amplitude! What could such a thing mean? The answer can be found in writing $r(t)$, for the case when $\alpha(t) < 0$, as

$$r(t) = \alpha(t) \cos(\omega_c t + \phi) = |\alpha(t)| \cos(\omega_c t + \phi + \pi).$$

That is, if $\alpha(t)$ goes negative (because $A_m > 1$) then the signal $r(t)$ is still the instantaneous value of a sinusoidal wave with a positive amplitude, but the negative sign appears as a *sudden phase shift* of π radians. This effect is called *phase-reversal distortion* because, as the bottom part of Figure 6.2 shows, the envelope of $r(t)$ is then *not* a mimic of the amplitude variation of $m(t)$, the message signal. Indeed, when $m(t)$ is decreasing (becoming

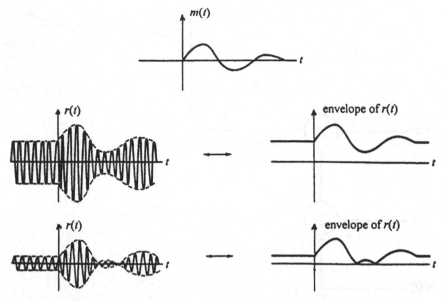

Figure 6.2   The envelopes of AM waves. For the given message signal $m(t)$, the middle curves show $r(t)$ and its envelope for less than maximum modulation, and the bottom curves show what happens in the case of overmodulation.

more negative), $|\alpha(t)|$ is increasing and the envelope of $r(t)$ is reversed from what it should be. The cure for this effect of overmodulation (as it is called) is exactly like what your doctor tells you to do when you say it hurts to jump up and down, i.e., don't jump up and down (don't overmodulate)!

Now, what *frequencies* are present in $r(t)$? (Be careful, this is not a trivial question!) Recall the trigonometric identity

$$\cos(\alpha)\cos(\beta) = \frac{1}{2}\left[\cos(\alpha+\beta)+\cos(\alpha-\beta)\right],$$

and make the associations of

$$\alpha = \omega_c t + \phi,$$
$$\beta = \omega_m t.$$

Thus,

$$r(t) = \cos(\omega_c t) + \frac{1}{2}A_m\left\{\cos\left(\left[\omega_c+\omega_m\right]t+\phi\right)+\cos\left(\left[\omega_c-\omega_m\right]t+\phi\right)\right\}.$$

That is, the transmitted signal consists of three frequencies. First, there is a carrier term, which itself contains no message information. Then, there are the so-called *sidetone* terms, at the sum (upper sidetone) frequency ω_c +

ω_m, and at the difference (lower sidetone) frequency $\omega_c - \omega_m$. These two sidetones frequencies carry the message information (e.g., the loudness of the whistler) in their peak amplitudes, $(\frac{1}{2})A_m$.

Notice that there is no component in $r(t)$ at frequency ω_m! This may seem puzzling (indeed, I think it *should* appear puzzling) because the envelope of $r(t)$ is so obviously varying at frequency ω_m. This is a good example of how pictures can be misleading—what appears *to the eye* to be a component at frequency ω_m is really an illusion, the net result of the two sidetones which are both at considerably higher frequencies than ω_m. We eventually *do* create a new signal at frequency ω_m, of course, by using the *nonlinear* envelope detector circuit discussed in the previous chapter. But what if the message signal is more complicated than a single pure tone? If the message signal is a *number* of distinct tones, each with its own amplitude, then what is true for the single tone in the previous analysis is true for each of the individual tones. Since human speech is a diverse collection of various frequencies, with various amplitudes, the resulting AM side*bands* (a *band* is an *interval* of frequency from the lowest sidetone frequency to the highest sidetone frequency) are equally diverse.

It is the shift of the baseband tone frequency up to the much higher sidetone frequencies that is the reason behind why we desire to implement this seemingly complicated process. The frequency shift simultaneously solves *two* fundamental problems. First, at baseband frequencies the physical size of an antenna would be enormous (recall Chapter 3 and the height of a one-quarter-wavelength vertical dipole). By shifting the message signal up in frequency, we shift the antenna down in size. Second, the shift allows multiple transmitters to geographically operate near each other without interference—each has its assigned carrier frequency with which it shifts the *same* baseband frequencies by *different* amounts. This makes it possible for a receiver to "tune in" a particular signal, and to reject all others.

The Radio Act of 1912 (essentially the legal seizure of the airwaves by the U.S. government, as a result of the *Titanic* disaster and the mass confusion caused by amateur radio chatter about the sinking) assigned all commercial stations to the same *two* carrier frequencies!— 833.3 kHz (360 meters) when broadcasting news, lectures and entertainment, and 618.6 kHz (485 meters) when transmitting government reports (e.g., the weather forecast). In August 1922 a third frequency was added (750 kHz, or 400 meters), but this did nothing to ease the vast mutual interference caused by hundreds of stations. It was not uncommon for a station, suffering interference, to simply change carrier frequency on its own initiative, a practice called "wavelength pirating." This bizarre state of affairs was finally recognized as a problem when the extended *band* of frequencies we have today was allocated for commercial AM use; eventually the 1912 Act was replaced with the far more regulatory Radio Act of 1927 (see the shaded box at the end of this chapter). For a contemporary view of the situation in the United States just before passage of that act, see H. V. Kaltenborn, "On the Air," *Century Magazine*, October 1926. For a modern scholarly assessment, see Marvin R. Bensman, "The Zenith–WJAZ Case and the chaos of 1926–7," *Journal of Broadcasting* 14, Fall 1970, pp. 423–440.

This method of achieving the desirable frequency up-shift doesn't come problem-free, however. For example, as shown in Appendix C (and used in the previous chapter), the average power in a sinusoidal waveform with peak value V_m is $(\frac{1}{2})V_m^2$. This tells us that the total average power in $r(t)$ is

$$\frac{1}{2}\left[1^2 + \left(\frac{1}{2}A_m\right)^2 + \left(\frac{1}{2}A_m\right)^2\right] = \frac{1}{2} + \frac{1}{4}A_m^2.$$

The fraction of this total average power that is in the carrier is

$$\frac{\frac{1}{2}}{\left(\frac{1}{2}\right) + \left(\frac{1}{4}\right)A_m^2} = \frac{2}{2 + A_m^2}.$$

This fraction decreases with increasing A_m, but even with A_m as large as possible without causing overmodulation distortion ($A_m = 1$) there is still two-thirds (!) of the total average power in the *non*-information-bearing carrier. The theoretical maximum efficiency for AM radio (as it is actually implemented) is therefore just 33%. When I discuss the nature of the AM sidebands in more mathematical detail in Section Four, you'll see that the upper and lower sidebands *each* contain all the message information. It is redundant to transmit both sidebands, i.e., transmitting just a single sideband, and no carrier, results in no loss of information. This means that AM radio (as it is actually implemented) is really only, at *most* about 17% efficient.

This is surely wasteful, you surely think, so why do radio engineers build such an inefficient system? The answer is, as will be developed as we go along, that the admitted inefficiency at the *lone* transmitter is compensated for, many-fold over, by the simplicity of the *millions* of receivers tuned to that transmitter. The presence of a strong (i.e., high-energy) carrier at the receiver is precisely what allows the simple, cheap envelope detector to work. Double-sideband AM transmission (DSB-AM) *without* a distinct carrier term is easy to achieve, yes, but as I'll show you in Section 4, the *receiver* for such signals is complicated (i.e., expensive). And single-sideband AM radio, which not only doesn't transmit the carrier but also suppresses one of the sidebands, is even more energy efficient but also *much* more complicated at both the transmitter *and* the receiver, compared to ordinary AM radio.

The evolution of the various radio acts in the United States can be summarized as follows. The first federal radio law was the nonregulatory Wireless Ship Act of 1910 which, for the safety of ships at sea, simply required the carrying of radio transmitters and receivers on all passenger ships licensed to carry 50 or more. On this score the U.S. was far behind the times: as far back as March 1899, when the British lightship *East Goodwin* sunk after colliding in the fog with the steamship *R. F. Matthews*, all hands survived by the lightship's radio call for help. Later, the world's first time trans-Atlantic superliner, the German *Kaiser Wilhelm der Grosse*, carried Marconi spark transmitters as of February 1900, and by the end of 1902 it was just one of seventy non-U.S. radio-equipped ships. Spark radio saved many lives from its earliest days with its electric calls for aid. In the *Goodwin's* case, it was most likely just a hurried, frantic call for help. In January 1904, however, the Marconi company declared the official distress signal to be "CQD," from "CQ" for the general call to all stations and "D" for distress (it did not stand for "Come Quick Danger" as was commonly believed!) This was changed to the famous "SOS" at the 1906 Berlin International Radio Conference; the new call did *not* stand for "Save Our Souls" (as was commonly believed) but rather was chosen simply because of its distinctive and easy-to-send Morse code (three dots, three dashes, three dots). Still it was a CQD that saved the passengers on the White Star liner (as was the *Titanic*) *Republic* in January 1909 when it collided in a fog with the Italian ship *Florida*. It was the first marine radio distress call in history. The *Republic* sank (it was this event, in fact, that prompted the passage of the Wireless Ship Act the following year) but, of the 1650 people on-board both ships, only six died (from the initial impact). The *Republic's* radio operator John ("Jack") Binns became an overnight hero, and when he died fifty years later in 1959, at age 75, his *New York Times* obituary was entitled "Pioneer in Radio . . . Was Hero of 1909 Rescue at Sea." Alas, it was both a CQD and an SOS that failed to save the *Titanic* in 1912 (the radio operator on the *Californian*, just twenty miles away, had gone off-duty for the night). The Radio Act of 1912, prompted by the *Titanic* disaster, had some regulatory powers, but only for interstate commerce and so provided no federal regulatory control over communications within an individual state. With the growth of the commercial broadcast industry in the 1920s this quickly led, as mentioned before, to a "free-for-all" among station owners until, finally, after four preliminary National Radio Conferences between February 1922 and November 1925, Congress passed the Radio Act of 1927; it created the Federal Radio Commission with power to regulate *all* of radio. The FRC, originally, was to exist only for a year, but it proved so effective that its charter was renewed annually until the Communications Act of 1934 replaced it with the permanent Federal Communications Commission (FCC), which continues to this day.

First Continuous Waves, Negative Resistance Oscillators and the Van der Pol Equation, and the Heterodyne Concept

In the two decades after the turn of the century and before the rise of commercial broadcast radio, the governments of the industrialized world were the prime movers behind the continued development of the new technology. Their interest was in the creation of strategic communication systems that, unlike the existing submarine telegraph cables, couldn't be disabled during time of war. In America, it was the U.S. Navy, with its need to communicate with a far-flung fleet at sea and remote outposts around the world, that funded much of the domestic development. Indeed, when on April 6, 1917 the American government seized all radio stations in the United States and its territories upon entering World War I, as required by national security measures contained in the Radio Act of 1912, it was the U.S. Navy that controlled all radio operations for the duration of the war. Spark radios were a step in the right direction for meeting the Navy's communication needs, but the inherent spreading of energy across the frequency spectrum by such radios made them both energy inefficient at the transmitter and unselective at the receiver.

It was recognized, quite early, that the solution at both ends of the communication link lay in the development of *continuous wave* radio, i.e., the transmission and reception of undamped waves. Such waves, precisely controlled in frequency, would solve the selective tuning problem, as well as concentrating the transmitter energy into well-defined frequency intervals, thereby increasing the effective power of the transmitter. When President Taft Signed the Radio Act of 1912, it became *almost* the law[1] to use only undamped waves.

It is important to realize that while constant amplitude, continuous wave radio is essential for voice transmission (and thus commercial radio), that was *not* the immediate goal in the 1900s and 1910s—selectivity and energy efficiency were. The ultimate means for achieving these goals would arrive with the vacuum tube (which will be discussed in the next chapter), but before the vacuum tube there were two other, now almost forgotten, interim technologies that had brief but important places on center stage. These were the radio-frequency alternator, and the negative resistance electric arc. While both have now been obsolete for many decades, their stories are testaments to the ingenuity of the early radio pioneers. (As I tell my own students—only partially in jest—the lack of a proper technology for the task at hand is but a trifling obstacle to a sufficiently motivated engineer!)

The first technology to be commercially developed to generate constant amplitude sinusoidal signals for radio communication was the *alternator*, similar to the alternators used in modern cars, and in power plants, to generate ac power. Their basic principle of operation is a direct application of Faraday's law of induction—if a nearly closed loop of wire surrounds a time-varying magnetic field, an electric field is created in the wire which then develops a potential difference (voltage drop) between the ends of the loop. In an alternator, *relative motion* between a magnetic field and an armature (essentially many loops of wire) generates an alternating or ac potential difference. The higher the relative rotation, the higher the rate of alternating (the frequency) of the ac voltage output.

It is a standard problem in freshman physics to show that a loop of wire rotating at constant speed in a uniform, constant magnetic field (produced, for example, by a permanent magnet) produces a *sinusoidally* alternating voltage with a frequency directly proportional to the loop rotation speed. (As the plane of the loop rotates, the magnetic flux penetrating the surface bounded by the loop varies, even though the field itself is constant). This is why even the very earliest ac power plants, run by water flowing over a paddlewheel which then rotated wire loops through a fixed, constant field, produced sine wave outputs *long before* the special mathematical properties of sine waves were appreciated by electrical engineers. It was all simply a fortuitous accident of geometry and physics! But whether it is the field or the conductor that actually does the rotating is immaterial (alternators are built both ways)—what matters is the *relative* rotation. Indeed, it was this specific example that led to Einstein's thinking that resulted in the special theory of relativity (Einstein's father and uncle jointly operated an electrical machinery manufacturing company in Munich when Einstein was a young boy in the early 1890s).

Figure 7.1 shows a simple transmitter circuit using an alternator, in which a stationary magnetic field is produced by an electromagnet powered by a dc source. The armature is the rotating part, and the ac output is directly fed to the antenna. To transmit Morse code, one simply operates the key in the field circuit—with the key up, there is no magnetic field (and so no ac), and with the key down, the field coils are energized and there is

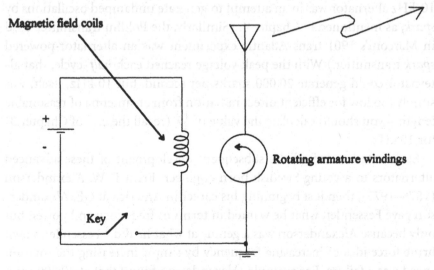

Figure 7.1 A radio-frequency alternator transmitter circuit.

an ac signal. Placing the key in the field coil circuit was not practical for very high-speed transmission, however, as the keying rate could be faster than the times required for the magnetic field to build up and collapse between individual dots and dashes. High-speed Morse code alternator transmitters (operated by prepunched paper tape) avoided this problem by placing the key in the antenna circuit. The field coils were then *always* energized. When the key was down, the alternator was connected to the antenna, and when the key was up, the alternator was connected to a "dummy load" that would absorb the alternator's output power. In practice, of course, the key actually controlled a high-power switching relay in the antenna circuit, because to run the alternator's output of perhaps hundreds of kilowatts directly through the key would have vaporized it (and its operator)! The dummy load was often simply a big tank of water—if the key was left up too long, with the alternator running, the water would literally boil.

This circuit's simplicity is deceptive, however. The dramatic difference between the radio frequency alternator and other forms of alternators is the rotation speed. The first such device,[2] ordered by Fessenden from General Electric in 1900, was designed by Charles Proteus Steinmetz (1865–1923) to run at 3750 rpm. It generated 1200 W at 10 kHz. Steinmetz, best known today to electrical engineers as a pioneer in the use of complex numbers in ac circuit analysis, was quite pleased at the performance of his alternator. But for Fessenden's purposes, it was just the beginning. He next wanted 100 kHz, and he even talked at one time of rotation speeds of 120,000 rpm! With no little irony, however, Fessenden's only use of the 10-kHz alternator was in an attempt to generate undamped oscillations by *spark*, as mentioned in Chapter 4. (Similarly, the Poldhu transmitter used in Marconi's 1901 trans-Atlantic experiment was an alternator-powered *spark* transmitter.) With the peak voltage reached each *half*-cycle, that alternator could generate 20,000 sparks per second, but 10 kHz, itself, was simply too low for efficient direct radiation from an antenna of reasonable length—you should calculate the value of $\frac{1}{4}\lambda$ (recall the end of Chapter 3) for 10 kHz.

Steinmetz handed off the subsequent development of these advanced alternators to a young Swedish born engineer, Ernst F. W. Alexanderson (1878–1975), then just beginning his career in America at GE. Alexanderson gave Fessenden what he wanted in terms of frequency and power, but only because Alexanderson was a genius at what he did.[3] Fessenden's own brute force idea of increasing frequency by simply increasing the rotation speed was a failure. For example, Alexanderson found that at 20,000 rpm, a 100-kHz alternator could be made to produce 2000 W of ac power—and 5000 W of waste heat due to the air friction. Such high-speed alternators also had a nasty tendency to suddenly disintegrate, or at least to spray the solder right off their electrical connections and into the face of anybody standing too close. Increasing the rotation speed from 3750 to 120,000 rpm increases the centrifugal forces on the rotating parts of the alternator by the *square* of the ratio of speeds, i.e., by a factor of 1024! It was Alexanderson's engineering skills that eventually overcame these problems.

On Christmas Eve of 1906, after alerting radio operators at sea of his intention, Fessenden used a 100-kHz alternator (although it probably never got above 50 kHz) to make what radio historians generally consider to be the first radio voice broadcast. He played phonograph records, spoke, and played a violin, but mostly he astonished the operators who heard music and speech in their headphones instead of the dots and dashes of Morse code. Amplitude modulation was achieved by directly inserting a water-cooled microphone in the high-power antenna circuit (modern radio transmitters introduce the modulation at the low power front end of the transmitter and *then* amplify to produce the high-power antenna signal).

Alexanderson's alternators were beautiful, highly precise pieces of machinery, with clearances of just one-tenth of a millimeter between relatively moving parts. For a machine weighing 30 *tons*, as did a typical 200,000-W alternator (delivering over 560 A to the antenna at a frequency of 22 kHz), that was high-tech indeed in 1918. It was with such an alternator that President Wilson's ultimatum for the Kaiser's abdication was transmitted directly to Germany in October 1918. But there was really no place for this technology to go. The radio frequency alternator is a good example of an idea pursued far beyond its natural limits. It worked, yes, but there was no possibility of extending it upward to even the lowest carrier frequency of commercial AM radio (540 kHz). Something different was needed, an entirely *new* approach to continuous wave radio. That new idea was the negative resistance electric arc.

Almost from the time of the invention of the electric battery at the beginning of the nineteenth century, it had been known that continuous low dc voltage (tens of volts) and high current (hundreds of amperes) produced electrical arcs emitting an extraordinarily intense light. Arcs are a phenomena totally different from the high-voltage, low-current discharge of a capacitor to produce a transient spark. If two electrodes carrying a large current are placed in contact and then slowly pulled apart, the current will continue to flow across the gap in the form of what can only be described as a *flame*. This flame is composed of ionized atmospheric gases and vaporized electrode material, and can reach temperatures as high as 9000° F.

During the 1870s, such arcs were used to light public streets and the interiors of large buildings, but the sheer brilliance of the light was far too overwhelming for arcs to be a competitor of the light bulb for use in private homes. Electric arcs were used in both world wars by all sides as defensive search lights during enemy bombing raids, as light sources for filming silent movies (the introduction of sound, the "talkies," made arcs obsolete for this purpose because of their noisy hiss), and today you still see them at Hollywood events and auto dealerships when the year's new models come out. And, of course, they are at the heart of electric arc welding. The electric arc was soon discovered to have an astonishing property, however, that gave it a future as far removed from the lightbulb as rocket flight is from gliding. By the 1890s it was known that the relationship between the voltage drop across the arc gap, and the gap current, was *not* the linear Ohm's law but rather is described by a curve of the general shape in Figure 7.2. The curious feature of this curve is, of course, that the middle section has a "kink" in which an increase in current is associated with a *decrease* in voltage drop! The total voltage to total current ratio is obviously always positive, but in the middle section the so-called *dynamic* ratio ($\frac{dv}{di}$) is negative, and for this reason the arc is said to have *negative* ac resistance. The electric arc was the first device discovered to have this remarkable property (the triode vacuum tube was next), and it is the key to the arc's ability to easily generate low-frequency oscillations.

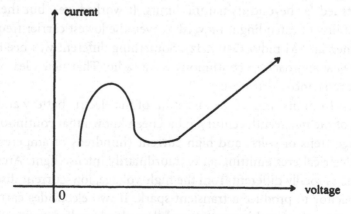

Figure 7.2   The electric arc's negative resistance voltage-current curve.

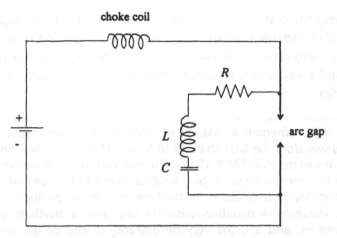

Figure 7.3   Duddell's oscillating arc circuit.

In Figure 7.3 a dc source is connected to an arc gap, which is in parallel with a series *LC* path. The figure also shows a resistance, *R*, in the shunt path to model the unavoidable energy loss mechanisms in any real circuit (as well as the *desired* radiation of energy). Such a resistance was the reason for why oscillations in a resonant circuit (shocked into existence by a spark) would quickly damp out in early radio transmitters (see Chapter 4). With the circuit of Figure 7.3, however, the negative dynamic resistance of the arc *cancelled* the positive resistance *R*!

To understand how this works, you have to visualize two distinct circuits in the figure. First, there is the dc circuit around the loop formed by the dc source, the "choke coil" (an inductor with only its ohmic resistance present at dc, but generating a relatively high ac impedance at nonzero frequencies), and the arc. Second, there is an ac loop formed by *R*, *L*, *C*, and, again, the arc. The total ac resistance in this second loop is the sum of *R* and the *dynamic* resistance of the arc (which, being negative, can result in a net ac resistance of zero). Because of the choke coil, the oscillations in the ac loop cannot "leak back" through the dc source (which would result in energy loss by simply heating the dc source via its internal resistance).

The arc current is, therefore, of two components—a steady dc current, on top of which is superimposed an oscillating (ac) component. This circuit, called the oscillating arc, was invented by the Englishman William Duddell (1872–1917) at the end of the nineteenth century. The oscillations could be produced only at audio frequency rates (one could actually *hear* the sound produced in air by the pulsating arc), however, with Duddell never getting above 10 kHz. Duddell knew the reason for this, too—at higher frequencies the curve of Figure 7.2 flattens out and the negative resistance characteristic is greatly degraded. Still, for the first time a circuit had been constructed which converted dc into *undamped* ac. The next step was to find a way to dramatically increase the oscillation rate up to radio frequencies.

> The first mathematical analysis of nonlinear negative resistance oscillators was done no later than 1920 by the Dutch electrical engineer Balthasar van der Pol (1889–1959). His goal was to calculate the amplitude of the nonlinear oscillations, a calculation you'll recall from Chapter 4 that depends crucially, in the case of a *linear* oscillator, on the initial conditions. A negative-resistance oscillator is anything but linear, however, and it is the very nonlinearity of the circuit itself (not the initial conditions) that limits the magnitude of the oscillations. The successful calculation of the nonlinear oscillation amplitude was one of Van der Pol's great achievements. His analysis was actually motivated by triode vacuum-tube oscillators (discussed in the next chapter) but it is equally valid for the case of the electric arc. In the next several boxes you'll see how to both derive the Van der Pol equation for negative resistance oscillators, and how to solve it using classical mathematics and MATLAB.

The key idea for this last crucial step had actually already been taken before Duddell, by the American Elihu Thomson (1853–1937), who added a *magnetic blowout* to an oscillating arc circuit he had patented in 1893 (but which was essentially unknown until Duddell's work). Using this idea, the Danish engineer Valdemar Poulsen (1869–1942) found that, by creating a transverse magnetic field through the arc gap, the arc could actually be extinguished (blown out) and then reignited during *each* cycle of oscillation. The charged particles in the arc gap experience the so-called *Lorentz force* by virtue of their motion through the magnetic field. This force is perpendicular to the directions of both the field and the particle motion, and so tends to bend the arc out sideways, to lengthen and to eventually break (blow out) the arc. Such arcs found their greatest popularity with the U.S. Navy, starting in 1913.

The final circuit for the arc transmitter, then, is shown in Figure 7.4. The arc current, itself, energizes two electromagnets (A and B) to produce a magnetic field through the gap. Duddell's shunt L and C are now simply the inductance and capacitance of the antenna at the far right of the circuit. As mentioned in the previous paragraph, the arc current has a dc value, with an ac component. Once each cycle of oscillation the current

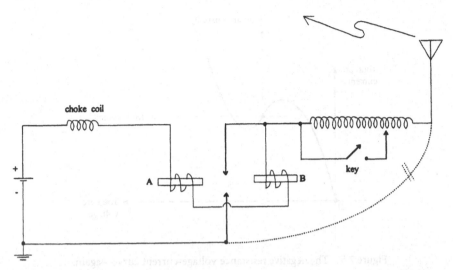

Figure 7.4   The radio frequency arc transmitter, with magnetic blowout for energy injection on *each rf* cycle.

would drop to sufficiently small values that the magnetic field could blow out the arc current. A short time later, the ac voltage in the resonant antenna circuit would become sufficiently large to reignite the arc (via the distributed antenna capacitance, shown in Figure 7.4 in dashed lines). This detail made the arc transmitter's relationship with its antenna system far more intimate than that between either a spark or (later) a vacuum tube transmitter, and their antennas. With those two technologies, oscillations are generated even if the antenna is a poor one (or even absent). With the arc, however, there were no oscillations (no periodic re-ignition of the arc) unless the antenna was very carefully integrated with the arc gap.

To see where Van der Pol's equation for negative resistance oscillators comes from, consider Figure 7.5, which is Figure 7.2 with a second set of axes centered on the middle of the negative resistance portion of the current-voltage curve. If we imagine that our device (*any* negative resistance device) operates at that midpoint $i = v = 0$ when there are no oscillations, then to say the device oscillates is to say we are interested in the current/voltage *deviations* around that midpoint, i.e., i and v denote the time-varying current and voltage, respectively, that *are* the oscillations.

Let's now place the device in parallel with an L and a C, as well as an R, to model the inherent ohmic losses in any real circuit, as shown in Figure 7.6. Applying Kirchhoff's current law to the top node, we can ignore the dc quantities in the circuit and write, for just the time-varying

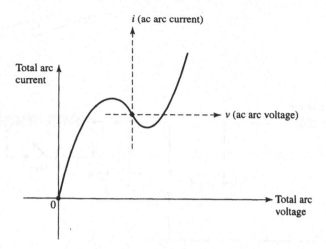

Figure 7.5   The negative resistance voltage-current curve—again.

currents,

$$i + \frac{v}{R} + C\frac{dv}{dt} + \frac{1}{L}\int v\, dt = 0.$$

We can do this because Kirchhoff's current law must be *separately* satisfied by the dc and the time-varying currents; think about this for a while and if you are still puzzled about *why* then read the last note[4] in this chapter. Van der Pol next wrote $i = -av + bv^3$ (with a and b both positive constants) to model the negative resistance "kink." Inserting this *i* into Kirchhoff's equation, differentiating once to remove the integral and multiplying through by L, we arrive at

$$LC\frac{d^2v}{dt^2} + \left[L\left(\frac{1}{R} - a\right) + 3bLv^2\right]\frac{dv}{dt} + v = 0.$$

This can be *greatly* simplified by remembering that we can treat differentials just like algebraic quantities. (The next few lines *are* just a

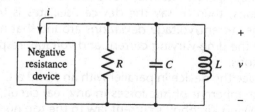

Figure 7.6   Negative resistance in parallel with a lossy resonant circuit.

bit tricky, so stay alert!) Start by writing a normalized time as $x = \omega_o t$ where $\omega_o = \frac{1}{\sqrt{LC}}$. Then, $dx = \omega_o\, dt$ and so $dt = \frac{dx}{\omega_o}$. Thus,

$$\frac{dv}{dt} = \frac{dv}{dx/\omega_o} = \omega_o \frac{dv}{dx}$$

and so

$$\frac{d^2v}{dt^2} = \frac{d}{dt}\left(\omega_o\frac{dv}{dx}\right) = \frac{d}{dx/\omega_o}\left(\omega_o\frac{dv}{dx}\right) = \omega_o\frac{d}{dx}\left(\omega_o\frac{dv}{dx}\right) = \omega_o^2\frac{d^2v}{dx^2}.$$

The Kirchhoff equation then becomes

$$\frac{d^2v}{dx^2} - \sqrt{\frac{L}{C}}\left(a - \frac{1}{R}\right)\frac{dv}{dx} + 3b\sqrt{\frac{L}{C}}v^2\frac{dv}{dx} + v = 0.$$

Next, define $\epsilon = \left(a - \frac{1}{R}\right)\sqrt{\frac{L}{C}}$ and make the change-of-variable $v = hu$ where h is the constant such that $h^2 = \dfrac{\epsilon}{\left(3b\sqrt{\frac{L}{C}}\right)}$. With a bit of algebra the Kirchhoff equation becomes

$$\frac{d^2u}{dx^2} - \epsilon(1 - u^2)\frac{du}{dx} + u = 0$$

which is Van der Pol's nonlinear equation for u (a normalized v) as a function of x (a normalized time). Notice how ϵ has completely absorbed the values of a, R, L and C, and that the value of b appears only in the amplitude scaling constant h. In the next box you'll see how, unlike the linear oscillator case, u does *not* depend on the initial conditions but *does* depend in a remarkable manner on ϵ (in addition, we will require $\epsilon > 0$ for *stable* solutions; see Problems 7.2 and 7.3 for more on stability). As a final comment, while Van der Pol had derived his equation by 1920 the English genius Lord Rayleigh had published an equivalent equation nearly *forty-five years* before; see Problem 7.1 for more on *Rayleigh's equation*.

Some final comments on the physics of the arc are important to make, because they get to the reason for *why* it could generate continuous, undamped high-frequency waves when spark couldn't. Since the arc was created anew on *each* rf cycle, it injected energy into the resonant antenna circuit during *each* cycle, something even Fessenden's 20,000 sparks per second alternator couldn't come close to doing. The ignition/blow out/re-ignition sequence put energy into the antenna circuit literally as fast as it was lost (via both internal heating and radiation). It was, of course, very important to make this sequence as stable as possible, with arc re-ignition occurring at the same relative time within each cycle. Slight variations in the relative re-ignition time translated into frequency "jitter," i.e., into a somewhat broadened frequency spectrum for the transmitter (rather than a pure, single frequency). It was found by early researchers that operating the arc in a hydrogen atmosphere facilitated stable re-ignition of the arc, because such an atmosphere replaced ionized nitrogen and oxygen atoms (relatively heavy ions of the major components of air) with the light ions of hydrogen. These light ions were more quickly removed from the arc gap by the transverse magnetic field than were the heavier ions, and this so-called scavaging of the arc made it possible to have consistent, reproducible conditions in the gap just before each re-ignition, as well as reducing the minimum time interval between successive re-ignitions (i.e., operation at higher frequencies).

The hydrogen atmosphere was achieved by placing the arc in a sealed chamber with a liquid hydrocarbon with a high vapor pressure, e.g., ethyl and methyl alcohol, gasoline, or kerosene. When an arc was first ignited, there could easily be some oxygen in the chamber, of course, and so there was often a start-up explosion! To periodically clear the gap of ionized debris, magnetic fields on the order of 15,000 Gauss (30,000 times the Earth's field) were used. The higher the operating frequency of the arc, the higher the necessary magnetic field strength, because the *shorter* the time available to clear the gap between successive re-ignitions. The physics of arcs is still of interest today, in the highly abstruse electrical engineering specialty of designing fast-response circuit breakers for commercial high-power transmission systems. For a discussion of how that is done (including the use of magnetic blowout), see Werner Rieder, "Circuit Breakers," *Scientific American*, January 1971. Power levels for arc transmitters were measured at the *input* and since they were at best only 50% efficient at producing a radiated signal, they got pretty hot and really high-power arc transmitters would typically be situated next to a cooling pond! Alternators, by contrast, were rated by power *output*. Thus, a 200-kW alternator was approximately equivalent to a 400-kW arc (which produced 200 kW of heat). Both transmitter types operated in the same general long wavelength frequency range (30–60 kHz).

Referring again to Figure 7.4, you can see how arc transmitters were keyed. With the key up, the full antenna coil was in the circuit, and the antenna resonated at a particular frequency. With the key down, part of the coil was shorted and the antenna resonated at a slightly different (higher) frequency. A receiver, sharply tuned to the transmitting frequency when the key was down, would not receive the frequency sent when the key was up (there is a subtle problem here, however, which is addressed at the end of this chapter). Arc transmitters reached truly impressive sizes, often in the many hundreds of kilowatts (with the largest I know of rated at well over a megawatt!). With the development of the vacuum tube, however, such transmitters went the same way as did spark and alternator. By the early 1920s, arc radio was commercially dead.

Van der Pol's equation, as a nonlinear differential equation, is not easy to solve. Indeed, Van der Pol himself was able to find analytic solutions only for the case of $\epsilon \ll 1$. He did develop a graphical method for finding solutions for large ϵ (see Problem 7.4), however, and others even found graphical solutions to generalizations of the Van der Pol equation. For example, the French mathematical physicist Alfred Liénard (1869–1958) studied $\ddot{u} + f(u)\dot{u} + g(u) = 0$ for $f(u)$ and $g(u)$ satisfying certain conditions (I'm using the dot notation for differentiation). For the special case of the Van der Pol equation, $f(u) = -\epsilon(1 - u^2)$ and $g(u) = u$ do satisfy Liénard's conditions. The generalized differential equation is called *Liénard's equation*. Today even very complicated nonlinear differential equations can be rapidly and accurately solved by computer. MATLAB, in particular, provides a selection of five differential equation "solvers." In preparation for invoking a solver, all one needs to do is to write the nth-order differential equation of interest as a set of n first-order differential equations. So, for the Van der Pol equation, define

$$u_1(x) = u(x) \quad \text{and} \quad u_2(x) = \dot{u}(x) = \dot{u}_1(x).$$

Then, our second-order differential equation becomes the two first-order differential equations

$$\dot{u}_1 = u_2$$
$$\dot{u}_2 = \epsilon(1 - u_1^2)u_2 - u_1.$$

Once the initial conditions of $u(0) = u_1(0)$ and $\dot{u}(0) = \dot{u}_1(0)$ are given then a solution is uniquely determined. The MATLAB program vdpol.m (which calls the subroutine function vdpolsub.m) implements the above first-order equations and the specified initial conditions, and plots the solution. The next box of this chapter shows some typical results and gives some general conclusions.

```
%vdpol.m/created by PJNahin for ''The Science of Radio 2e''(6/14/98)
xinterval = [0, 100];          %normalized time interval
uo = [0;0.01];                 %initial conditions, with uo(1)=u(x=0)
                               %and uo(2)=u'(x)=u-dot(x)
[x,u] = ode15s('vdpolsub',xinterval,uo);
                               %call Ordinary Differential
                               %Equation solve, with
                               %epsilon defined in vdpolsub
plot(x,u(:,1))                 %plot u(1) = u(x)
grid
xlabel('normalized time, x')
ylabel('normalized amplitude, u')
title('Fig.7.9 - One Possible Solution to the Van Der Pol Equation')
legend(u'(0)=0','u-dot=0.01','epsilon=10')
figure(1)


%vdpol.subm/created by PJNahin for ''The Science of
%Radio 2e''(6/14/98)
function        uvector=vdpolsub(x,u);
epsilon=10;
uvector = [u(2); epsilon*(1-u(1)\^2-u(1)];       %uvector is a column
%vector, i.e., notice the semicolon separation of elements, where
%the first column u(:,1) = u(x) and the second column u(:,2) =
%u'(x) = u-dot(x)
```

Even though alternator and arc radio transmitters were based on very different physical principles, they shared the characteristic of producing signals with no amplitude modulation. For the alternator, it either transmitted a signal at one frequency (key down) or no signal at all (key up). In a certain sense, of course, one might consider this as a limiting case of amplitude modulation (the signal amplitude is either full-on or full-off), but the information in the alternator's signal was actually carried by the sheer presence of the signal and not by the signal's particular amplitude (as in true AM). And for the arc, it either transmitted a signal at one frequency (key down) or at another frequency (key up). This sort of signal is today called FSK (frequency shift keying) and it is often used for the transmission of digital data—there is *no* AM present.

The following plots illustrate how the nature of the solutions to Van der Pol's equation change as a function of ϵ, and of the initial conditions. There are four general conclusions that can be drawn from the plots: (1) the solutions oscillate, appearing sinusoidal for $\epsilon \ll 1$ but becoming less so as ϵ increases; (2) the final normalized amplitude of the oscillations is very nearly 2, *independent* of both ϵ and the initial conditions (the next box shows how to derive this result for the case[5] of $\epsilon \ll 1$); (3) the frequency of the oscillations decreases as ϵ increases; (4) the larger ϵ is, the faster the steady-state oscillation amplitude is reached.

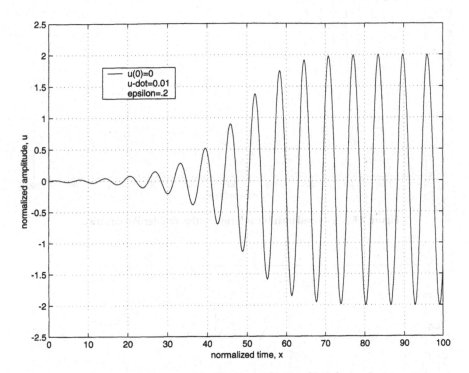

Figure 7.7   One possible solution to the Van der Pol equation.

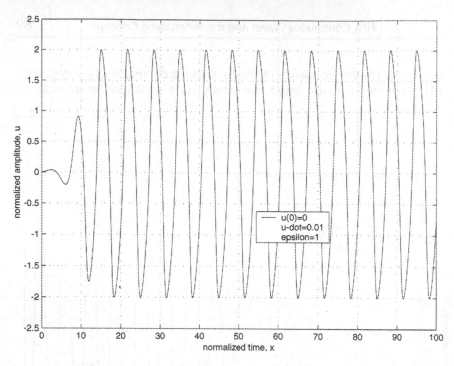

Figure 7.8   A second possible solution to the Van der Pol equation.

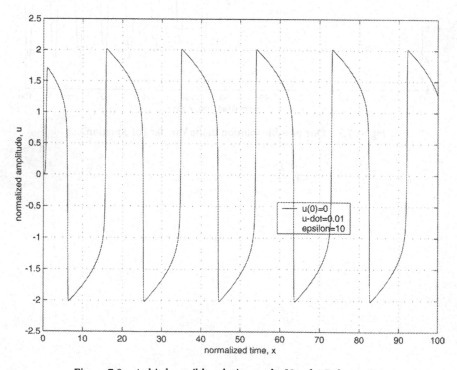

Figure 7.9   A third possible solution to the Van der Pol equation.

It is not possible to detect FSK signals using the envelope detection process described in Chapter 5. The constant amplitude FSK signal of the arc would produce no output at all from an envelope detector, and the on-off signal of the alternator would (at best) produce short, pulselike "clicks" in the headphones of Figure 5.4. The problem of how to receive alternator and arc signals resulted, in fact, in a solution that is one of the fundamental concepts of modern radio—the concept of *heterodyning* (from the Greek *heteros* or external, and *dynamics* or force). Of all his contributions to early radio, probably Reginald Fessenden's most important (and lasting) is the *heterodyne detector* (an example is shown in Figure 7.10, dating from 1907, but the original patent dates from 1901). The received signal in the antenna is nonlinearly "mixed" with the output of a local oscillator (i.e., the "external force") via the magnetic coupling and, as discussed in Chapter 6, new signals will then appear in the headphones at the sum and difference frequencies. By adjusting the frequency at the local oscillator, the headphone difference frequency could be placed in the audio range. Fessenden's circuit was ahead of its time, however, as there simply was no technology available then with which to build the required local oscillator with the necessary frequency stability. The invention of the triode vacuum tube would change that situation—it was the tube that made AM broadcast radio a commercial possibility.

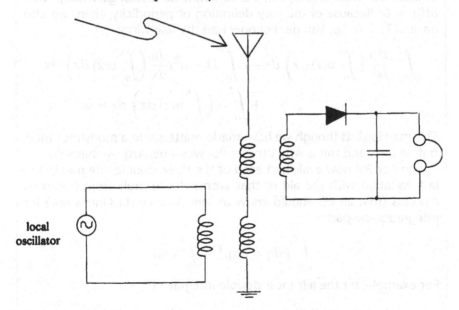

local oscillator

Figure 7.10   Fessenden's heterodyne detector circuit, using a local oscillator, for the non-AM waves produced by alternator and arc transmitters.

In a 1926 paper[6] Van der Pol gave an ingenious derivation for why the steady-state amplitude of his equation's oscillatory solutions, for the case of $\epsilon \ll 1$, is 2. His analysis also shows why this amplitude, at least for small ϵ, is independent of ϵ. Every electrical engineering student should study Van der Pol's analysis simply to admire an extraordinarily clever mind at work. Here's what he did (although I've expanded on his original, somewhat terse remarks).

Starting with his equation, Van der Pol multiplied through by the common factor $\int_o^x u(z)\,dz$ to get

$$\frac{d^2u}{dx^2}\int_o^x u(z)\,dz - \epsilon(1-u^2)\frac{du}{dx}\int_o^x u(z)\,dz + u\int_o^x u(z)\,dz = 0.$$

He then assumed that the steady state $u(x)$ is *periodic*, with zero average value, independent of ϵ. This is a good assumption; after all, Van der Pol had experimental solutions such as shown in Figures 7.7, 7.8, and 7.9 on the desktop in front of him with just these properties. Let's call the period T. You should note carefully that in all that follows we will *not* have to know the actual value of T (but a variation of the analysis that follows will let you calculate, as did Van der Pol, the value of T; see Problem 7.6). This is important because, as Figures 7.7, 7.8, and 7.9 show, T *does* depend on ϵ.

Van der Pol next integrated the above equation, term-by-term, over a period. Now, we can arbitrarily define the time $x = 0$ as the start of a period when $u = 0$, i.e., we can with no loss of generality write $u(0) = 0$. Because of the very definition of periodicity, then, we also have $u(T) = 0$. So, Van der Pol now had the following:

$$\int_o^T \frac{d^2u}{dx^2}\left(\int_o^x u(z)\,dz\right)dx - \epsilon\int_o^T (1-u^2)\frac{du}{dx}\left(\int_o^x u(z)\,dz\right)dx$$
$$+ \int_o^T u\left(\int_o^x u(z)\,dz\right)dx = 0.$$

This may *look* as though we have made matters into a monstrous mess but, in fact (and this is why Van der Pol was a genius), we haven't.

Van der Pol now evaluated each of the three double integrals in the last equation with the aid of that wonderful formula from freshman calculus (that all EEs should know as well as they do Ohm's law) for integration-by-parts:

$$\int_o^T p\,dq = (pq|_o^T - \int_o^T q\,dp.$$

For example, for the left-most double integral let

$$dq = \frac{d^2u}{dx^2}\,dx \quad \text{and} \quad p = \int_o^x u(z)\,dz.$$

Then,

$$q = \frac{du}{dx} \quad \text{and} \quad dp = u(x)\,dx.$$

So, our wonderful formula tells us that

$$\int_o^T \frac{d^2u}{dx^2}\left(\int_o^x u(z)\,dz\right)dx = \left(\frac{du}{dx}\int_o^x u(z)\,dz\Big|_o^T\right) - \int_o^T \frac{du}{dx}u(x)\,dx$$

$$= \left(\frac{du}{dx}\right)_{x=T}\int_o^T u(z)\,dz - \int_o^T u\,du.$$

The first of the last two integrals is zero since the area under the $u(x)$ curve over a period is zero (because $u(x)$ has zero average value). And the last integral is zero because it is equal to

$$\frac{1}{2}u^2\Big|_o^T = \frac{1}{2}\left[u^2(T) - u^2(0)\right] = \frac{1}{2}[0 - 0] = 0.$$

If you repeat this process, i.e., if you integrate the right-most double integral by parts, you'll see that the same result is reached:

$$\int_o^T u\left(\int_o^x u(z)\,dz\right)dx = 0.$$

These first two results can only mean that the middle double integral must also vanish, i.e., that (notice that ϵ has vanished from the equation)

$$\int_o^T \frac{du}{dx}\left(\int_o^x u(z)\,dz\right)dx = \int_o^T u^2\frac{du}{dx}\left(\int_o^x u(z)\,dz\right)dx.$$

Integrating each side by parts, this statement quickly reduces to

$$\int_o^T u^2(x)\,dx = \frac{1}{3}\int_o^T u^4(x)\,dx.$$

All we have assumed to this point is that $u(x)$ is periodic (with unspecified period T) and has a zero average value. If we now impose the additional observation that, for ϵ "small," $u(x)$ appears to be very nearly sinusoidal, then we can write

$$u(x) = A\sin\left(\frac{2\pi x}{T}\right)$$

where A is the amplitude value we want to calculate. Thus,

$$A^2\int_o^T \sin^2\left(\frac{2\pi x}{T}\right)dx = \frac{1}{3}A^4\int_o^T \sin^4\left(\frac{2\pi x}{T}\right)dx$$

or,

$$A = \sqrt{3\frac{\frac{1}{T}\int_o^T \sin^2\left(\frac{2\pi x}{T}\right)\,dx}{\frac{1}{T}\int_o^T \sin^4\left(\frac{2\pi x}{T}\right)\,dx}}.$$

The two integrals are simply the average values of $\sin^2$ and $\sin^4$ over a period and can be looked up in any good table of integrals; their values are $\frac{1}{2}$ and $\frac{3}{8}$, respectively. So,

$$A = \sqrt{3\frac{\frac{1}{2}}{\frac{3}{8}}} = \sqrt{4} = 2,$$

and this is just what is observed.

Notes

1. While not specifically banned by the Radio Act of 1912, the following language in the Act makes it clear that the days of spark radio were numbered: "At all stations if the sending apparatus, to be referred to hereinafter as the 'transmitter,' is of such a character that the energy is radiated in two or more wavelengths, more or less sharply defined, or indicated by a sensitive wave meter, the energy in no one of the lesser waves shall exceed ten per centum of that in the greatest."

2. Or perhaps I should say the second, as in the early 1890s the eccentric electrical genius Nikola Tesla had built a 15-kHz alternator for ac experimentation. But he never applied it to radio.

3. The definitive story of Alexanderson's life and work is in James Brittain's *Alexanderson*, John Hopkins, 1992. The development of the radio-frequency alternator was really just a small part of this amazing man's creativity. In an engineering career that spanned nearly three-quarters of a century, he received over 340 patents (the first in 1903, the last in 1968 at age 90).

4. If there are no oscillations when the negative-resistance device is at the midpoint of the voltage-current "kink" then there are only dc currents in the circuit, which of course satisfy Kirchhoff's current law at the top node. When the circuit begins to oscillate there are then time varying currents superimposed on the dc currents, and the total currents (dc plus time-varying) in the components connected to the top node also of course satisfy Kirchhoff's current law. But that means the time-varying components, alone, must also satisfy Kirchhoff's current law.

5. This result was established for arbitrary ϵ by the English mathematicians Mary Cartwright (1900–1998) and John Littlewood (1885–1977), in response to a January 1938 request from the British Radio Research Board for aid in helping engineers understand the behavior of "certain types of non-linear differential equations involved in the technique of radio engineering." They were able to show, in fact, that the amplitude of the oscillations tends to 2 only in the case of ϵ "large," while in the case of ϵ "small" the steady-state amplitude is *just* a bit different from 2. This work was of great importance in electronic radio engineering during the Second World War. See Shawnee L. McMurran and James J. Tattersall, "The Mathematical Collaboration of M. L. Cartwright and J. E. Littlewood," *The American Mathematical Monthly* 103, December 1996, pp. 833–845, and Cartwright's own paper "Some Points in the History of the Theory of Nonlinear Oscillations," *Bulletin of the Institute of Mathematics and Its Applications* 10, 1974, pp. 329–333.

6. Balthazar Van der Pol, "On Relaxation-Oscillations," *Philosophical Magazine* 2, November 1926, pp. 978–992.

Problems

7.1. In an 1883 article ("On Maintained Vibrations") in *Philosophical Magazine*, Lord Rayleigh discussed the nonlinear differential equation

$$\frac{d^2\theta}{dt^2} + K\frac{d\theta}{dt} + K'\left(\frac{d\theta}{dt}\right)^3 + n^2\theta = 0.$$

Indeed, he had even earlier (1874) encountered essentially the same equation in connection with the vibrations of a violin string. By appropriately selecting the values of the constants K, K' and n^2, it is evident that this equation includes the following special case (now called *Rayleigh's equation*), where I've made the trivial notational changes of y for θ and x for t:

$$\frac{d^2y}{dx^2} - \epsilon\left[\frac{dy}{dx} - \frac{1}{3}\left(\frac{dy}{dx}\right)^3\right] + y = 0.$$

It can be shown that this equation is mathematically equivalent to Van der Pol's equation and, in a certain sense, it is "simpler" because the nonlinear term involves only $\frac{dy}{dx}$ while the nonlinear term in Van der Pol's equation involves both u and $\frac{du}{dx}$. To show the equivalence, perform the following steps:

1. make the change-of-variable $u = \frac{dy}{dx}$ in the Van der Pol equation.
2. indefinitely integrate the result from (1), term-by-term, and so derive Rayleigh's equation.

Discuss the detail of the constant-of-integration that, in general, appears in the integration operation of (2).

7.2. As shown by the plots in Figures 7.7, 7.8, and 7.9, solutions to the Van der Pol equation are oscillations that result from essentially *any* choice of initial conditions. One particular choice, however, seems to be the exception, i.e., $u(0) = 0$, $\dot{u}(0) = 0$. That is, if a negative resistance circuit starts off at time $x = 0$ with zero amplitude *and* with zero rate-of-change, then the time-varying voltage *remains* at zero. While this is mathematically correct, it can be shown that this particular solution is also an unstable one; the slightest perturbation in u—for example, a passing cosmic ray that knocks an electron out of a valence orbit in a copper atom in the circuit's wire and thereby makes $\dot{u} \neq 0$—will trigger the circuit into oscillation and so Van der Pol's

equation (and any circuit it describes) will "almost certainly" spon-
taneously burst into oscillation when energized. To show that this is
so, fill-in the details for the following argument. We can always define
the origin of normalized time ($x = 0$) to be the instant of the perturbation that results in the initial conditions $u(0) = 0$, $\dot{u}(0) \neq 0$. Then,
over a following very short time interval when u is still very small the
solution to the Van der Pol equation should differ only slightly from
the solution to the *linear* differential equation

$$\frac{d^2u}{dx^2} - \epsilon\frac{du}{dx} + u = 0$$

with the same initial conditions. By the method of Chapter 4 (i.e.,
assume $u(x) = Ue^{sx}$) you should be able to calculate the two resulting
values of s (call them s_1 and s_2) as function of ϵ, and so

$$u(x) = U_1e^{s_1x} + U_2e^{s_2x}.$$

Since $u(0) = 0$ then $U_1 = -U_2$ or, dropping the subscripts,

$$u(x) = U(e^{s_1x} - e^{s_2x}).$$

Now, since $\dot{u}(0) \neq 0$ then $U \neq 0$ (why?) and so $u(x)$ is *not* zero.
Indeed, since *if $\epsilon > 0$* both s_1 and s_2 can be shown to be either real and
positive, or to be complex with positive real parts (this is to be done
by *you*), then $u(x)$ will grow in amplitude as x increases. Whether $u(x)$
grows more positive or more negative depends on the algebraic sign
of $\dot{u}(0)$, but in either case the instability of the $u(x) = 0$ solution is
shown. As a final part of your analysis, discuss what happens if $\epsilon < 0$
(but don't forget, this analysis is for the linear approximation to the
Van der Pol equation and not for the true Van der Pol, itself).

7.3. In the derivation of Van der Pol's equation it was stated that the condition $\epsilon > 0$ is required for a stable solution. What is the *physical*
reason behind this requirement? Hint: recall that $i = -av + bv^3$, calculate the dynamic resistance of the "kink" at $v = 0$, and then observe
how your result connects to the definition of ϵ.

7.4. In the days before computers and MATLAB, how *were* non-linear differential equations like Van der Pol's studied? One clever technique,
called *phase-plane analysis*, is based on the idea of eliminating time
from the problem. That is, define the variable $y = \frac{du}{dx}$, and so

$$\frac{d^2u}{dx^2} = \frac{dy}{dx} = \frac{dy}{du}\frac{du}{dx} = y\frac{dy}{du}.$$

Then the Van der Pol equation becomes

$$y\frac{dy}{du} - \epsilon(1 - u^2)y + u = 0$$

or,

$$\frac{dy}{du} = \epsilon(1 - u^2) - \frac{u}{y}$$

and so the normalized time variable (x) no longer explicitly appears. To illustrate the details of this method for a specific circuit, let's begin with the arc circuit of Figure 7.11 where i_g and v_g are the arc current and voltage drop, respectively.

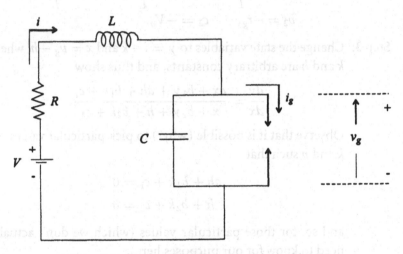

Figure 7.11   The oscillating arc circuit for problem 7.4.

Step 1: Write Kirchhoff's laws for the circuit, eliminate the time variable, and thus show that

$$\frac{L}{C}\frac{di}{dv_g} = \frac{V - v_g - iR}{i - i_g}.$$

The variables i (the battery current) and v_g are what electrical engineers call *state variables*, i.e., their instantaneous values define the instantaneous state of the circuit. These, we (you) will show, oscillate.

Step 2: Assume the state of the circuit is always confined to the "negative resistance" region of the arc, the "kink," which we'll model as

$$v_g = V_0 - r_g i_g$$

where V_0 and r_g are positive constants (that is, rather than Van der Pol's "cubic kink" I have *linearized* the negative-resistance "kink" region for mathematical simplicity). Show then that

$$\frac{di}{dv_g} = \frac{a v_g + b_1 i + c_1}{v_g + b_2 i + c_2}$$

where

$$a = -\frac{r_g C}{L}, \quad b_1 = -\frac{r_g RC}{L}, \quad c_1 = \frac{r_g CV}{L}$$

$$b_2 = -r_g, \quad c_2 = -V_0.$$

Step 3: Change the state variables to $y = i - k$ and $x = v_g - h$, where k and h are arbitrary constants, and thus show

$$\frac{dy}{dx} = \frac{ax + b_1 y + ah + b_1 k + c_1}{x + b_2 y + h + b_2 k + c_2}.$$

Observe that it is possible (why?) to pick particular values for k and h such that

$$ah + b_1 k + c_1 = 0$$
$$h + b_2 k + c_2 = 0$$

and so, for those particular values (which we don't actually need to know for our purposes here),

$$\frac{dy}{dx} = \frac{ax + b_1 y}{x + b_2 y}.$$

Step 4: Despite initial (horrifying?) appearances, this equation can be exactly integrated (try the change of variable $y = zx$ and watch the variables separate), but for our purposes here let's be far less general. Suppose we adjust R, C, and/or L such that

$$\frac{r_g RC}{L} = 1.$$

Then the differential equation takes on the particularly simple form

$$x\,dy + y\,dx + r_g y\,dy + \frac{1}{R} x\,dx = 0.$$

Since the first two terms are the exact differential $d(xy)$, then we can integrate directly to

$$xy + \frac{1}{2}r_g y^2 + \frac{1}{2}\frac{1}{R}x^2 = \text{constant}.$$

Step 5: We are done! All we need to do is observe that the integrated equation for the state variables x and y is an ellipse (actually a *family* of ellipses, as the constant on the right-hand side is still undetermined). That is, the state variables (which of course are functions of time) are such that the circuit state moves along a *closed loop path*, which means the state is *periodic*. One complete orbit of the state around the elliptical path is one period. The state variables x and y (or i and v_g) are therefore oscillating, and in fact they both execute undamped sinusoidal oscillation (*undamped* because the state orbit is closed rather than an inward spiral, and *sinusoidal* because the orbit is elliptical—question: what would the state orbits be for a circuit generating damped, and then undamped, *square* wave oscillations?). These conclusions have been made on the basis of the very special assumption that $r_g RC = L$, but in fact a more detailed analysis would show they hold under much more general conditions. The point here was simply to *mathematically* demonstrate that undamped sinusoidal oscillations are indeed possible in negative resistance circuits, and we did so *without* explicitly solving for anything!

Step 6: You may have noticed that Figure 7.11 doesn't look quite like Figure 7.3, which was used in describing Duddell's arc circuit. Try repeating this analysis for Duddell's circuit.

7.5. Oscillatory circuits that obey Van der Pol's equation were referred to by him in his 1926 paper (see Note 6) as *relaxation oscillators*, since their frequency is determined by the decay (or "relaxation") of some physical process inherent in the circuit. Another famous example of a relaxation oscillator, one mentioned in passing at the end of the 1926 paper, is shown in Figure 7.12a, in which the component at the far left is a gas-filled bulb (the gas is typically neon). The behavior of the bulb is highly nonlinear (the resistor r is not a separate component, but rather represents the resistance of the gas), and is as follows:

(1) When the voltage drop, v, across the bulb terminals is less than V_1 the gas is a nonconductor and the bulb is an open circuit; $r = \infty$;

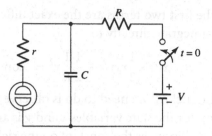

Figure 7.12a   A relaxation oscillator.

(2) when v reaches V_1 the gas ionizes (valence electrons in the neon atoms are ripped free and the gas becomes a mixture of free negative charges and positively charged ions) and so becomes a relatively good conductor (r is finite); (3) Once ionized, the gas remains a conductor until v drops to V_2, where $V_2 < V_1$, and then the gas recombines and once again becomes nonconducting. So, here is what happens in the circuit of Figure 7.12a, assuming C is initially uncharged at time $t = 0$ when the switch is closed. Since $v(0) = 0$ the bulb is an open circuit and C is charged by the battery through R. The voltage drop across C, v, approaches the battery voltage V in an exponentially decaying manner but, assuming $V > V_1$, at some finite time $t = T$ the capacitor voltage reaches V_1 and the bulb gas ionizes. This presents C with a relatively low resistance discharge path, assuming $r \ll R$. After some further time interval (which we'll assume is short compared to T) the voltage drop across C falls to V_2 and so the bulb gas recombines. Thus, C starts to recharge. This charge/discharge process repeats over and over. With each discharge a pulse of current flows in the bulb and the gas glows, briefly. Find an expression for the time interval between successive blinks of the bulb.

Note: EWB does not have a gas bulb in its components bins, but it does have an equivalent component, called a "voltage-controlled switch." In the EWB schematic, Figure 7.12b, such a switch has been set to close at 90 volts ($= V_1$) and to open at 10 volts ($= V_2$). For a battery voltage of 100 volts ($= V$) and an RC time constant of one second, the oscilloscope clearly displays the oscillatory nature of the voltage drop across C. The time interval between successive switch closings (bulb ionizations) is about 2.2 seconds. Does this agree with your expression?

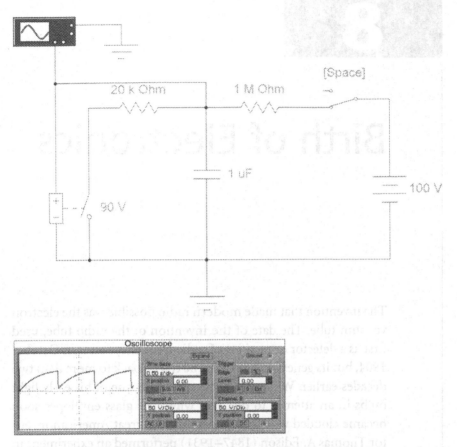

Figure 7.12b The circuit is activated when the switch in series with the battery is closed (by depressing the computer keyboard space key). The voltage-controlled switch *AND* the 20k-ohm resistor, together, model the gas-filled bulb. The switch is set to close (i.e., the bulb-gas ionizes) when there is a 90 volt drop across the resistor/switch, and to open (i.e., the bulb-gas recombines) when the voltage drop decreases to 10 volts across the resistor/switch. The oscilloscope screen is displaying the signal on channel A.

7.6. Examination of Figure 7.7 shows that the period of the oscillatory solution to Van der Pol's equation, for ϵ "small," is about $T = 6$. Use the following approach, described in Van der Pol's 1926 paper, to show how this value can be *derived*: multiply through the Van der Pol equation by u, write the integrals term-by-term over a period (T), and then integrate by parts. This will give you a simple equation for T if you make the assumption $u(x) = A \sin\left(2\pi \frac{x}{T}\right)$.

Birth of Electronics

The invention that made modern radio possible was the electron vacuum tube. The date of the invention of the radio tube, used first as a detector or rectifier of radio frequency waves, is October 1904, but its genesis can actually be traced back to more than two decades earlier. While experimenting with one of his early light bulbs in an attempt to discover why their glass envelopes soon became clouded with a dark deposit, the great American inventor Thomas A. Edison (1847–1931) performed an experiment in February 1880 that was far ahead of its time. Using the circuit of Figure 8.1, he found that if a wire probe (p) was sealed inside a glass envelope along with the filament (f), then there was a small current (measured in microamperes) flowing in the evacuated space between the probe and the filament. That is, there was a current *if* the current meter (m) was connected to the positive terminal of the filament. Moving the connection to the negative terminal resulted in *no* detected current.

This observation became known as the *Edison effect*, but because it failed to help Edison solve his immediate problem with lightbulbs, he failed to pursue it. What was happening is easy to understand today, with more than a century of progress in physics, but for Edison and his contemporaries it was truly a great mystery. So what *is* going on in Figure 8.1? The purpose of the battery is simply to heat the filament to incandescence (to temperatures between 1800°F to over 4500°F), to produce light.

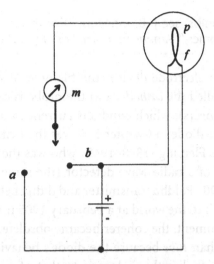

Figure 8.1   The Edison effect circuit. When the switch is in position "b" there is a current through the current meter m, but in position "a" there is no current.

Unknown to Edison, however, was that so great is the thermal agitation in such an intensely hot filament that electrons are literally "boiled" out of the filament and into the vacuum of the bulb. These electrons form a negatively charged electron "cloud" around the filament. The size of the cloud depends on the temperature of the filament, but for any given temperature the "space charge" cloud quickly self-limits. That is, the negative charge of the cloud quickly reaches a level where Coulomb repulsion prevents any further electrons from joining it (because like charges repel). When the meter switch in Figure 8.1 was moved to position b (the positive terminal of the filament battery), then electrode p attracted the negative electrons in the filament space charge of Edison's lightbulb. The motion of these electrons is the Edison effect current. Putting the switch in position a, however (connecting the negative battery terminal to electrode p) did not attract the space charge electrons.

The Edison effect wouldn't be understood until after the turn of the century, in large part because of the work of Sir Owen W. Richardson (1879–1959), who received the 1928 Nobel Prize in physics for his theoretical studies of electron emission in vacuum tubes. The electron itself wasn't discovered until 1897, for which Sir J. J. Thomson (1856–1940) received the 1906 physics Nobel Prize. For more on the Edison effect experiments (and what was actually causing the dark deposit on Edison's lightbulbs) see J. B. Johnson, "Contributions of Thomas A. Edison to Thermionics," *American Journal of Physics*, December 1960.

This simple, two-electrode device (the filament or electron emitter is more technically called the *cathode*, and the probe is called the *anode*, or *plate* by radio engineers), which conducts current in only one direction, behaves just like the diode in Chapter 5. It was the British electrical engineer John Ambrose Fleming (1849–1945) who was the first to appreciate this in the context of a radio wave detector (the same Fleming who designed Marconi's 1901 Poldhu transmitter and didn't believe in sidebands), and he announced it to the world at a February 1905 meeting of the Royal Society.[1] At that moment, the coherer became obsolete. Fleming used the term *valve* rather than *tube* because the diode's behavior with respect to current struck him as directly analogous to that of a plumber's valve in controlling the flow of water.

If the circuit of Figure 8.1 is slightly modified by having the filament indirectly heat the cathode (which can be made from a copious electron-emitting material), and also by inserting an adjustable dc voltage source between cathode and anode (see Figure 8.2), then Figure 8.3 shows how the tube current *I* varies as a function of *V* (the anode-to-cathode voltage). When *V* is "low" the current increases rapidly with increases in *V*, as more of the available space charge electrons are attracted to the *plate* (as I'll call the anode from now on). This current is called the *space charge current* and within very weak assumptions it is described by the so-called Child–

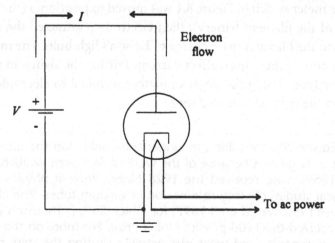

Figure 8.2   A forward-biased vacuum diode with indirectly heated cathode.

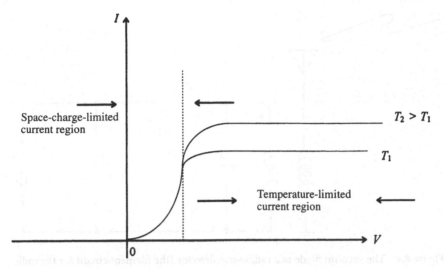

Figure 8.3   Voltage-current characteristic curves for a forward-biased vacuum diode.

Langmuir law

$$I = KV^{3/2},$$

where K is a *constant* dependent on the details of the physical geometry of the tube. As V is increased, however, it eventually becomes sufficiently large that *all* of the space charge electrons are attracted to the plate as fast as they are boiled off the cathode, and so additional increases in V result in no increase in the tube current. The tube current is then said to be *temperature limited*. That is, the only way to further increase I is to increase the size of the space charge by increasing the temperature of the cathode. When used as an rf detector, the vacuum diode can be connected as shown in Figure 8.4.

These comments are valid only for the case where the filament has reached operating temperature, i.e., only after the tube has 'warmed up'. Until then, for a time lapse that could be as long as ten to fifteen seconds, the circuitry the tube is part of will not properly operate. For the same reason, a tube circuit may continue to operate for a few seconds *after* power is disconnected, until the filament has cooled. By comparison, modern solid-state radio circuits turn on instantly, although they, too, may play for a few seconds after power is turned off—*but for a different reason* (the power supply capacitors may retain sufficient charge, for a second or two, to continue operation). An amusing illustration of the first case, for vacuum tubes, occurs in the classic

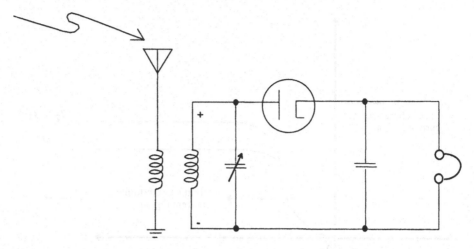

Figure 8.4   The vacuum diode as a radio wave detector (the filament circuit for the indirectly heated cathode is now shown). When the polarity of the arriving signal is as indicated, the diode conducts. The circuit is equivalent to that in Figure 5.1.

1939 film *Another Thin Man,* in which Nick Charles (William Powell) is about to be shot by two gangsters. To cover the noise of their guns, they turn-on a nearby radio but are forced to wait "until it warms up." That's enough time, of course, for the police to arrive and save our hero.

Indirectly heated cathodes are convenient for a variety of reasons. A practical one is that the filament (or *heater,* as it is called by electrical engineers in its electronic, non-lightbulb application) can be powered by the ac power out of a wall socket instead of by an expensive battery, and yet, because of the relative massiveness of the cathode and its thermal inertia, there is no ac frequency "hum" introduced on the tube current. Another result of indirect heating, that theoreticians like, is that the entire cathode is at the same potential, while if a cathode filament is directly powered by a battery, then different parts of the cathode are at different potentials (because of the ohmic voltage drop across the cathode structure from one battery terminal to the other). Such potential variations cause theoretical analysis problems.

In 1906–1907 the American inventor Lee De Forest (1873–1961) inserted a third electrode between the cathode and the plate. De Forest called his device the "audion," but it is now called the *triode.* The triode extended the capability of the vacuum tube from that of a mere detector of signals to that of an *amplifier* of them and, perhaps even more importantly, as the *generator* of rf signals (even the earliest of commercial triodes could operate over a frequency interval from fractions of a hertz to several hundreds of kilohertz). (As discussed in Chapter 21, however, these great circuit *applications* of the triode were the work of Edwin Armstrong, *not* De Forest. Although he had a PhD from Yale, there is no doubt among modern historians that De Forest had, at best, only a confused idea of just how and why his three-element vacuum tube actually worked.)[2] Sparks, arcs, and alternators became engineering history; the triode revolutionized radio, and started an electronics boom that hasn't stopped since.

The Child–Langmuir equation can be derived from fundamental principles, and has been known since 1911. It was derived that year by C. D. Child, and again in 1913 by Irving Langmuir (1881–1957) who in 1932 won the Nobel Prize in chemistry. See, for a theoretical analysis, any "older" book on advanced electronics, e.g., Jacob Millman, *Vacuum-Tube and Semiconductor Electronics*, McGraw-Hill, 1958, pp. 99–103. ["Older" because most authors of modern advanced electronics texts seem to have concluded that vacuum tubes are obsolete, and thus devote themselves exclusively to discussing solid state electronics. For electronic applications requiring operation at extraordinarily high frequencies (thousands of megahertz) and high power (hundreds of kilowatts, and even megawatts), however, the vacuum tube and its descendants still reign supreme over solid state devices.] The Child–Langmuir law is very general, holding for a wide variety of cathode/plate geometries (each with its own particular value for the constant K).

De Forest's third electrode, called the *grid* because it often takes the form of a fine-meshed screen, is normally operated at a voltage negative with respect to the cathode (and so there is normally no current in the grid circuit). That is, in the notation of Figure 8.5, $V_{gk} < 0$ and the plate-to-cathode potential difference $V_{pk} > 0$. Since the grid is physically closer to the cathode (and to its space charge) than is the plate, the tube current is more responsive to changes in the grid potential than it is to equal changes in the plate potential. I should mention that a zero grid current is a theoretically nice assumption that isn't always actually true. It is most closely approximated when triodes are used as amplifiers, although even then some of the electrons on their way from cathode to plate can still run into the grid despite its negative, repulsive voltage. The grid current then, however, is still very small, typically only a fraction of a microampere. In more sophisticated circuits, beyond what I'll discuss in this book, the grid is intentionally driven positive (briefly) and this can result (briefly) in substantial grid currents.

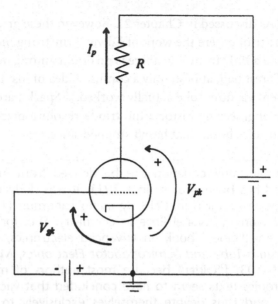

Figure 8.5   A triode tube with its electrode voltages in their "usual" state.

Following tradition, I am using the letter k for cathode, rather than the more obvious c. Pay close attention to this notation! We could just as well write $V_{kg} > 0$ (the cathode-to-grid potential difference), and $V_{kp} < 0$. The order of the subscripts has meaning. I am using capital letters to denote dc quantities, and will use lowercase for time-varying ones. When we come to differentials (like dI_p), we will think of them as small amplitude ac variations around some particular dc value.

It is found experimentally that the Child–Langmuir law can be extended from diodes to triodes, as long as there is no grid current, by writing

$$I_p = K_1(V_{gk} + K_2 V_{pk})^{3/2}$$

where K_1 and K_2 are constants (this reduces to the diode case when $V_{gk} = 0$ and $K_1 K_2^{3/2} = K$). Notice, in particular, that K_2 must be dimensionless and, since I_p is more responsive to changes in V_{gk} than it is to equal changes in V_{pk}, then $K_2 < 1$. We can write the total change in I_p due to changes in V_{gk} and V_{pk} as

$$dI_p = \left.\frac{\partial I_p}{\partial V_{gk}}\right|_{\substack{V_{pk}\ \text{held} \\ \text{constant}}} dV_{gk} + \left.\frac{\partial I_p}{\partial V_{pk}}\right|_{\substack{V_{pk}\ \text{held} \\ \text{constant}}} dV_{pk}.$$

The two partial derivatives have the units, respectively, of ohms and conductance (reciprocal ohms, or what used to be amusingly called mhos!), i.e., denoting the two partial derivatives by g_m and $\frac{1}{r_p}$ we have

$$dI_p = g_m dV_{gk} + \frac{1}{r_p} dV_{pk},$$

where g_m and r_p are called, respectively, the tube's *transconductance* and *plate resistance*. The *trans* is used because the g_m current-to-voltage ratio is not calculated from quantities measured at the same location, but rather one quantity is *transferred* (in the mathematics) to the location of the other.

Now, suppose $dI_p = 0$. Then,

$$\frac{dV_{pk}}{-dV_{gk}} = r_p g_m.$$

That is, if we increase V_{pk} we tend to increase I_p, and to "undo" that increase we must *decrease* (make more negative) V_{gk}. The ratio of these two voltage changes, equal to $r_p g_m$ as just shown, is usually written as μ and, as you'll see next, is a *constant*. Thus, from the Child–Langmuir law for triodes,

$$\frac{1}{r_p} = \left.\frac{\partial I_p}{\partial V_{pk}}\right|_{\substack{V_{gk} \text{ held} \\ \text{constant}}} = \frac{3}{2}K_1(V_{gk} + K_2 V_{pk})^{1/2}K_2$$

$$g_m = \left.\frac{\partial I_p}{\partial V_{gk}}\right|_{\substack{V_{pk} \text{ held} \\ \text{constant}}} = \frac{3}{2}K_1(V_{gk} + K_2 V_{pk})^{1/2}$$

and so $g_m r_p = \mu = \frac{1}{K_2} > 1$ (as $K_2 < 1$).

More specifically, for the 12AX7 triode operated at a dc plate-to-cathode voltage of 175 volts and a dc grid-to-cathode voltage of -1 volt (these dc operating voltages are called the *bias* voltages and they determine what electrical engineers call the quiescent or *Q point* of the tube), we find directly from the curves of Figure 8.6 that

$$r_p \approx \left.\frac{\Delta V_{pk}}{\Delta I_p}\right|_{V_{gk}=-1} = \frac{250 - 100}{3.3 \times 10^{-3} - 0.5 \times 10^{-3}} = 53.5 \text{ k}\Omega$$

$$g_m \approx \left.\frac{\Delta I_p}{\Delta V_{gk}}\right|_{V_{pk}=175} = \frac{1 \times 10^{-3} - 3 \times 10^{-3}}{-1.5 - (-0.4)}$$

$$= 1.8 \times 10^{-3} \text{ siemens (mhos)}.$$

Thus, for the 12AX7 over a wide range of Q points, $\mu \approx 97$.

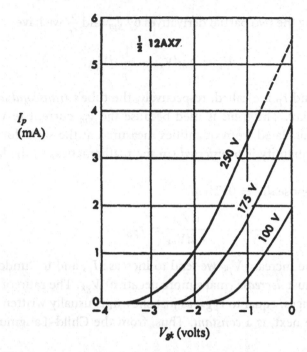

Figure 8.6   The plate current of a triode versus the grid-to-cathode potential difference, for three different plate-to-cathode potential differences. The "$\frac{1}{2}$" refers to the fact that a single glass envelope contains two independent electrode assemblies, i.e., the 12AX7 is a "dual triode."

The application of an input ac signal to the grid causes the instantaneous state of the tube to deviate from the Q point, with the result being the generation of an *output* ac signal at the plate. If the circuit is designed properly, then the ac output signal can be *much* larger than the input. That is, we'll have an amplifier. Figure 8.7 shows a simple amplifier circuit, called a *common cathode* amplifier because the cathode is part of both the input (grid) circuit and the output (plate) circuit. As the grid voltage becomes more negative, the tube current decreases. This results in a decreased voltage drop across R, and so the plate voltage increases, i.e., becomes more positive. Since the plate voltage increases in response to a grid voltage decrease, there is a 180° phase shift in this circuit from grid to plate. To understand in detail how this circuit works, we start by doing a dc analysis to determine the Q point. Thus, we take $v_i = 0$ and so the dc voltage on the grid *with respect to ground* is zero. This is *not*, as you'll see, V_{gk}, which is the grid voltage with respect *to the cathode*. The cathode is *not* at dc ground potential (because of the dc voltage drop across R_k).

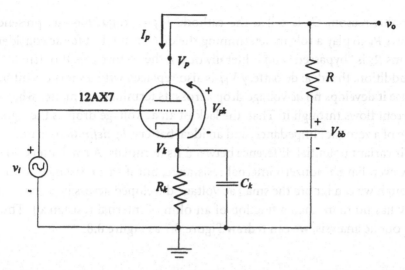

Figure 8.7 A simple common cathode, self-biased triode amplifier. The "self-bias" refers to the automatic generation of a negative grid-to-cathode potential difference (across R_k), thus avoiding the use of an expensive battery.

The battery (V_{bb}) in the plate circuit establishes some dc current I_p in the tube, and this current results in a voltage drop across R_k, the cathode resistor. (In the dc case we are doing now, the cathode capacitor C_k simply charges to the dc voltage drop across R_k *and plays no role at this point*.) The polarity of the voltage drop across R_k is such that the cathode potential V_k is positive with respect to ground, and so $V_{gk} < 0$. To see how all this works out in the numbers, suppose we've decided to put the Q point at $V_{pk} = 250$ V and $I_p = 2$ milliamperes. The curves in Figure 8.6 then tell us $V_{gk} = -1.5$ volts. Since $V_k = I_p R_k$, then we have $2 \times 10^{-3} R_k = 1.5$ which gives $R_k = 750$ ohms. Since there is a 250-volt drop across the tube, then the plate voltage *with respect to ground* is 251.5 volts. Finally, we need to select R and V_{bb} so that $V_{bb} - V_p$ is just the right voltage drop across R to give $I_p = 2$ milliamperes. Generally, one picks V_{bb} and calculates R. So, suppose $V_{bb} = 300$ volts, which gives $R = (300 - 251.5)/2 \times 10^{-3} = 24.7$ kΩ.

Now, what happens if $v_i \neq 0$, i.e., if we apply an ac input signal to the grid? We begin an answer to this question by assuming that C_k is sufficiently large that, even at the lowest frequency of interest, it has an impedance that is "small" compared to R_k (electrical engineers usually take this to mean $\leq 0.1 R_k$). That is, C_k can be effectively replaced with a short circuit (for the ac case), and so the cathode is at ac ground potential.

For this reason, C_k is called the cathode *bypass capacitor*—its presence
allows R_k to play a role in determining the dc Q point, but for ac consider-
ations R_k is "bypassed" and is literally out of the picture (see Problem 8.1).
In addition, the ideal dc battery V_{bb} is also replaced with a short circuit because it develops no *ac* voltage drop across its terminals no matter what ac
current flows through it. That, the lack of an ac voltage drop, is the signa-
ture of a zero (ac) impedance, and an ideal battery, *by definition*, maintains
an invariant potential difference between its terminals. A *real* battery does,
however, have a nonzero internal resistance, but if this resistance is small
enough we can ignore the small ac voltage developed across it (a *good* battery has no more than a fraction of an ohm of internal resistance). Thus,
for our ac analysis, we can redraw Figure 8.7 as Figure 8.8.

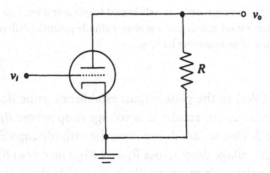

Figure 8.8   The ac version of Figure 8.7 (all capacitors and dc voltage sources have been
replaced with short circuits).

Our final step in developing an ac analysis for our simple amplifier is to
replace the triode, itself, with an *equivalent* circuit. Then we can analyze
the entire amplifier using just Kirchhoff's laws. The equivalent circuit for
the triode follows immediately from the equation I wrote earlier for the
total change in the tube current I_p (reproduced below):

$$dI_p = g_m dV_{gK} + \frac{1}{r_p} dV_{pk}.$$

As mentioned earlier, the differentials represent *small* ac variations, i.e., we
can write

$$i_p = g_m v_{gk} + \frac{1}{r_p} v_{pk},$$

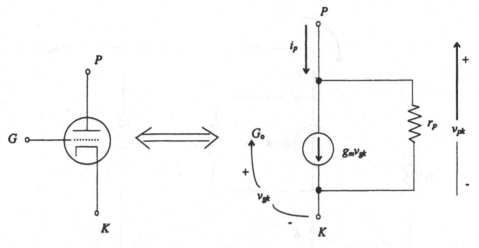

Figure 8.9   The small signal, ac equivalent circuit model for the triode.

where the lowercase letters represent just the time-varying (ac) components of the variables. We can then, as shown in Figure 8.9, replace the triode tube with a small-signal equivalent circuit consisting of just a *dependent, voltage-controlled* current source, in parallel with a resistor.

The current source in Figure 8.9 is what electrical engineers call the *dual* of a voltage source. Just as an ideal voltage source (e.g., a battery) maintains the same voltage drop across its terminals independent of the current through it, an ideal current source maintains the same current through itself independent of the voltage drop across its terminals. The current source in the triode small signal equivalent circuit is not an independent source (like a battery), but is *voltage controlled* (by v_{gk}). Under extreme situations, both sources can cause nonphysical happenings (short circuiting an ideal voltage source results in an infinite current, and open circuiting an ideal current source results in an infinite potential difference), which simply means more sophisticated models are needed. We won't need to be that sophisticated in this book. (As an aside, if you change the G (grid), P (plate) and K (cathode) nomenclature to G (gate), D (drain) and S (source) respectively, and v_{gk} to v_{gs}, you have the small-signal ac equivalent circuit model for the modern junction field-effect transistor, or JFET. The curves shown in Figure 8.3 of I versus V for the tube are identical in shape with the analogous curves for the JFET, with a quiet salute to the past being the term used for the JFET's 'space-charge limited' region—it is called the *triode region*.)

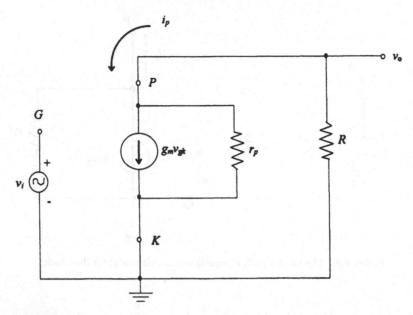

Figure 8.10   The ac equivalent circuit for the amplifier of Figure 8.7.

With the triode circuit model in hand we can then draw Figure 8.10, which is a completely reduced circuit equivalent of the original electronics in Figure 8.7. Since $v_{gk} = v_i$ and $v_{pk} = v_0$, we have

$$i_p = g_m v_i + \frac{1}{r_p} v_0$$

and

$$v_0 = -i_p R.$$

Thus, writing A as the voltage gain of the circuit, we have

$$A = \frac{v_0}{v_i} = -g_m \frac{R r_p}{R + r_p}$$

or, as $\mu = r_p g_m$,

$$A = -\mu \frac{R}{R + r_p}.$$

There are two immediate conclusions that we can draw from this result. First, the minus sign indicates there is a 180° phase shift from input to output, a result we already arrived at in a less mathematical way. Second, the voltage gain is less than μ, reaching μ only in the limit as $R \to \infty$. For our particular circuit elements, for example, with $\mu = 97$, $r_p = 53.5$ kΩ, and $R = 24.7$ kΩ, we have $A = -30.6$. Increasing R to increase A comes with a penalty, however. Increasing R requires that the dc plate supply voltage (V_{bb}) also be increased, or otherwise the designed Q point value for I_p will change (decrease). Eventually, V_{bb} will become unreasonably large.

The invention of electronic amplification was of enormous importance for the commercial development of radio. Indeed, it was crucial. Before the triode, the only energy available to the listener at a receiver was the energy intercepted by the antenna. This is not much energy, and so the need to use headphones. With the availability of amplification, however, the puny signal out of an antenna wire could be magnified by truly heroic proportions and applied to a loudspeaker, which could then entertain a roomful of listeners (none of whom had to be literally tethered to the radio with a set of headphones). If the gain of our simple one-tube amplifier is insufficient for the required amplification, the output can be applied to the input of a second amplifier. If A is the voltage gain of each amplifier, then n such identical amplifiers in sequence (or *cascade*) would have an overall gain of A^n. For $|A| = 30$ and $n = 3$, for example, a gain of $30^3 = 27,000$ is easy to achieve, and so even a tiny 10-μV antenna signal will emerge from the third amplifier stage as 0.27 V. Since the input stage is being driven by such a tiny signal, its grid will always remain negative and so, theoretically, essentially zero energy would be required from the antenna.

But of course a roomful of people would not be well entertained by nothing but amplified Morse code! To listen to broadcasts of human speech and music required AM radio, and that in turn requires constant amplitude oscillations. The triode solved that problem, too, and the lengthy quest for such oscillations by Fessenden and others was at last over. Triode oscillators had no moving parts (goodbye to the alternator), and no fiery discharges along with occasional explosions (adios to the electric arc). You could hold a triode oscillator in your hand, whereas it took a crane to move an alternator or an arc. Triode oscillators were cheap, too, as a triode is essentially just an enhanced lightbulb. And, best of all, it is *easy* to get a triode to oscillate. In essence, one simply connects the output of an amplifier back to its own input terminals.

Figure 8.11 shows in more detail how an "oscillating amplifier" is constructed. Imagine that a tiny input signal, somehow created, is present at the input terminals to an amplifier with voltage gain A, with a $180°$ phase shift through the amplifier. That is, $A < 0$. The output of the amplifier is fed into a network of passive components that has a frequency-dependent phase shift. At some frequency, in particular, the network is designed to have a phase shift of $180°$. Thus, at that frequency (call it ω_0), the total

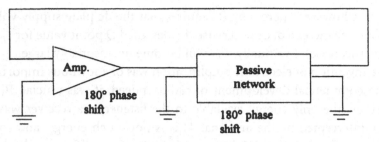

Figure 8.11   An oscillator is an amplifier with positive feedback.

phase shift around the loop from amplifier input back to amplifier input is 360°. That is, at ω_0 the output of the network is *in phase* with the original, "starting" signal.

Now, suppose that at frequency ω_0 the network's output amplitude is the input amplitude attenuated by the factor $\beta < 1$. Then, *if* $|A\beta| = 1$ we see that the signal originally present at the amplifier input terminals will sustain itself "around the loop." Indeed, if $|A\beta| > 1$ the signal will *grow* as it circulates round and round the loop. This is only true at frequency ω_0, however, as at any other frequency the output of the phase-shifting network will be out of phase with the initial signal at the amplifier input and so the circulating signal will destructively interfere with itself. The signal growth will cease once the signal becomes so large that nonlinear effects limit and finally cut off any further growth. Thus, the closed loop system of Figure 8.11 oscillates only at frequency ω_0. A simple resistor/capacitor network that has the appropriate frequency dependent phase shift is given in problem C.2 of Appendix C. If an amplifier has a positive (zero phase shift) gain, one can still build an oscillator by using a network with a frequency-dependent phase shift that has zero phase shift, too, at some particular frequency.

> The statement that $|A\beta| \geq 1$ for oscillations to exist is called the *Barkhausen condition*, after the German electrical engineer Heinrich Barkhausen (1881–1956). While the "circulating-in-a-loop" idea is intuitive, it is *fundamentally* flawed, which wasn't understood until the late 1920s (see Chapter 21). Now, a question: where does the *initial* signal at frequency ω_0 come from, the signal we assumed is present at the amplifier input that starts the oscillations? In fact, in a "perfect" world of no noise, there would be no such signal! We do seem to have a "chicken-and-the-egg" problem. In the *real* world, however, there are always tiny energy fluctuations present (e.g., thermal agitation of

electrical charges in the circuit components, and cosmic rays smashing through the oscillator circuit, will result in random voltage fluctuations). As you will see in the next chapter, such fluctuations invariably have their energy spread over an enormous interval of frequency, from dc up to many hundreds of megahertz. Try as you might, you couldn't *avoid* having some energy at frequency ω_0 present, whatever ω_0 might be.

This kind of oscillator, called a *phase-shift oscillator*, can be easily built to oscillate over the interval, roughly, from 0.1 Hz to a megahertz or so. To build oscillators that work at even higher frequencies, more intricate circuits are required but all generally depend on the idea of feeding a signal back from the output of an amplifier to its input. And so, with a handful of resistors and an enhanced lightbulb, the massive continuous wave machines of Alexanderson and Poulsen were rendered obsolete at a stroke. The technical basis for AM broadcast radio was therefore completed well before 1914. Indeed, by 1913 the American Telephone and Telegraph Company had acquired all of De Forest's audion patents (for just $50,000) and then quickly put their new technology to spectacular use. In October 1915 AT&T used a very complicated transmitter (with 550 tubes in parallel to achieve a power level of about 11,000 watts) to send *voice* signals from the U.S. Navy's station NAA in Arlington, Virginia to receivers located at both Pearl Harbor, Hawaii and on the Eiffel Tower in Paris. It took several more years, however, before the final inventions of how to *use* the new electronic technology in practical radio circuits were finally developed. To understand how those circuits work, you now need to learn some more mathematics.

Notes

1. Fleming's seminal paper, "On the Conversion of Electric Oscillations into Continuous Currents by Means of a Vacuum Value," was published the following month in the *Proceedings of the Royal Society*.

2. See Robert A. Chipman, "De Forest and the Triode Detector," *Scientific American*, March 1965. It was Edwin Armstrong who gave the first scientific explanation (in 1915) for the triode's operation in radio circuits.

Problems

8.1. Redo the amplifier analysis when there is no cathode bypass capacitor. This won't affect the dc Q point, but now the cathode will no longer be at ac ground potential. You should find that the voltage gain decreases from that of the bypassed case (that's why C_k is usually included!). The cause of this is the ac signal generated across R_k which has a polarity that *opposes* v_i. This is called *ac degeneration*, and it is a special case of negative feedback (which has some good points, which is why C_k is not *always* included!).

8.2. In the two-tube amplifier of Figure 8.12, assume the tubes are identical. They then clearly have the same dc Q points (if $v_i = 0$ then the tubes are in identical situations, with both grids connected to ground). Once, $v_i \neq 0$, however, the ac situations are different for the two tubes since the grid-to-cathode voltages are different). Use the triode small-signal equivalent circuit model to find the voltage gain as a function of μ and r_p. Assume that V_{bb} is whatever it should be so that the small-signal model is valid (essentially that there is no grid current at any time). (Partial answer: for $\mu = 19$ and $r_p = 10$ kΩ, $A = +4.75$, i.e., there is zero phase shift from input to output.)

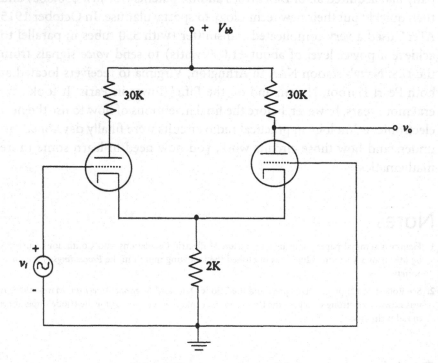

Figure 8.12   A two-tube amplifier.

Mostly Math and a Little History

All of the next Section is mathematical. There are, from time to time, comments directly related to radio, but not many. Have faith, however! The mathematics is important! Don't slide over it. The circuitry in Sections 3 and 4 will simply make no sense without a proper understanding of Section 2. The math, *itself* is beautiful, but it is also most practical as well.

Fourier Series and Their Physical Meaning

In a 1744 letter to a friend, Euler stated that

$$x(t) = \frac{\pi - t}{2} = \sum_{n=1}^{\infty} \frac{\sin(nt)}{n} = \sin(t) + \frac{\sin(2t)}{2} + \frac{\sin(3t)}{3} + \cdots.$$

This was probably (almost certainly) the first "Fourier series," although of course Euler didn't call it that since Fourier wouldn't be born until twenty-four years later. Where in the world, you may now be wondering, does such a wonderful formula come from? And, what does it *mean*? There certainly seems to be a problem at $t = 0$, for example, as Euler's formula then appears to claim that $\frac{\pi}{2} = 0$! Our start at answering these questions is what this chapter is all about, and at the end of it I'll show you how, with MATLAB, to visualize what Euler wrote to his friend. And in the next chapter you'll learn how modern electrical engineers derive the above expression. This will all prove to be of such value to understanding radio that it is impossible for me to exaggerate the importance of what follows. But—it will require that we take a temporary journey away from radio circuits and into mathematics. Not everybody appreciates this!

For example, Edwin Armstrong, the inventor of the superheterodyne radio receiver, is one of my heroes. But that doesn't mean I am blindly uncritical of him. Indeed, one of his friends remembered him saying (in January 1936, at a meeting of the Institute of Radio Engineers (IRE), after listening to a theoretician's explanation of the theory of FM radio), "You don't make inventions by fancy mathematics. You make them by jackassing storage batteries around the laboratory."[1] The President of the IRE, Armstrong's friend Professor Louis Hazeltine, took exception to that rash assertion and, of course, so do I. Just look at this book!

During this chapter, and the next, I will be particularly careful in properly establishing the theoretical underpinnings of the Fourier series. From the series for a periodic signal comes the Fourier integral for *any* signal, periodic or otherwise, and from that integral comes *everything in radio*. We don't want to build our "House of Radio" on a loose pile of sand, and so we must be quite careful at the start. We begin.

Suppose $x(t)$ is a periodic, real-valued time signal. The signal $x(t)$ is periodic with period T if it satisfies the condition

$$x(t) = x(t + T), \qquad -\infty < t < \infty$$

and we call the smallest possible positive T the *fundamental period* of $x(t)$. A little thought should convince you that there is *no such signal*! That is because for a signal to be periodic it must have been on forever, and it must be immortal, i.e., it *will* stay on forever. So, from a pure engineering view, there is no such signal. Mathematically, of course, we can at least *imagine* such signals; and from a practical point of view, if a signal has been on for a million periods (1 sec for a 1-MHz sine wave) it may as well have been on forever.

The concept of periodicity, so seemingly obvious at first glance, has some subtle issues. For example, is the sum of two periodic functions, *itself* always periodic? If you said *yes* (and most people do), then what is the fundamental period of $x(t) = \cos(t) + \cos(t\sqrt{2})$? What ever it might be, let's call it T. Now, I'll show you that this assumption that there is such a T leads to a contradiction. We have, for all t,

$$x(t) = x(t + T) = \cos(t + T) + \cos\left\{(t + T)\sqrt{2}\right\}.$$

For $t = 0$, in particular, this means

$$x(0) = \cos(0) + \cos(0) = 2 = \cos(T) + \cos(T\sqrt{2}).$$

Since the maximum value of the cosine function is 1, then both $\cos(T)$ and $\cos(T\sqrt{2})$ must be 1, i.e., $T = 2\pi n$ and $T\sqrt{2} = 2\pi m$ where n and m are different integers. Thus,

$$\frac{T\sqrt{2}}{T} = \frac{2\pi m}{2\pi n} = \sqrt{2} = \frac{m}{n}.$$

But, since $\sqrt{2}$ is irrational then *by definition* there are no integer m and n with a ratio of $\sqrt{2}$. So, our original assumption that T exists must be false!

Here's another question for you: if two periodic functions *do* sum to give a periodic function, can the period of the sum be *less* than each of the original two periods? When I've asked electrical engineers this, even other professors, most say *no*. In fact, the answer is *yes*. Consider, for example, $x_1(t) = 2\sin(\pi t)$, and $x_2(t)$ defined as

$$x_2(t) = \begin{cases} -\sin(\pi t), & 0 \le t \le 1 \\ -3\sin(\pi t), & 1 < t < 2 \\ x_2(t-2), & -\infty < t < \infty. \end{cases}$$

Both $x_1(t)$ and $x_2(t)$ are periodic with fundamental period $T = 2$, while their sum is equal to $|\sin(\pi t)|$—sketch x_1, x_2 and $x_1 + x_2$ and see it for yourself—with a fundamental period of *one*! For a surprising extension of this example, see Problem 9.1.

And finally, how about this question—can a function be periodic without having a *fundamental* period? Sure—just ponder $x(t) = 2$ or any other constant. Then $x(t + T) = x(t)$ for all t *and* for all T, i.e., there is no *smallest* $T > 0$, because for any T you pick, I can always pick a smaller one, e.g., $\frac{1}{2}T$! If you don't like this argument because a *constant* isn't varying with time, then how about the following time signals where things are varying *very, very* fast indeed!: define $x_1(t)$ and $x_2(t)$ to be

$$x_1(t) = \begin{cases} 1 \text{ if } t \text{ is an integer} \\ 0 \text{ if } t \text{ is } not \text{ an integer} \end{cases}$$

$$x_2(t) = \begin{cases} 1 \text{ if } t \text{ is rational but not an integer} \\ 0 \text{ if } t \text{ is irrational } \underline{or} \text{ is an integer}. \end{cases}$$

$x_1(t)$ is a "nice" function, in that we can draw a sketch of it, but $x_2(t)$ is obviously pretty wild—you *cannot* sketch it because between any two rationals there are an infinite number of irrationals (and vice versa). Both $x_1(t)$ and $x_2(t)$, however, have the same fundamental period: *one*. And yet, their sum satisfies the rule

$$x_1(t) + x_2(t) = \begin{cases} 1 \text{ if } t \text{ is rational} \\ 0 \text{ if } t \text{ is irrational}, \end{cases}$$

which is a periodic function (every rational number is a period). But, since there is no smallest nonzero positive rational then there is *no* fundamental period!

Periodicity—it *is* a deeper concept than you perhaps thought at first glance.

Our discussion of period functions will assume very little about the specific nature of $x(t)$, but one restriction we will impose is that $x(t)$ has *finite energy* in a period. This is a very weak restriction (such signals are said to be of *finite power*), and any "real world" time function will satisfy it. What is meant by the energy of a signal is quite straightforward. Suppose, to be specific to electrical engineering, that $x(t)$ is a periodic *voltage* signal. If $x(t)$ is applied to a resistor, R, then the instantaneous power is

$$p(t) = \frac{x^2(t)}{R} = x^2(t) \qquad \text{if } R = 1 \text{ ohm.}$$

The energy per period is, thus,

$$W = \int_{\text{period}} p(t)\, dt = \int_{\text{period}} x^2(t)\, dt$$

which we assume to be finite; that is, $0 \leq W < \infty$. We then simply *define* this integral to be the energy (*per period*) of any real periodic signal $x(t)$, whether it is a voltage or not (or if $R = 1$ ohm or not). Even the mathematicians do this! Now, assume (for the present) that we can write

$$x(t) = \sum_{k=-\infty}^{\infty} c_k e^{jk\omega_0 t} \qquad \text{where} \qquad \omega_0 = \frac{2\pi}{T}$$

and the c_k are yet-to-be-determined constants.

There is absolutely no reason for you to either understand what motivates writing this astonishing expression [called the *Fourier series expansion* of $x(t)$] or to even believe it is possible to so express $x(t)$. Just accept it for the present as a possibility, and by the time the next chapter is over you will see that it is possible to write $x(t)$ this way, as well as what is the physical meaning of the Fourier series. The use of complex exponentials is not essential (real-valued sines and cosines can be used), but the exponential form is mathematically very convenient as you will see when we get to the Fourier integral.

Expressing functions as trigonometric sums has a long history. The French mathematician Charles Bossut (1730–1814) wrote a number of such sums with a *finite* number (n) of terms, and in 1773 the Swiss mathematician Daniel Bernoulli (1700–1782) took Bossut's formulas and simply let $n \to \infty$. Bernoulli played fast and loose with his math, however, even more than I sometimes do in this book; he arrived at correct results only by making the unjustified (and meaningless) argument that $\sin(\infty)$ and $\cos(\infty)$ are zero! As mentioned at the start of this chapter Euler had, as early as 1744, written down a specific expansion of a function using an infinite number of sine terms, and in 1777 he actually found general formulas for the coefficients of a given function expanded as a sum of cosines. These were the same formulas deduced thirty years later (1807) by the French mathematical physicist Joseph Fourier (1768–1830). Still, even Euler wasn't the *first*; in 1759 similar results (unknown to Euler) had been obtained by the French mathematician Alexis-Claude Clairaut (1713–1765). Despite this lengthy history, however, when Fourier published his famous treatise *The Mathematical Theory of Heat* (1822) it still generated (appropriately!) a lot of heat among mathematicians on whether Fourier was right or wrong.

Fourier was in fact mostly right, but mostly from hazy (at best) reasoning. The central puzzle in Fourier's results, for the mathematicians of his day, was that they believed a sum of continuous functions (sines and cosines) should also be continuous. That is true for a finite number of terms, but it is *not* necessarily true for Fourier series that have an infinite number of terms. Even *discontinuous* periodic functions have Fourier series, and that was not understood by mathematicians until Fourier's work. You can find more on the history of these developments in the essay "Fourier Analysis" in the book by Philip J. Davis and Reuben Hersh, *The Mathematical Experience*, Birkhäuser 1981, and I'll say more about such developments in the next chapter.

The energy of $x(t)$ is (note, in particular, that $\omega_0 T = 2\pi$)

$$W = \int_{\text{period}} \left\{ \sum_{k=-\infty}^{\infty} c_k e^{jk\omega_0 t} \right\}^2 dt.$$

The c_k coefficients are as yet undetermined constants but, in general, because of the complex-valued exponentials, the c_k are complex valued as well [because $x(t)$ is real valued]. Again, because $x(t)$ is real valued, we also have

$$x(t) = \sum_{k=-\infty}^{\infty} c_k e^{jk\omega_0 t} = x^*(t) = \sum_{k=-\infty}^{\infty} c_k^* e^{-jk\omega_0 t}.$$

Writing out the two sums and matching both sides, term-by-term by frequency, we conclude that

$$c_{-k} = c_k^*.$$

(Note carefully: this is true, in general, only if $x(t)$ is *real*. It is not generally true if $x(t)$ is complex.) And since $c_0 = c_0^*$ then, in particular, c_0 is *real*-valued for *any* real-valued $x(t)$. The physical significance of this is discussed next.

$\frac{1}{T}$, with units of sec^{-1}, is called the *fundamental frequency* of the periodic signal $x(t)$. Thus, a Fourier series represents the signal as the sum of a dc term (c_0, which is indeed real valued, as a dc level better be!) and of sinusoids of various frequencies which are integer multiples of the fundamental frequency. These frequencies are often called *harmonics* of the fundamental (the fundamental, itself, is called the *first harmonic*). Each harmonic has its own real-valued amplitude, which we can calculate as follows: if, for $k \geq 1$ we combine the c_{-k} and c_k terms using Euler's identity, we obtain

$$c_{-k}e^{-jk\omega_0 t} + c_k e^{jk\omega_0 t} = c_k^*(e^{jk\omega_0 t})* + c_k e^{jk\omega_0 t}$$
$$= (c_k e^{jk\omega_0 t})* + c_k e^{jk\omega_0 t} = 2\text{Re}(c_k e^{jk\omega_0 t}).$$

If we then express c_k in rectangular form

$$c_k = a_k + jb_k, \quad k \neq 0$$

then

$$2\text{Re}(c_k e^{jk\omega_0 t}) = 2[a_k \cos(k\omega_0 t) - b_k \sin(k\omega_0 t)],$$

which by elementary trigonometry is a sinusoid with frequency $k\omega_0$, a phase angle, and a peak amplitude of

$$2\sqrt{a_k^2 + b_k^2} = 2|c_k|.$$

Specifically,

$$a_k \cos(k\omega_0 t) - b_k \sin(k\omega_0 t) = \sqrt{a_k^2 + b_k^2} \cos\left\{k\omega_0 t + \tan^{-1}(b_k/a_k)\right\},$$

and so the *phase* of the kth harmonic is the angle

$$\tan^{-1}(b_k/a_k).$$

Now, recalling the energy integral for W, expand the integrand and write

$$\left\{\sum_{k=-\infty}^{\infty} c_k e^{jk\omega_0 t}\right\}^2 = \sum_{m=-\infty}^{\infty} \sum_{n=-\infty}^{\infty} c_m c_n e^{j(m+n)\omega_0 t}$$

and so

$$W = \int_{\text{period}} \sum_{m=-\infty}^{\infty} \sum_{n=-\infty}^{\infty} c_m c_n e^{j(m+n)\omega_0 t} \, dt$$
$$= \sum_{m=-\infty}^{\infty} \sum_{n=-\infty}^{\infty} c_m c_n \int_{\text{period}} e^{j(m+n)\omega_0 t} \, dt.$$

This integral is particularly easy to evaluate. For any integration interval with a one-period duration, beginning at the arbitrary time t',

$$\int_{t'}^{t'+T} e^{j(m+n)\omega_0 t}\, dt = \frac{e^{j(m+n)\omega_0 t}}{j(m+n)}\bigg|_{t'}^{t'+T} = \frac{e^{j(m+n)\omega_0(t'+T)} - e^{j(m+n)\omega_0 t'}}{j(m+n)}$$

$$= \frac{e^{j(m+n)\omega_0 t'}(e^{j(m+n)\omega_0 T} - 1)}{j(m+n)}.$$

Now, recall that $\omega_0 T = 2\pi$, and that $e^{j(m+n)2\pi} = 1$ for all m and n (which are both integers), and so this integral is *almost* always equal to zero. But not *always*, because for the special cases where $m = -n$ the last expression becomes the indeterminate $\frac{0}{0}$. For those special cases, set $m = -n$ in the energy integral *first*, and *then* do the integral:

$$\int_{\text{period}} e^0\, dt = T.$$

Thus,

$$W = \sum_{k=-\infty}^{\infty} c_k c_{-k} T = T \sum_{k=-\infty}^{\infty} c_k c_k^* = T \sum_{k=-\infty}^{\infty} |c_k|^2 = T\left[c_0^2 + 2\sum_{k=1}^{\infty} |c_k|^2\right].$$

Dividing through by the period we arrive at Parseval's theorem:

$$\frac{1}{T} W = \frac{1}{T} \int_{\text{period}} x^2(t)\, dt = c_0^2 + 2\sum_{k=1}^{\infty} |c_k|^2.$$

Marc Antoine Parseval des Chenes (1755–1836) was a French mathematician who lived on the edges of scientific life. In 1792 he was imprisoned for his Royalist beliefs, and later had to flee France when Napoleon ordered his arrest for writing anti-establishment poetry. There are literally dozens of equations in mathematical physics called "Parseval's equation," and in fact the theorem Parseval presented to the Paris Academy of Sciences in 1799 bears only the most superficial resemblance to the theorem stated here. One such result, sometimes called "Parseval's integral," is important in the theory of FM radio. Starting with the power series expansion of the exponential, we have

$$e^{jx\sin(\theta)} = 1 + j\frac{x\sin(\theta)}{1!} + \frac{[jx\sin(\theta)]^2}{2!} + \frac{[jx\sin(\theta)]^3}{3!} + \cdots$$

$$= 1 - jx\sin(\theta) - x^2\frac{\sin^2(\theta)}{2!} - jx^3\frac{\sin^3(\theta)}{3!} + \cdots.$$

Integrating on both sides from $-\pi$ to π, and using the integration formula

$$\int_{-\pi}^{\pi} \sin^n(\theta)\,d\theta = \begin{cases} 0 \text{ if } n \text{ is odd} \\ \dfrac{(n-1)(n-3)\cdots 3\cdot 1}{n(n-2)\cdots 4\cdot 2}\, 2\pi & \text{if } n \text{ is even} \end{cases}$$

we then have from Euler's identity that

$$\int_{-\pi}^{\pi} e^{jx\sin(\theta)}\,d\theta = \int_{-\pi}^{\pi} \cos\{x\sin(\theta)\}\,d\theta + j\int_{-\pi}^{\pi} \sin\{x\sin(\theta)\}\,d\theta$$

$$= 2\pi \left[1 - \frac{x^2}{2^2} + \frac{x^4}{2^2\cdot 4^2} - \frac{x^6}{2^2\cdot 4^2\cdot 6^2} + \cdots \right].$$

The imaginary part (the last integral) must be zero since the power series expression is strictly real (also, of course, that integral must vanish since the sine is an odd function), and so our result reduces to the exotic (and *very* useful)

$$\frac{1}{2\pi}\int_{-\pi}^{\pi} \cos\{x\sin(\theta)\}\,d\theta = 1 - \frac{x^2}{2^2} + \frac{x^4}{2^2\cdot 4^2} - \frac{x^6}{2^2\cdot 4^2\cdot 6^2} + \cdots.$$

This formula was derived by Parseval in 1799 (published in 1805), but today it is commonly called the "Bessel function of the first kind, of order zero" and written as $J_0(x)$. It is named after the German astronomer Friedrich Wilhelm Bessel (1784–1846) who encountered it as a solution to a differential equation he studied to explain gravitational perturbations in the motions of stars and planets. It is a so-called nonelementary integral, which is why it is *not* usually discussed in freshman calculus. See the final shaded box in Chapter 10 for a related calculation.

Parseval's theorem has an elegant physical interpretation. The integral on the left is the energy of $x(t)$ *per period*, or the *total average power*. On the right, c_0^2 is the dc power. And each $2|c_k|^2$ term represents the average power in the kth harmonic. That is, the total average power of $x(t)$ is simply the sum of the average powers in the Fourier series components of $x(t)$. Recall that we showed $2|c_k|$ is the peak amplitude of the kth harmonic. You should be able to show that the average power of a sinusoid with this peak value, over a period, is indeed equal to $2|c_k|^2$. (If this presents some difficulty, review the average power discussion in Appendix C.)

This physical interpretation is encouraging, hinting strongly that we are on to something interesting. But, to return to the question raised at the beginning of this chapter, how do we know we *really can* write a periodic $x(t)$ as

$$x(t) = \sum_{k=-\infty}^{\infty} c_k e^{jk\omega_0 t}?$$

And even if $|c_k|^2$ is related to the power in the kth harmonic, what *are* the c_k, i.e., how do we calculate them? In the next chapter both of these questions are answered.

This *does* present us with an interesting historical question: how did Euler, in 1744, arrive at his "Fourier series" for $\frac{\pi - t}{2}$ given at the start of this chapter? After all, he did not know how to calculate c_k either! Well, Euler was a spectacularly inventive genius, and he got his result in the following spectacular (if mathematically wild) way. Writing the geometric series

$$S = e^{jt} + e^{j2t} + e^{j3t} + \cdots$$

he summed it in the usual manner (see Appendix A):

$$e^{jt}S = e^{j2t} + e^{j3t} + \cdots$$

and so $S - e^{jt}S = e^{jt}$ or,

$$S = \frac{e^{jt}}{1 - e^{jt}} = \frac{e^{jt}(1 - e^{-jt})}{(1 - e^{jt})(1 - e^{-jt})} = \frac{e^{jt} - 1}{1 - e^{jt} - e^{-jt} + 1}$$

$$= \frac{\cos(t) + j\sin(t) - 1}{2 - 2\cos(t)} = \frac{-[1 - \cos(t)] + j\sin(t)}{2[1 - \cos(t)]}$$

$$= -\frac{1}{2} + j\frac{1}{2} \cdot \frac{\sin(t)}{1 - \cos(t)}.$$

Now, from the original definition of S, Euler also wrote

$$S = \cos(t) + \cos(2t) + \cdots + j[\sin(t) + \sin(2t) + \cdots]$$

or, upon equating the real parts of his two expressions for S,

$$\cos(t) + \cos(2t) + \cos(3t) + \cdots = -\frac{1}{2}.$$

Then, integrating indefinitely term by term,

$$\sin(t) + \frac{\sin(2t)}{2} + \frac{\sin(3t)}{3} - \cdots = -\frac{1}{2}t + C$$

where C is the constant of integration. In particular, for $t = \frac{\pi}{2}$ Euler could write

$$1 - \frac{1}{3} + \frac{1}{5} - \frac{1}{7} + \cdots = -\frac{\pi}{4} + C.$$

By 1744 it had already been known for over 70 years that the sum on the left is equal to $\frac{\pi}{4}$ (see Problem 10.1), and so $\frac{\pi}{4} = -\frac{\pi}{4} + C$ or, $C = \frac{\pi}{2}$.

Therefore, Euler could immediately write to his friend that

$$\sin(t) + \frac{\sin(2t)}{2} + \frac{\sin(3t)}{3} + \cdots = \frac{\pi - t}{2} = x(t).$$

Euler's "derivation" is, despite the devil-may-care manipulations, correct *if* properly interpreted. In Figures 9.1, 9.2, 9.3, and 9.4 I've plotted (using the MATLAB program figure9.m) the left-hand side for the first 5, 10, 20, and 40 terms in the series, respectively, and you can see that the curves *do* look more and more like $\frac{\pi-t}{2}$ as more and more terms are included; but the plots do *not* equal $\frac{\pi-t}{2}$ for all t. Rather, the series is periodic, and repeats with period 2π. The series is said to be the *periodic extension* of $x(t)$, where one period is $0 < t < 2\pi$. Between the periods (at $t = 0, \pm2\pi, \pm4\pi$, etc.) there are discontinuities, i.e., *jumps*. What the Fourier series for $x(t)$ converges to *at* a discontinuity of $x(t)$ may seem a mystery—what *is* $x(t)$ at a discontinuity? In fact, as will be discussed in the next chapter, the series converges to the *average* of the values of $x(t)$ on each side of the discontinuity. Since $t = 0$ is the location of a discontinuity in Euler's series, we can now understand why simply plugging $t = 0$ into

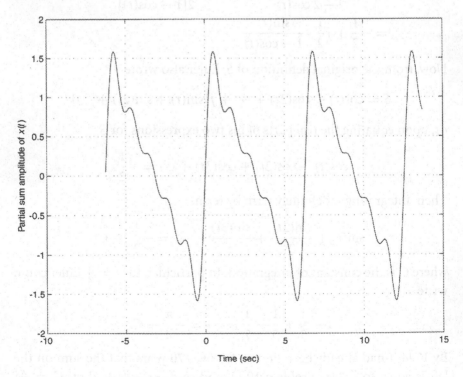

Figure 9.1   First five terms in the Euler series.

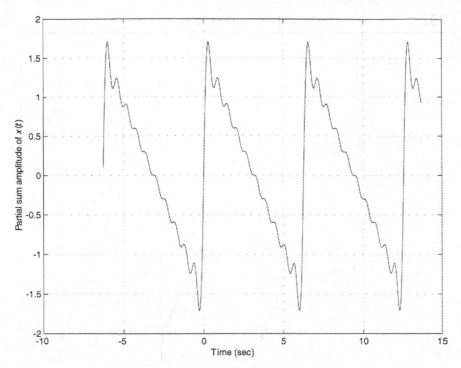

Figure 9.2   First ten terms in the Euler series.

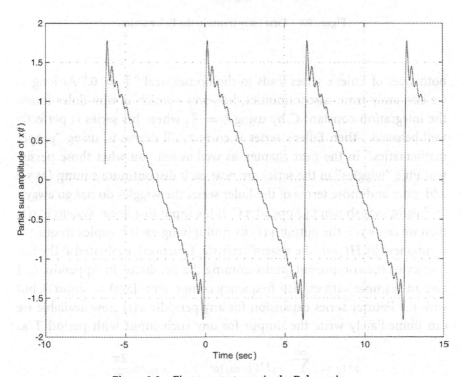

Figure 9.3   First twenty terms in the Euler series.

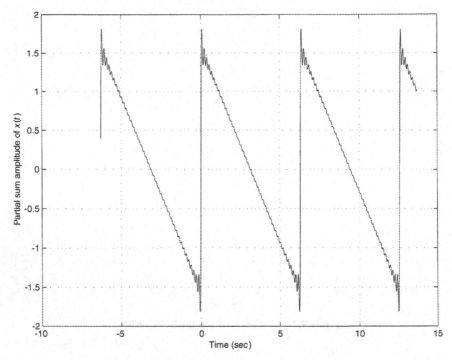

Figure 9.4   First forty terms in the Euler series.

both sides of Euler's series leads to the nonsensical "$\frac{\pi}{2} = 0$." As long as we stay away from discontinuities, however—which is why Euler found the integration constant C by using $t = \frac{\pi}{2}$, where his series is perfectly well behaved—then Euler's series is correct. I'll derive it, using "proper mathematics," in the next chapter, as well as tell you what those persistent little "wiggles" in the series are near each discontinuous jump (as we add more and more terms of the Euler series the wiggles do *not* go away!)

Finally, as is shown in Appendix C, if the input to a linear system is $x(t)$ then we can write the output $y(t)$ by multiplying each complex frequency component by $H(j\omega)$ (the system's transfer function), evaluated at the frequency of the component, and summing the products. In Appendix C, I give an example with explicit frequency components [$x(t) = \sin(at)$], but with the Fourier series expansion for any periodic $x(t)$ now available we can immediately write the output for any such input with period T as

$$y(t) = \sum_{k=-\infty}^{\infty} c_k H(jk\omega_0)e^{jk\omega_0 t}, \quad \omega_0 = \frac{2\pi}{T}.$$

```
%figure9.m/created by PJNahin for "Science of Radio"(6/15/99)
%Euler's 'Fourier series'
time=.01:.01:20;
timeshift=time-2*pi;
x=zeros(1,2000);
for r=1:40
    for c=1:2000
        t=c/100;
        matrix(r,c)=sin(r*t)/r;
    end
end
for c=1:2000
    for r=1:5
        x(c)=x(c)+matrix(r,c);
    end
end
figure(1)
plot(timeshift, x)
grid
xlabel('time (seconds)')
ylabel('partial sum amplitude of x(t)')
title('Figure 9.1 - First Five Terms in the Euler Series')
for c=1:2000
    for r=6:10
        x(c)=x(c)+matrix(r,c);
    end
end
figure(2)
plot(timeshift, x)
grid
xlabel('time (seconds)')
ylabel('partial sum amplitude of x(t)')
title('Figure 9.2 - First Ten Terms in the Euler Series')
for c=1:2000
    for r=11:20
        x(c)=x(c)+matrix(r,c);
    end
end
figure(3)
plot(timeshift,x)
grid
xlabel('time (seconds)')
ylabel('partial sum amplitude of x(t)')
title('Figure 9.3 - First Twenty Terms in the Euler Series')
for c=1:2000
    for r=21:40
        x(c)=x(c)+matrix(r,c);
    end
end
figure(4)
plot(timeshift,x)
grid
xlabel('time (seconds)')
ylabel('partial sum amplitude of x(t)')
title('Figure 9.4 - First Forty Terms in the Euler Series')
```

Figure 9

At the end of the next chapter I'll use this result to solve for the steady-state output of a series RLC-circuit with a periodic (square-wave) input. It will then be interesting to compare this theoretical result (plotted using MAT-LAB) with an Electronics Workbench simulation (we would, of course, expect the two answers to agree).

Note

1. See the remembrances of Harold H. Beverage in *IEEE Potentials*, February 1992, pp. 42–44.

Problems

9.1. Define $x_1(t)$ and $x_2(t)$ as follows, where N is any positive integer:

$$
x_1(t) = \begin{cases} \sin(2N\pi t), & 0 \le t \le \frac{1}{N} \\ 0, & \frac{1}{N} < t < 1 \\ x_1(t-1), & -\infty < t < \infty \end{cases}
$$

$$
x_2(t) = \begin{cases} 0, & 0 \le t \le \frac{1}{N} \\ \sin(2N\pi t), & \frac{1}{N} < t < 1 \\ x_2(t-1), & -\infty < t < \infty. \end{cases}
$$

It should be clear that both $x_1(t)$ and $x_2(t)$ have fundamental periods of one, independent of the value of N. Explain why their *sum*, however, has a fundamental period of $\frac{1}{N}$ (and so can be made as small as desired by simply making N larger and larger).

9.2. Suppose $x(t)$ is any real periodic function with zero average value from which the Fourier series of $x'(t)$ can be found by differentiating the series for $x(t)$ term by term. Show that

$$
\int_{\text{period}} x'^2(t)\, dt \ge \int_{\text{period}} x^2(t)\, dt
$$

if the period is no larger than 2π. Hint: use Parseval's theorem.

9.3. If $x(t)$ is real with period T then, as shown in the text, the energy per period of $x(t)$ is

$$
W = \int_{\text{period}} x^2(t)\, dt = T \sum_{k=-\infty}^{\infty} |c_k|^2.
$$

Show that this is still true even if $x(t)$ is complex, with the definition

$$W = \int_{\text{period}} |x(t)|^2 \, dt.$$

9.4. Consider the voltage $v(t) = 3 \cos(2\pi f_1 t - 1.3) + 5 \cos(2\pi f_2 t + 0.5)$, where of course the arguments of the cosines are in radians. The maximum value of $v(t)$ is obviously no larger than 8, but in fact it can be significantly smaller. Use a computer (e.g., MATLAB) to determine the maximum value of $v(t)$ for $f_1 = 2$ Hz and $f_2 = 1$ Hz. Hint: It is less than 6. Another hint (if you *are* using MATLAB): consider using the function $f$min on $-v(t)$, since the minimum of $-v(t)$ is the maximum of $v(t)$.

Convergence in Energy of the Fourier Series

To show you that it is indeed possible to write a periodic $x(t)$ as the infinite-sum Fourier series of Chapter 9, I'll begin with the truncated sum with a finite number of terms:

$$x_N(t) = \sum_{k=-N}^{N} c_k e^{jk\omega_0 t}.$$

I'll next define the *integrated-squared-error* that $x_N(t)$ makes as an approximation to $x(t)$:

$$E_N = \int_{\text{period}} [x(t) - x_N(t)]^2 \, dt = \int_{\text{period}} \left[x(t) - \sum_{k=-N}^{N} c_k e^{jk\omega_0 t} \right]^2 dt.$$

Notice that the physical significance of E_N is that it represents the energy of the difference signal between $x(t)$ and $x_N(t)$, over a period. This definition of E_N counts with equal weight the cases of $x(t) > x_N(t)$ and $x(t) < x_N(t)$. This avoids the problem that would occur if we defined E_N as, for example, simply the integral of the difference between $x(t)$ and $x_N(t)$; E_N could be zero even with *wildly* different $x(t)$ and $x_N(t)$. This cannot happen with a *squared* integrand. But why a *squared* integrand? Why not some other even power of the difference between $x(t)$ and $x_N(t)$? The answer is a pragmatic one; for the above definition of E_N we can actually calculate what E_N is, explicitly, while for other powers we would find this very hard (or even impossible) to do. And *physically*, of course, squared voltages lead one to naturally think of power and energy.

The analysis in this section will now proceed as follows:

1. I will find the c_k, for a given finite N, that minimize E_N. Interestingly, the results are *independent* of N.

2. If we denote the minimized E_N by min E_N, then I will show you that

$$\lim_{N \to \infty} \min E_N = 0.$$

That is, the Fourier series expansion of a periodic function has zero integrated-squared-error (i.e., *zero* energy in the error signal), and it is this result that justifies the use of Fourier series to represent periodic signals.

To minimize E_N, we must choose the $N + 1$ independent c_k (remember, $c_{-k} = c_k^*$) coefficients "properly." Let's suppose we've done that for all but one; call that last independent coefficient to be determined c_n, where $0 \le n \le N$. Then,

$$\frac{dE_N}{dc_n} = 0.$$

Thus,

$$\int_{\text{period}} 2 \left[x(t) - \sum_{k=-N}^{N} c_k e^{jk\omega_0 t} \right] (-e^{-jn\omega_0 t}) \, dt = 0$$

or,

$$\int_{\text{period}} x(t) e^{jn\omega_0 t} \, dt = \sum_{k=-N}^{N} \int_{\text{period}} e^{j(k-n)\omega_0 t} \, dt.$$

As shown in the previous chapter,

$$\int_{\text{period}} e^{j(k-n)\omega_0 t} \, dt = \begin{cases} T, & \text{for } k = n \\ 0, & \text{for } k \ne n. \end{cases}$$

Thus,

$$\int_{\text{period}} x(t) e^{jn\omega_0 t} \, dt = T c_n$$

and so

$$c_n = \frac{1}{T} \int\limits_{\text{period}} x(t)e^{-jn\omega_0 t} \, dt, \; -N \le n \le N.$$

You should notice *two* characteristics of this expression for c_n:

1. c_n depends only on $x(t)$ and n, and is independent of all the other c_k. Thus, our expression for c_n is valid for *any* particular value of n, $-N \le n \le N$.

2. c_n is independent of N. Thus, if we determine c_n for some particular value of N, and then increase N to M, we will find the first $2N+1 c_k$ for $x_M(t)$ are the $2N+1 c_k$ for $x_N(t)$.

A mathematician would say the c_k are *robust*, i.e., the coefficients c_k are *insensitive* [to everything but $x(t)$!]. But now we come to the crux of the matter—do these choices for the coefficients (the Fourier coefficients) lead to a *convergent* series, i.e., does

$$\lim_{N \to \infty} \min E_N = 0?$$

I'll now show you that the answer is yes. Inserting our newly found c_k into the expression for E_N, we have

$$\min E_N = \int\limits_{\text{period}} \left[x(t) - \sum_{k=-N}^{N} \left\{ \left(\frac{1}{T} \int\limits_{\text{period}} x(t)e^{-jk\omega_0 t} \, dt \right) e^{jk\omega_0 t} \right\} \right]^2 \, dt$$

$$= \int\limits_{\text{period}} x(t)^2 \, dt - \frac{2}{T} \int\limits_{\text{period}} x(t)$$

$$\times \left[\sum_{k=-N}^{N} \left(\int\limits_{\text{period}} x(t)e^{-jk\omega_0 t} \, dt \right) e^{jk\omega_0 t} \right] dt$$

$$+ \frac{1}{T^2} \int\limits_{\text{period}} \left\{ \sum_{k=-N}^{N} \left(\int\limits_{\text{period}} x(t)e^{-jk\omega_0 t} \, dt \right) e^{jk\omega_0 t} \right\}^2 \, dt.$$

Since we earlier found that

$$\int\limits_{\text{period}} x(t)e^{-jn\omega_0 t} \, dt = Tc_n$$

then we can write the expression for $\min E_N$ as

$$\min E_N = \int_{\text{period}} x^2(t)\,dt - \frac{2}{T} \int_{\text{period}} x(t) \sum_{k=-N}^{N} T c_k e^{jk\omega_0 t}\,dt$$

$$+ \frac{1}{T^2} \int_{\text{period}} \left\{ \sum_{k=-N}^{N} T c_k e^{jk\omega_0 t} \right\}^2 dt.$$

Now, concentrating on the second term, and cancelling the Ts, we see it is

$$2 \int_{\text{period}} x(t) \sum_{k=-N}^{N} c_k e^{jk\omega_0 t}\,dt = 2 \sum_{k=-N}^{N} c_k \int_{\text{period}} x(t) e^{jk\omega_0 t}\,dt$$

$$= 2 \sum_{k=-N}^{N} c_k T c_{-k} = 2T \sum_{k=-N}^{N} c_k c_k^*$$

$$= 2T \sum_{k=-N}^{N} |c_k|^2.$$

Similarly, for the third term of $\min E_N$ (and cancelling the T^2's),

$$\int_{\text{period}} \left\{ \sum_{k=-N}^{N} c_k e^{jk\omega_0 t} \right\}^2 dt = \int_{\text{period}} \left(\sum_{k=-N}^{N} c_k e^{jk\omega_0 t} \right) \left(\sum_{\ell=-N}^{N} c_\ell e^{j\ell\omega_0 t} \right) dt.$$

When the two sums are multiplied out term by term and integrated over a period, all the integrals vanish except for the cases where $\ell = -k$. In those special cases, of which there are $2N + 1$ in number, each integral equals T (as shown in the previous chapter). Thus, the third term in $\min E_N$ is

$$\sum_{k=-N}^{N} T c_k c_{-k} = T \sum_{k=-N}^{N} |c_k|^2.$$

Therefore, combining these partial results we arrive at

$$\min E_N = \int_{\text{period}} x^2(t)\,dt - 2T \sum_{k=-N}^{N} |c_k|^2 + T \sum_{k=-N}^{N} |c_k|^2$$

$$= \int_{\text{period}} x^2(t)\,dt - T \sum_{k=-N}^{N} |c_k|^2.$$

Now, as $|c_k| = |c_{-k}|$, we can rewrite this last result as

$$\min E_N = \int\limits_{\text{period}} x^2(t)\, dt - T\left(c_0^2 + 2\sum_{k=1}^{N} |c_k|^2\right).$$

Since $\min E_N$ is never negative—it is an integrated-*squared* error—then it must be true that for *any* value of N

$$T\left(c_0^2 + 2\sum_{k=1}^{N} |c_k|^2\right) < \infty$$

because the expression on the left is bounded from above by the energy of $x(t)$ in a period (which we've assumed to be finite). This requires (since N, although finite can be arbitrarily large) that

$$\lim_{k\to\infty} |c_k| = 0.$$

This is simply the mathematics making the physically obvious statement that the power in the kth harmonic of a periodic $x(t)$ must decrease to zero as k increases. Suppose that was *not* true. Then, as k goes to infinity, we would find the total harmonic power would be arbitrarily large, contrary to our initial, fundamental restriction that $x(t)$ has *finite* energy per period.

Now, recall Parseval's theorem from the previous chapter, which says

$$\int\limits_{\text{period}} x^2(t)\, dt = T\left(c_0^2 + 2\sum_{k=1}^{\infty} |c_k|^2\right).$$

Thus,

$$\min E_N = T\left(c_0^2 + 2\sum_{k=1}^{\infty} |c_k|^2\right) - T\left(c_0^2 + 2\sum_{k=1}^{N} |c_k|^2\right)$$

or,

$$\min E_N = 2T\sum_{k=N+1}^{\infty} |c_k|^2.$$

From this we immediately have (remembering that $\lim_{k\to\infty} |c_k| = 0$) that

$$\lim_{N\to\infty} \min E_N = 0.$$

Thus, the Fourier series for $x(t)$ has *zero* integrated-squared-error when compared to $x(t)$, and it is this result that gives physical meaning to the Fourier series of $x(t)$. This conclusion does not mean that the series for

$x(t)$, and $x(t)$, are equal for *every* value of t. If there is a finite discontinuity in $x(t)$ at $t = t_0$, for example, it can be shown that the series converges to the *average* value, i.e., that the series converges to

$$\frac{1}{2}\left[x(t_0^-) + x(t_0^+)\right].$$

Most electrical engineers and physicists are surprised to learn that this result wasn't proven until nearly a hundred years after Fourier's work, in 1906, by the American mathematician Maxime Bôcher (1867–1918), who taught at Harvard. An illustration of this sort of behavior at a discontinuity is shown in Figures 9.1 through 9.4.

What our convergence result does say, however, is that the Fourier series for $x(t)$, and $x(t)$, are *almost always* equal. Indeed, the convergence of the Fourier series of a periodic $x(t)$ is taken for granted by engineers as long as $x(t)$ satisfies the so-called *Dirichlet conditions*—after the French mathematician Peter Lejeune Dirichlet (1805–1859)—which require a *finite* number of both discontinuities and exrema in a period. For any "real world" $x(t)$, of course, we are *guaranteed* there will not be an infinite number of either of those two cases.

Now, before going any further, let me show you the actual details of calculating a Fourier series. That is, I'll use our wonderful formula for the Fourier coefficients c_k that we now know leads to a *convergent* result. In particular, we saw in Figures 9.1 through 9.4 how Euler's 1744 "Fourier series" appears to converge, as more and more terms are included, to the *periodic extension* of $x(t) = \frac{\pi-t}{2}$ where one period is $0 < t < 2\pi$. Euler got his series by what I called, in Chapter 9, "wild mathematics." What does our "proper mathematics" give us? (To keep the suspense from becoming overwhelming, I'll tell you right here—the same thing!) Here's how to do it. We have

$$x(t) = \sum_{k=-\infty}^{\infty} c_k e^{j\omega_0 t} = \sum_{k=-\infty}^{\infty} c_k e^{jt}$$

since $\omega_0 = \frac{2\pi}{T} = 1$ as $T = 2\pi$. Also,

$$c_k = \frac{1}{2\pi} \int_0^{2\pi} x(t) e^{-jkt}\, dt.$$

Thus,

$$c_k = \frac{1}{2\pi} \int_0^{2\pi} \frac{\pi - t}{2} e^{-jkt}\, dt = \frac{1}{4} \int_0^{2\pi} e^{-jkt}\, dt - \frac{1}{4\pi} \int_0^{2\pi} t e^{-jkt}\, dt.$$

For $k = 0$, this reduces to

$$c_0 = \frac{1}{4} \int_0^{2\pi} dt - \frac{1}{4\pi} \int_0^{2\pi} t \, dt = \frac{1}{4} \left(t \Big|_0^{2\pi} - \frac{1}{4\pi} \left(\frac{t^2}{2} \Big|_0^{2\pi} \right) \right)$$

$$= \frac{2\pi}{4} - \frac{4\pi^2}{8\pi} = \frac{\pi}{2} - \frac{\pi}{2} = 0.$$

This is actually physically obvious, of course, since c_0 is the average value of $x(t)$ over a period and Euler's $x(t)$ clearly has zero average value.

For $k \neq 0$, we have (if you recall that $\int te^{at} \, dt = \frac{ate^{at} - e^{at}}{a^2}$ and set $a = -jk$)

$$c_k = \frac{1}{4} \left(\frac{e^{-jkt}}{-jk} \Big|_0^{2\pi} - \frac{1}{4\pi} \left(\frac{-jkte^{-jkt} - e^{-jkt}}{-k^2} \Big|_0^{2\pi} \right) \right)$$

$$= \frac{e^{-jk2\pi} - 1}{-j4k} - \frac{-jk2\pi e^{-jk2\pi} - e^{-jk2\pi} + 1}{-4\pi k^2}.$$

Since $e^{jk2\pi} = 1$ for *all* k, then this seemingly awful expression for c_k reduces to

$$c_k = -j\frac{1}{2k}, \qquad k \neq 0.$$

Notice carefully that $c_{-k} = c_k^*$, just as it should since $x(t)$ is a *real* function of time.

So, what we have arrived at is that, over a period of $x(t)$,

$$x(t) = \sum_{k=-\infty}^{\infty} c_k e^{jkt} = \sum_{k=-\infty}^{-1} -j\frac{e^{jkt}}{2k} + 0 + \sum_{k=1}^{\infty} -j\frac{e^{jkt}}{2k}.$$

In the first sum let $n = -k$, and so

$$\sum_{k=-\infty}^{-1} -j\frac{e^{jkt}}{2k} = \sum_{n=\infty}^{1} -j\frac{e^{-jnt}}{-2n} = \sum_{n=1}^{\infty} j\frac{e^{-jnt}}{2n} = \sum_{k=1}^{\infty} j\frac{e^{-jt}}{2k}.$$

So, for $0 < t < 2\pi$,

$$x(t) = \frac{\pi - t}{2} = \sum_{k=1}^{\infty} j\frac{e^{-jkt}}{2k} + \sum_{k=1}^{\infty} -j\frac{e^{jkt}}{2k}$$

$$= \sum_{k=1}^{\infty} j\frac{e^{-jkt} - e^{jkt}}{2k} = \sum_{k=1}^{\infty} j\frac{-2j\sin(kt)}{2k}$$

$$= \sum_{k=1}^{\infty} \frac{\sin(kt)}{k} = \sin(t) + \frac{\sin(2t)}{2} + \frac{\sin(3t)}{3} + \cdots,$$

which is precisely Euler's series.

The convergence of a Fourier series expansion of a time function is best appreciated, I think, with a picture. As you will show in Problem 10.1, the Fourier series for a periodic "square wave" with period 2π [as shown in Figure 10.1 with little circles on the time axis, at multiples of π, to indicate

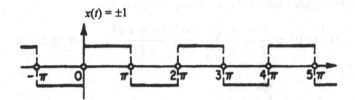

Figure 10.1   A "square wave."

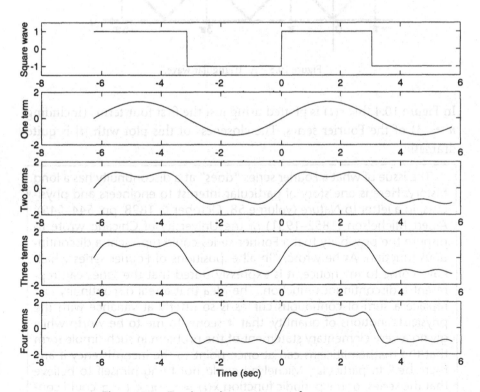

Figure 10.2   Fourier series convergence for a square wave.

the value of $x(t)$ at its discontinuities], with unit amplitude, is

$$x(t) = \pm 1 = \frac{4}{\pi} \left[\sin(t) + \frac{1}{3} \sin(3t) + \frac{1}{5} \sin(5t) + \cdots \right], \quad -\pi < t < \pi.$$

In Figure 10.2 a single period (centered on $t = 0$) of $x(t)$ is shown at the top, and beneath it are shown the partial Fourier series approximations using just the first one, two, three, and four terms, in succession. This is pretty impressive, especially so as this $x(t)$ is discontinuous.

If $x(t)$ is a continuous function, then the convergence is even more dramatic. For example, in Problem 10.2 you will show that the Fourier series for the "triangular wave" shown in Figure 10.3 is

$$x(t) = |t| = \frac{\pi}{2} - \frac{4}{\pi} \sum_{n=1}^{\infty} \frac{\cos\{(2n - 1)t\}}{(2n - 1)^2}, \quad -\pi \leq t \leq \pi.$$

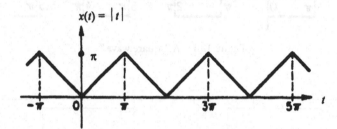

Figure 10.3   A "triangular wave."

In Figure 10.4 this $x(t)$ is plotted using just the first four terms (including $n = 3$) of the Fourier series. The closeness of this plot with $|t|$ is quite dramatic.

The issue of what a Fourier series "does" at a discontinuity has a long history. Here is one story of particular interest to engineers and physicists. In a letter to Nature (volume 58, October 6, 1898, pp. 544–545), Albert Michelson (1852–1931) of the University of Chicago wrote to dispute the possibility that a Fourier series could represent a discontinuous function. As he wrote, "In all expositions of Fourier series which have come to my notice, it is expressly stated that the series can represent a discontinuous function. The idea that a real discontinuity can replace a sum of continuous curves is so utterly at variance with the physicists' notions of quantity, that it seems to me to be worth while giving a very elementary statement of the problem in such simple form that the mathematicians can at once point to the inconsistency if any there be." In particular, Michelson could not bring himself to believe that the series for the periodic function $x(t) = t$, $-\pi < t < \pi$ could con-

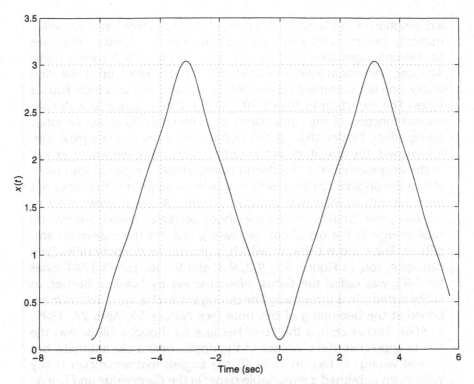

Figure 10.4   First four terms of the Fourier series of absolute value of t.

verge to zero at the times of the discontinuities (odd multiples of π). Michaelson's reasoning was, in fact, in error, as was pointed out in a reply from a Cambridge math professor (*Nature* 58, September 13, 1898, pp. 569–570). Michelson later replied that he had not been convinced (*Nature* 59, December 29, 1898, p. 200), in letter printed along with two others. One was from J. Willard Gibbs (1839–1903) at Yale (more on Gibbs at the end of this box), which attempted to answer Michelson's original confusion. The second was from the Cambridge math professor who pointed out that Gibbs had some misunderstandings in his letter, too! It is important to understand that Michelson was not being dense, although he was mathematically naive in this particular case. Indeed, at the time of his writing, it had been twelve years since he had performed the famous 1887 experiment (the "Michelson–Morley experiment") that would win him the 1907 Nobel Prize in physics. While Michelson is remembered today only for that and other optical experiments, he was a gadget builder extraordinaire. In particular, his interest in Fourier series was experimental, not theoretical. Before his first letter to *Nature* he had already reported on his construction of a mechanical device to sum and plot up to 80 (!) terms of any given Fourier series. For a description of this fantastic machine, along with some beautiful, automatically generated ink-pen plots of a large number of exotic series, see Michelson's paper (co-authored with S. W. Stratton) "A New Harmonic Analyzer," *American Journal of Science* 5, January 1898, pp. 1–14. And finally, one last comment on discontinuous functions and their Fourier series. The depiction in Figure 10.2 of the convergence to a discontinuous function of the partial sums of a Fourier series shows an interesting effect. Notice that, as the number of terms in the partial sum is increased, the amplitude of the ripples decreases everywhere *except* in the neighborhood of the discontinuity. There the partial sum overshoots the original function, with a maximum amplitude that does not decrease with an increasing number of terms. As the number of terms increases, the *duration* of the overshoot decreases (which means the total *energy* in the overshoot decreases), but not the maximum amplitude. This curious behavior, which is generic for discontinuities (you can see it, too, in Figures 9.1, 9.2, 9.3, and 9.4 for Euler's 1744 series for $\frac{\pi-t}{2}$), was called the *Gibbs' phenomenon* by Maxime Bôcher, as Gibbs stated it in a throwaway line during the exchange of letters mentioned at the beginning of this note (see *Nature* 59, April 27, 1899, p. 606). Bôcher coined this term because he thought Gibbs was the first to appreciate the existence of the ripples near a discontinuity, but he was wrong. In fact, in 1848 (!) the English mathematician Henry Wilbraham published a remarkable paper in the *Cambridge and Dublin Mathematical Journal*[1] that derives the precise nature of the Fourier series near the discontinuity of a square wave. His paper (titled "On a Certain Periodic Function"), was soon forgotten, but it even includes a carefully plotted graph of the yet-to-be-named Gibbs' phenomenon!

The physical interpretation of the distribution of the power of a periodic $x(t)$ over the Fourier harmonics is, of course, of great interest to electrical engineers. Indeed, it was of paramount importance (and frustration!) to the early radio engineers who ran head-on into Parseval's theorem with their spark-gap transmitters. The next chapter shows exactly what these pioneers were up against, as we put all the mathematical machinery just developed into action.

Before leaving the pure theory of Fourier series, let me show you two absolutely stunning results in pure mathematics that come directly out of Parseval's theorem. Imagine, for this box, that we have the *complex* periodic signal $x(t)$, defined over a period as

$$x(t) = e^{jAt}, \quad -\pi < t < \pi$$

where A is any real constant *not* equal to an integer (you'll see why this restriction is important, soon). Now, we have so far in this book always thought of $x(t)$ as real, but that was simply a concession to the physical reality of the signals we can actually generate. On paper, however, we can indeed discuss complex-valued, periodic functions as well. So, since the period is $T = 2\pi$, we have $\omega_0 = 1$ and we can write $x(t)$ in a Fourier series as

$$x(t) = \sum_{k=-\infty}^{\infty} c_k e^{jkt}.$$

Then

$$c_k = \frac{1}{2\pi} \int_{-\pi}^{\pi} x(t) e^{-jkt}\, dt = \frac{1}{2\pi} \int_{-\pi}^{\pi} e^{jAt} e^{-jkt}\, dt$$

$$= \frac{1}{2\pi} \int_{-\pi}^{\pi} e^{j(A-k)t}\, dt = \frac{1}{2\pi} \left(\frac{e^{j(A-k)t}}{j(A-k)} \Big|_{-\pi}^{\pi} \right)$$

$$= \frac{e^{j(A-k)\pi} - e^{-j(A-k)\pi}}{j2\pi(A-k)} = \frac{j2\sin\left\{(A-k)\pi\right\}}{j2\pi(A-k)}$$

or, at last,

$$c_k = \frac{\sin\left\{\pi(A-k)\right\}}{\pi(A-k)}.$$

This result holds for *all* k because the denominator is never equal to zero (because of our initial assumption that A is not equal to an integer). Notice that $c_{-k} \neq c_k^*$ because $x(t)$ is *not* real.

In the last chapter we defined the energy of a *real* periodic signal to be

$$W = \int_{\text{period}} x^2(t)\, dt.$$

If we wish to keep the energy as a real quantity, even for complex-valued signals, then we must generalize the energy definition; this is done by writing

$$W = \int_{\text{period}} x(t) x^*(t)\, dt = \int_{\text{period}} |x(t)|^2\, dt$$

where $x^*(t) = e^{-jAt}$ is, as usual, the conjugate of $x(t)$. This extended definition reduces (as it must!) to the original definition for the case of $x(t)$ real. Thus, for our particular case we have

$$W = \int_{-\pi}^{\pi} e^{jAt} e^{-jAt}\, dt = \int_{-\pi}^{\pi} dt = 2\pi.$$

But Parseval's theorem tells us that even with $x(t)$ complex (see Problem 9.3) the energy per period is $W = T\sum_{k=-\infty}^{\infty} |c_k|^2$ and so (since

$T = 2\pi$) we have

$$\sum_{k=-\infty}^{\infty} |c_k|^2 = 1.$$

That is,

$$\sum_{k=-\infty}^{\infty} \frac{\sin^2\{\pi(A-k)\}}{\pi^2(A-k)^2} = 1.$$

Now, let $A = \frac{u}{\pi}$. Then,

$$1 = \sum_{k=-\infty}^{\infty} \frac{\sin^2\left\{\pi\left(\frac{u}{\pi}-k\right)\right\}}{\pi^2\left(\frac{u}{\pi}-k\right)^2} = \sum_{k=-\infty}^{\infty} \frac{\sin^2(u-k\pi)}{(u-k\pi)^2}.$$

Since $\sin^2(u - k\pi) = \sin^2(u)$, this reduces to one of the most famous identities in mathematics, the beautiful

$$\frac{1}{\sin^2(u)} = \sum_{k=-\infty}^{\infty} \frac{1}{(u-k\pi)^2}.$$

This basic identity, known to Euler (by other means)[2] no later than 1740, takes on special forms when specific values of u are used. For example if $u = \frac{\pi}{3}$ then $\sin(u) = \frac{1}{2}\sqrt{3}$ and the basic identity reduces to

$$\sum_{k=-\infty}^{\infty} \frac{1}{(1-3k)^2} = \frac{4\pi^2}{27} = 1.4621636\cdots.$$

Who would have guessed it? This result is easily verified on a computer; when I wrote a simple MATLAB loop to evaluate the sum for $-50{,}000 \le k \le 50{,}000$ the result was $1.46215917\cdots$.

Here's another example of the value of discussing complex-valued time functions (you'll see yet another when we get to single-sideband radio). Suppose $x(t)$ is defined as

$$x(t) = e^{(1/2)ae^{jt}} = e^{(1/2)a\cos(t)}e^{j(1/2)a\sin(t)}.$$

Since $|e^{j(\text{anything real})}| = 1$, and as the first exponential on the right-hand side is always nonnegative, then

$$|x(t)|^2 = e^{a\cos(t)}.$$

Also, since $x(t)$ is periodic with period $T = 2\pi$, then once again the energy per period is

$$W = \int_{-\pi}^{\pi} |x(t)|^2 \, dt = \int_{-\pi}^{\pi} e^{a\cos(t)} \, dt = 2\pi \sum_{k=-\infty}^{\infty} |c_k|^2.$$

This gives us the important integral (it appears in the theory of FM radio)

$$\frac{1}{2\pi} \int_{-\pi}^{\pi} e^{a\cos(t)} \, dt = \sum_{k=-\infty}^{\infty} |c_k|^2.$$

The integral is called the "modified Bessel function of the first kind, of order zero," and is usually denoted by $I_o(a)$—all we need to do is find the c_k.

To find the c_k is, despite initial appearances (perhaps), not hard to do. We have the standard Fourier series, with $\omega_o = 1$, of

$$x(t) = \sum_{k=-\infty}^{\infty} c_k e^{jkt}.$$

But we also have

$$x(t) = e^{(1/2)ae^{jt}} = \sum_{k=o}^{\infty} \frac{\left(\frac{1}{2}ae^{jt}\right)^k}{k!} = \sum_{k=0}^{\infty} \frac{a^k}{2^k k!} e^{jkt},$$

where I've used the power-series expansion of the exponential. Comparing coefficients, term by term, we see by inspection that

$$c_k = 0, \quad k < 0$$

$$c_k = \frac{a^k}{2^k k!}, \quad k \geq 0,$$

and you should notice that, again, $c_{-k} \neq c_k^*$ because $x(t)$ is complex, not real. In any case,

$$I_o(a) = \frac{1}{2\pi} \int_{-\pi}^{\pi} e^{a\cos(t)} \, dt = \sum_{k=-\infty}^{\infty} |c_k|^2 = \sum_{k=0}^{\infty} \frac{a^{2k}}{2^{2k}(k!)^2}.$$

This is *almost* the same result that we got back in the final shaded box of Chapter 9, for J_o, except that now all the terms in the sum have the same sign while J_o had alternating signs.

An important use of Fourier series is the determination of the steady-state response of linear filters to *non*sinusoidal (yet still periodic) inputs. Recall the last equation of Chapter 9 for the filter output;

$$y(t) = \sum_{k=-\infty}^{\infty} c_k H(jk\omega_o) e^{jk\omega_o t}, \quad \omega_o = \frac{2\pi}{T}$$

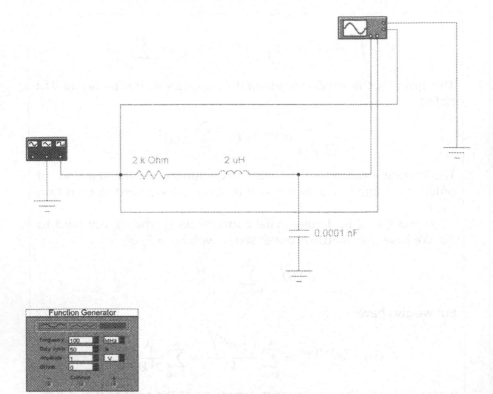

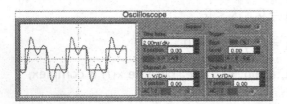

Figure 10.5   This series RLC circuit has an input of a 100 MHz square wave (−1 to
+1 volts), with the output taken across the capacitor. Notice the "ringing," i.e., that the
circuit oscillates, and also that the peak of the output voltage exceeds the input voltage. To
obtain an easily observed display, the oscilloscope is triggered by leading edge of the square
wave (on Channel B).

where $H(j\omega)$ is the transfer function of the filter. For the filter in the Electronics Workbench simulation of Figure 10.5 you should be able to show
that

$$H(j\omega) = \frac{10^{19}}{10^{19} - 2\omega^2 + j2 \times 10^9\omega}.$$

In addition, if the input is the 100 MHz square-wave

$$x(t) = -1, \quad -\frac{T}{2} < t < 0$$

$$+1, \quad 0 < t < \frac{T}{2}, T = 10 \times 10^{-9} \text{ sec},$$

then you should also be able to show that $c_k = -j\frac{2}{\pi k}$ for k odd, and zero for k even. Thus,

$$y(t) = -j\frac{2}{\pi} \sum_{k=-\infty}^{\infty} \frac{1}{k} \cdot \frac{10^{19} e^{jk\omega_o t}}{10^{19} - 2(k\omega_o)^2 + j2 \times 10^9 k\omega_o}.$$

With a bit of algebra this reduces to

$$y(t) = \frac{4}{\pi} \times 10^{19} \sum_{k=1,3,5,\ldots}^{\infty} \frac{1}{k}$$

$$\cdot \frac{(10^{19} - 2k^2\omega_o^2)\sin(k\omega_o t) - 2 \times 10^9 k\omega_o \cos(k\omega_o t)}{4k^4\omega_o^4 - 3.6 \times 10^{19} k^2\omega_o^2 + 10^{38}}$$

which is plotted in Figure 10.6. Notice the excellent agreement with the Electronics Workbench simulation.

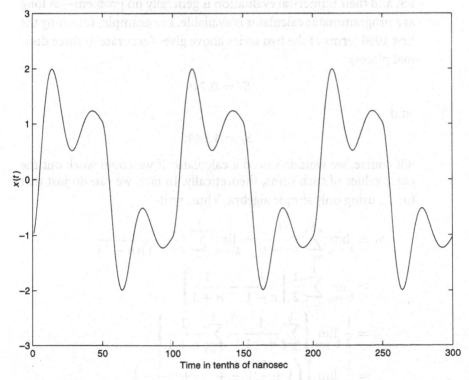

Figure 10.6   Steady-state output of RLC-filter with square-wave input.

Notes

1. For just what Wilbraham did, see the scholarly paper by Edwin Hewitt and Robert E. Hewitt, "The Gibbs–Wilbraham Phenomenon: An Episode in Fourier Analysis," *Archive for History of Exact Sciences* 21, 1979, pp. 129–160.

2. For these "other means" see my *An Imaginary Tale: the story of $\sqrt{-1}$*, Princeton 1998, pp. 156–157.

3. See, for example, my book, *An Imaginary Tale* Princeton 1998, 155–156.

Problems

10.1. As an example of the power of Fourier series, consider the numerical problem of summing an infinite series, such as

$$S_1 = \sum_{n=2}^{\infty} \frac{1}{n^2 - 1} = \frac{1}{3} + \frac{1}{8} + \frac{1}{15} + \cdots$$

and

$$S_2 = \sum_{n=1}^{\infty} \frac{1}{n^2} = 1 + \frac{1}{4} + \frac{1}{9} + \cdots.$$

Such expressions often occur in engineering and theoretical analyses, and their numerical evaluation is generally no problem—as long as a programmable calculator is available. For example, summing the first 1000 terms of the two series above gives (accurate to three decimal places),

$$S_1 = 0.749$$

and

$$S_2 = 1.644.$$

Of course, we wouldn't need a calculator if we could work out the exact values of such sums, theoretically. In fact, we can do just that for S_1, using only simple algebra. Thus, write

$$S_1 = \lim_{k \to \infty} \sum_{n=2}^{k} \frac{1}{n^2 - 1} = \lim_{k \to \infty} \sum_{n=2}^{k} \frac{1}{(n+1)(n-1)}$$

$$= \lim_{k \to \infty} \sum_{n=2}^{k} \frac{1}{2} \left\{ \frac{1}{n-1} - \frac{1}{n+1} \right\}$$

$$= \frac{1}{2} \lim_{k \to \infty} \left\{ \sum_{n=2}^{k} \frac{1}{n-1} - \sum_{n=2}^{k} \frac{1}{n+1} \right\}$$

$$= \frac{1}{2} \lim_{k \to \infty} \left[\left(1 + \frac{1}{2} + \frac{1}{3} + \cdots + \frac{1}{k-1} \right) \right.$$

$$-\left(\frac{1}{3}+\cdots+\frac{1}{k-1}+\frac{1}{k}+\frac{1}{k+1}\right)\Bigg]$$

$$=\frac{1}{2}\lim_{k\to\infty}\left[1+\frac{1}{2}-\frac{1}{k}-\frac{1}{k+1}\right]=\frac{1}{2}\cdot\frac{3}{2}=\frac{3}{4}=0.75.$$

This approach (called *telescoping* the series because of the internal cancellations) will obviously not work for S_2, however. In fact, even though the two series superficially look very similar, the second series requires a much more sophisticated approach. Because S_2 stumped so many mathematicians for a very long time, the problem of summing the reciprocals of the squares of the positive integers became a famous problem in the history of mathematics. It was first solved by Euler in 1736, using an extremely clever and deep method which, however, is understandable to anyone who has had freshman calculus.[3] You can do it today with Fourier analysis (which, while known to Euler, was also believed by him to apply only in very special circumstances).

(a) For the function shown in Figure 10.1 (called a "square wave" for obvious reasons), verify that its Fourier series is

$$x(t)=\frac{4}{\pi}\left[\sin(t)+\frac{1}{3}\sin(3t)+\frac{1}{5}\sin(5t)+\frac{1}{7}\sin(7t)+\cdots\right].$$

Notice that at $t=\frac{\pi}{2}$ we have $x(\frac{\pi}{2})=1$, and so

$$1=\frac{4}{\pi}\left[\sin\left(\frac{\pi}{2}\right)+\frac{1}{3}\sin\left(\frac{3\pi}{2}\right)+\frac{1}{5}\sin\left(\frac{5\pi}{2}\right)+\cdots\right]$$

or

$$1=\frac{4}{\pi}\left[1-\frac{1}{3}+\frac{1}{5}-\frac{1}{7}+\cdots\right]$$

or, finally, that

$$\frac{\pi}{4}=1-\frac{1}{3}+\frac{1}{5}-\frac{1}{7}+\cdots.$$

This is called Leibniz's series [Leibniz discovered it in 1673, by an entirely different method, but it is known that the Scottish mathematician James Gregory (1638–1675) knew it earlier, in 1671]. Of this expression, Leibniz is reported to have said "God loves the odd integers." It is not of any use in finding S_2, but we get it along the way—for free, so to speak! Recall from Chapter 9, however, how Euler used it in 1744 to evaluate the infinite constant of integration in his development of the "Fourier Series" for $\frac{\pi-t}{2}$, $0 < t < 2\pi$.

(b) Next, observe that $x^2(t) = 1$ for any t. Thus,

$$\frac{16}{\pi^2} \left[\sin(t) + \frac{1}{3}\sin(3t) + \frac{1}{5}\sin(5t) + \cdots \right]^2 = 1.$$

Expand the right-hand side, integrate term by term from 0 to 2π, and thereby show that

$$1 + \frac{1}{3^2} + \frac{1}{5^2} + \frac{1}{7^2} + \cdots = \frac{\pi^2}{8}.$$

When doing the integrals you'll find it helpful to use the so-called *orthogonality property* of the sine function, i.e.,

$$\int_0^{2\pi} \sin(mt)\sin(nt)\, dt = \begin{cases} 0, & m \neq n \\ \pi, & m = n \end{cases}$$

where m and n are any integers, positive or negative.

(c) The result in part b isn't quite S_2, of course, being simply the sum of the reciprocals of the squares of the odd integers. This is, part b says

$$\sum_{n=0}^{\infty} \frac{1}{(2n+1)^2} = \frac{\pi^2}{8}.$$

To complete the calculation of S_2, show that

$$S_2 = \sum_{n=1}^{\infty} \frac{1}{n^2} = \frac{4}{3}\sum_{n=0}^{\infty} \frac{1}{(2n+1)^2} = \frac{\pi^2}{6} \quad (= 1.6449\ldots).$$

Hint: Use the (almost trivial) observation that all the integers can be separated into two sets—the evens and the odds. Thus, we can immediately write

$$S_2 = \sum_{n=1}^{\infty} \frac{1}{n^2} = \sum_{n=1}^{\infty} \frac{1}{(2n)^2} + \sum_{n=0}^{\infty} \frac{1}{(2n+1)^2}.$$

See Problem G6 for yet another fantastic way to calculate S_2.

10.2. For an even more direct solution to the previous problem, find the Fourier series for the periodic function shown in Figure 10.3, where $x(t) = |t|$, $-\pi \leq t \leq \pi$. Then, set $t = \pi$ and use the hint at the end of the previous problem to again evaluate Euler's sum.

10.3. As another example of the power of Fourier series to calculate sums, find the Fourier expansion of the function shown in Figure 10.7,

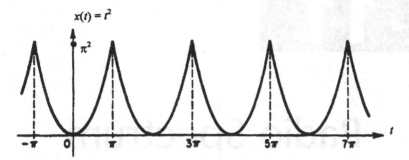

Figure 10.7   A "parabolic wave."

where $x(t) = t^2$, $-\pi \leq t \leq \pi$. Then, set $t = 0$ in the result and derive the sum $1 - \frac{1}{2^2} + \frac{1}{3^2} - \frac{1}{4^2} + \cdots = \frac{\pi^2}{12}$.

10.4. Starting with the periodic function given by $x(t) = e^{-t}$ over the interval $0 < t < 2\pi$, use Parseval's theorem to derive the formula

$$\sum_{k=-\infty}^{\infty} \frac{1}{1+k^2} = \pi \left(\frac{1 + e^{-2\pi}}{1 - e^{-2\pi}} \right).$$

Write a simple computer program to numerically check the truth of this result.

10.5. What is the value of

$$\sum_{k=-\infty}^{\infty} \frac{1}{(1-6k)^2}?$$

Hint: Remember the identity derived at the end of the chapter with the aid of Parseval's theorem. Write a computer program to numerically check the truth of your answer.

10.6. Recall the claim (verified with MATLAB) made at the very beginning of the book that

$$\int_0^1 \frac{1}{x} \ln \left[\left(\frac{1+x}{1-x} \right)^2 \right] dx = \frac{1}{2}\pi^2.$$

Prove that this is so. *Hint:* Write

$$\ln \left[\left(\frac{1+x}{1-x} \right)^2 \right] = 2 \left[\ln(1+x) - \ln(1-x) \right],$$

expand the log functions in power series, integrate term by term, and use the result derived in Problem 10.1 for the sum of the reciprocals of the squares of the odd integers.

11

Radio Spectrum of a Spark-Gap Transmitter

In Chapter 4, I stated (without proof) that the earliest radio transmitters, the spark-gap transmitters dating from Hertz's original late 1880s experiments, scatter electromagnetic energy all across the frequency spectrum. Such transmitters are very wasteful of their available energy, and rather than concentrating it in a narrow band of frequencies, they literally splatter energy everywhere. You are now in a position to mathematically study the radio signal generated by these kinds of transmitters, and to substantiate my earlier statements.

Economists talk of the "tragedy of the commons" when discussing what happens when a limited resource is made freely available to the public. The phrase comes from the old English village tradition of setting aside a certain amount of public grazing land (a "commons") on which anyone in the village could graze sheep for free. What typically happened is that so many people would attempt to do this that none of the sheep could find enough to eat, and so the commons was ruined for all (that's the "tragedy," of course). The finite width of the broadcast spectrum is a commons, too, and it also can support only a finite number of users at any given instant. It is of great importance, then, that each user (each radio transmitter) not "consume" too much spectrum—and spark-gap transmitters are not such considerate users. Each spark-gap transmitter was, in fact, like a great, fat sheep lurching about the commons while devouring huge chunks of spectrum. Spark-gap transmitters were eventually forbidden to be used.

As stated in Chapter 4, a spark-gap transmitter's signal is a periodic, exponentially damped sinusoid. That is, we can write it, for the particular period $0 < t < T$ and where N is some constant, as

$$x(t) = e^{-at} \sin(\omega_c t), \quad T = N \frac{2\pi}{\omega_c}.$$

N is related to the spark-rate frequency (how fast sparks are generated), and ω_c is the *resonant* frequency of the spark transmitter's antenna circuit (see Figure 4.3 again). Without loss of very much generality, I have made the following assumptions:

1. The initial signal amplitude factor is unity (this is simply an arbitrary scaling).

2. The signal period is an *integer* (N) number of complete cycles at frequency ω_c. This is almost certainly *not* exactly true, but it keeps the math simple and it won't affect the essential nature of the final conclusions.

In the notation of the previous chapters, then, we write

$$x(t) = \sum_{k=-\infty}^{\infty} c_k e^{jk\omega_0 t}$$

where

$$\omega_0 = 2\pi f_0 = 2\pi \frac{1}{T}.$$

Thus,

$$\omega_0 = \frac{\omega_c}{N}$$

and

$$x(t) = \sum_{k=-\infty}^{\infty} c_k e^{jk(\omega_c/N)t}, \quad 0 < t < T = \frac{2\pi N}{\omega_c}.$$

The Fourier coefficients are given by

$$c_k = \frac{\omega_c}{2\pi N} \int_0^{2\pi N/\omega_c} e^{-at} \sin(\omega_c t) e^{-jk(\omega_c/N)t} \, dt,$$

a somewhat formidable-looking integral which is actually not too difficult to do. If you replace $\sin(\omega_c t)$ with its complex exponential equivalent, and are careful with the algebra, you will find that

$$c_k = \frac{\omega_c^2}{2\pi N} \left[\frac{1 - e^{-a2\pi N/\omega_c}}{a^2 + \omega_c^2 (1 - k^2/N^2) + j2a\omega_c k/N} \right].$$

While doing the integral you will find it *very* helpful to remember that both k and N are integers; when you find factors of the form $e^{\pm j2p(N\pm k)}$ you will of course remember they are unity for all k and N.

Now, what is a? It is, of course, the reciprocal of the time constant in the amplitude exponential decay factor, and thus has the units of frequency. The only specific frequency parameters we have in the analysis, however, are N and ω_c. It will be convenient to retain N as an explicit parameter, and if we write $a = \eta\omega_c$ (where η is a dimensionless, nonnegative parameter), then all explicit appearances of both a and ω_c disappear and we obtain

$$c_k = \frac{1 - e^{-2\pi N\eta}}{2\pi N} \frac{1}{\eta^2 + (1 - k^2/N^2) + j2\eta k/N}.$$

$\eta = 0$ means no amplitude decay, and $\eta > 0$ means decay. $\eta < 0$ would mean amplitude growth, which implies an energy source mechanism in the spark (which does not exist). The energy lost in the spark appears in several forms; the most obvious, literally by definition, is the visible radiation by which you *see* the spark. Sparks were also very noisy (the rapid sparking of some transmitters was likened to machine gunfire), and energy had to be dissipated as well to create sound waves.

Now, notice that $c_{-k} = c_k^*$, just as we require, since $x(t)$ is real-valued. Also, notice that $c_0 = 0$ if $\eta = 0$, which is physically correct as $\eta = 0$ means no amplitude damping, which in turn means $x(t)$ is a pure, single-frequency, constant-amplitude sinusoid with zero dc; and c_0 is the dc value. Specifically, setting $k = 0$ for arbitrary η, we obtain

$$c_0 = \frac{1 - e^{-2\pi N\eta}}{2\pi N} \frac{1}{1 + \eta^2}, \quad \eta \geq 0.$$

The energy per period of $x(t)$ is

$$W = \int_{period} x^2(t)\, dt = \int_0^{2\pi N/\omega_c} e^{-2at} \sin^2(\omega_c t)\, dt.$$

Using $a = \eta\omega_c$ and again being careful with the algebra, you can evaluate this integral, and then divide it by the period, to get the average power of

$x(t)$:

$$\frac{1}{T}W = \frac{1 - e^{-4\pi N\eta}}{8\pi N\eta(1 + \eta^2)}.$$

However, recall Parseval's theorem from Chapter 9, where it was shown that

$$\frac{1}{T}W = c_0^2 + 2\sum_{k=1}^{\infty}|c_k|^2.$$

So, using the expressions for c_0, and for c_k for $k > 1$, we arrive (after some more algebra) at the following absolutely astonishing conclusion:

$$\underbrace{\left(\frac{1 - e^{-2\pi N\eta}}{2\pi N}\frac{1}{1 + \eta^2}\right)^2}_{\text{dc power}} + \underbrace{\frac{1}{2}\sum_{k=1}^{\infty}\frac{\left(\frac{1}{\pi N}\right)^2(1 - e^{-2\pi N\eta})^2}{\left(\eta^2 + 1 - \frac{k^2}{N^2}\right)^2 + \frac{4\eta^2 k^2}{N^2}}}_{\substack{\text{average ac power, by} \\ \text{frequency harmonic (by } k)}} = \underbrace{\frac{1 - e^{-4\pi N\eta}}{8\pi N\eta(1 + \eta^2)}}_{\substack{\text{total average} \\ \text{power}}}.$$

Be clear in your mind about the frequencies of the harmonics of $x(t)$. They are integer multiples of $\omega_0 = \frac{\omega_c}{N}$, i.e., frequencies that are *below* ω_c (for $k < N$), as well as above ω_c (for $k > N$), where ω_c is the resonant frequency of the transmitter's antenna circuitry.

How do we know this incredible identity is correct? It certainly isn't obvious, at least not to me! I don't know what a mathematician might do to check this purely mathematical statement, but to the engineering mind there is a simple approach to enhance one's confidence that a factor of pi hasn't been lost somewhere (or even worse). The statement is supposed to be true for any integer $N > 1$ and any $\eta > 0$. So, simply write a computer program that calculates each side and see if the answers agree, for a wide range of choices of N and η. It seems quite unlikely that such an agreement would occur simply by chance. I've done this and (no surprise, since you see it here in print!) it checks. This isn't a *proof*, of course, but it is convincing. For example, with $N = 1000$ and $\eta = 0.041$, the left side calculates (using the first six thousand terms of the sum) as 0.968756×10^{-3}, while the right side is equal to 0.968828×10^{-3}. If we increase η to 0.41, then the left side is 0.82995×10^{-4} and the right side is 0.8308×10^{-4}. For engineering work, this is pretty good agreement!

These particular values for N and η were not picked "at random." Here's how I arrived at them, in an attempt to realistically model the transmitted signal of a typical turn-of-the-century spark-gap Morse code transmitter:

a. The radio historian Hugh Aitken has presented convincing evidence that the spark rate frequency varied from perhaps 8 sparks per second to up to 20,000 sparks per second. I use 500 sparks per second.[1] The value of 500 produced an audible tone in the headphones of crystal sets receiving Morse code signals (see Chapter 5 again), while the much higher sparking rates were an attempt (failed) to approximate continuous waves by generating damp oscillations that had relatively little time to damp out, compared to oscillations created by the lower sparking rates.

b. The resonant frequencies of the antenna circuits in the early spark transmitters were in the tens of kilohertz to hundreds of kilohertz range. The lower frequencies were used by long-range stations that used the radio wave reflection property of the ionosphere to "bounce" signals back down to the earth, around the curvature of the planet. Such low frequencies (25 kHz was popular) required enormous antenna structures, as discussed in Chapter 3. In this analysis, I use 500 kHz, the international maritime distress call frequency.

c. Thus, $N = (\frac{1}{500})$ seconds/spark $\times 5 \times 10^5$ cycles/second $= 10^3$ cycles/spark $= 1000$ cycles for each period of the damped sinusoid.

d. Professor Aitken says the typical damping factor of a spark transmitter was a factor of ten after nine cycles.[2] Since the amplitude-damping factor is $e^{-\eta \omega_c t}$, then for t equal to the duration of nine cycles (at frequency ω_c), we have $t = 9(2\pi/\omega_c)$ and so $e^{-\eta 18\pi} = \frac{1}{10}$ or, $\eta = \frac{\ln(10)}{18\pi} = 0.041$. Arbitrarily multiplying by ten is then used to represent even more severe damping (that is, $\eta = 0.41$).

We can now complete our study of spark-gap transmitters by returning to the "incredible identity," and in particular concentrating our attention on the ac harmonics. If $f_c = 500$ kHz and if $N = 1000$ (500 sparks per second), then each k represents a frequency step of 0.5 kHz (e.g., $k = N = 1000$ corresponds to 500 kHz). Again using a computer to perform the odious numerical work, we can calculate how the ac power of the signal $x(t)$ is distributed over the frequency spectrum. Thus, by first subtracting off the dc power (which, from Chapter 3, doesn't radiate) from the total power of $x(t)$, and then calculating the individual ac powers, we can construct a table like the following one, where, for the two damping factors of $\eta = 0.041$ and $\eta = 0.41$, the power distribution is given.

This table immediately shows us why spark-gap transmitters are called "spectrum polluters." With $\eta = 0.041$, for example, only 10% of the total ac power is located within a 6.5-kHz band of frequencies centered on 500 kHz [50% of the ac power is below 497 kHz ($= 994 \times 0.5$ kHz), and 60% is below 503.5 kHz ($= 1007 \times 0.5$ kHz)]. Another 10% of the ac power is below 416 kHz ($= 832 \times 0.5$ kHz), and another 5% is above 583 kHz ($= 1{,}166 \times 0.5$ kHz). With more severe damping ($\eta = 0.41$) the frequency-spreading of power is even more pronounced. For example, with a high-power spark of many kilowatts,[3] even just 1% of the total ac power could be a significant signal—and the table shows us that 1% of the ac power, for such a heavily damped transmitter, is above 1.422 MHz ($= 2845 \times 0.5$ kHz), almost *three times* the nominal signal frequency of 500 kHz! This analysis does, I must point out, ignore how much power at each frequency is actually radiated, i.e., how efficient is the coupling, by the antenna, of the spark-gap energy to space. This coupling is not independent of frequency, but the broad range of frequencies in the antenna circuitry is quite suggestive of the range of frequencies over which radiated energy occurs.

% of ac power at or below $k\,\omega_c/N$	k $\eta = 0.041$	$\eta = 0.41$
5	654	112
10	832	220
20	928	417
30	962	582
40	980	719
50	994	838
60	1007	949
70	1022	1066
80	1044	1212
90	1092	1467
95	1166	1773
99	1509	2845

By 1914–1915, the spark-gap radio transmitter was commercially on its way out. Spark-gap transmitters are outlawed today (since 1923), because no one else could broadcast and still hope to be received without interference if even just one such "dirty" transmitter turned on anywhere nearby.[4] Indeed, such transmitters are intentionally used today by military organizations as electromagnetic signal *jammers*.

The low frequencies of the early radio transmitters are still used today by the U.S. Navy for communication and/or navigation purposes by submerged submarines. One such transmitter (station NDT in Yosami, Japan), in fact, is a Poulsen arc (!) operating at 17.4 kHz. The reason for the low frequencies for transmission to subs at great depths is particularly interesting—as frequency increases, conductive sea water increasingly shorts out radio waves. The lower the frequency, then, the better. Indeed, in the late 1960s and early 1970s the U.S. Navy seriously proposed building a HUGE radio system (Project Seafarer/Sanquine) to operate at 45 Hz (!) and 500 megawatts (!), with a buried 10,000 square mile (!) antenna. See the entire April 1974 issue of the *IEEE Transactions on Communications*. The month of publication *does* seem appropriate.

Parseval's theorem about the energy of a periodic $x(t)$ and the c_k Fourier series coefficients of $x(t)$, has played a big role in the previous two chapters, and in this one, too. We have derived important *general* results based on assuming no more than $x(t)$ is periodic. If we make additional assumptions, then of course our conclusions can be correspondingly sharper, e.g., as in the analysis of this chapter of just how the energy is distributed over frequency for the specific $x(t)$ of an amplitude damped sinusoid. If we make assumptions that are "in-between" these two extreme cases, i.e., assumptions that are not as weak as simply taking $x(t)$ as periodic, but also not as strong as defining a specific $x(t)$, *then* what sort of conclusions can we make? In this final technical analysis of the chapter, I'll show you an elegant illustration of how to "play" with the mathematics to get an answer to that question.

Let's suppose that not only is $x(t)$ periodic (with period 2π), but that it is real *and* non-negative, i.e., $x(t) \geq 0$ for all t. Then, as I'll show you, the Fourier series coefficients satisfy the following inequality:

$$c_o^2 + c_o|c_{2k}| \geq 2|c_k|^2 \text{ for all } k.$$

This statement is obviously true for the case of $k = 0$ because, since c_o is real and nonnegative (because c_o is the average value of a real, nonnegative $x(t)$), the inequality reduces to the unarguably true identity $2c_o^2 = 2c_o^2$. For all other k, however, the inequality is nontrivial. For $k = 1$, for example, it says $c_o^2 + c_o|c_2| > 2|c_1|^2$. (See Problem 11.2 for more on this particular case.)

As the first step in proving the above general inequality, I'll establish the following preliminary result: for a given k, there always exists a real number a such that $e^{jk\alpha}c_k$ is real. From Euler's identity we have

$$e^{jk\alpha}c_k = c_k \cos(k\alpha) + jc_k \sin(k\alpha),$$

and since we can write the generally complex-valued c_k in rectangular form as $c_k = a_k + jb_k$ (where a_k and b_k are real), then we have

$$e^{jk\alpha}c_k = a_k \cos(k\alpha) + jb_k \cos(k\alpha) + ja_k \sin(k\alpha) - b_k \sin(k\alpha).$$

This *is* real if we set

$$b_k \cos(k\alpha) + a_k \sin(k\alpha) = 0,$$

a condition that lets us solve for α. That is,

$$\frac{\sin(k\alpha)}{\cos(k\alpha)} = \tan(k\alpha) = -\frac{b_k}{a_k}$$

and so we have established the existence of α by actually finding α, i.e., $\alpha = -\tan^{-1}(\frac{b_k}{a_k})$.

Thus, using our well-known formula for c_k (with $\omega_o = \frac{2\pi}{T} = \frac{2\pi}{2\pi} = 1$), and the α just derived, we have

$$e^{jk\alpha}c_k = \frac{1}{2\pi} \int_{-\pi}^{\pi} x(t)e^{-jk(t-\alpha)} \, dt.$$

(In particular, for $k = 0$, we have $c_o = \frac{1}{2\pi} \int_{-\pi}^{\pi} x(t) \, dt$.) Now, since $e^{jk\alpha}c_k$ is real, then Euler's identity tells us the imaginary part of the integral is zero and, since $x(t)$ is real, then

$$e^{jk\alpha}c_k = \frac{1}{2\pi} \int_{-\pi}^{\pi} x(t) \cos\{k(t - \alpha)\} \, dt.$$

Taking absolute values, and remembering that $|e^{jk\alpha}| = 1$, we have

$$|e^{jk\alpha}c_k| = |e^{jk\alpha}||c_k| = |c_k| = \left| \frac{1}{2\pi} \int_{-\infty}^{\infty} x(t) \cos\{k(t - \alpha)\} \, dt \right|.$$

Since we can drop the absolute value sign around the integral if we *square* both sides of the equality, we can write

$$|c_k|^2 = \left\{ \frac{1}{2\pi} \int_{-\pi}^{\pi} x(t) \cos\{k(t - \alpha)\} \, dt \right\}^2.$$

If we don't square the integral, then we can only write

$$|c_k| \geq \frac{1}{2\pi} \int_{-\pi}^{\pi} x(t) \cos\{k(t - \alpha)\} \, dt.$$

Finally, if we make the trivial notational change of replacing every "k" with "$2k$" in this last inequality, then

$$|c_{2k}| \geq \frac{1}{2\pi} \int_{-\pi}^{\pi} x(t) \cos\{2k(t - \alpha)\}\, dt.$$

Now, let's consider the quantity Q, defined as

$$Q = c_o^2 + c_o|c_{2k}| - 2|c_k|^2.$$

Our last step will be to simply show that $Q > 0$ and that will be the inequality that we wish to show is true. To do this, I'll substitute what c_o is *equal* to, and I'll substitute the *lower* bound on $|c_{2k}|$, and I'll put in what $|c_k|^2$ is *equal* to. That will give a *lower* bound on Q—let's call it L—and the result will turn out to be $L \geq 0$. Thus, $Q \geq 0$, too. So, following that plan,

$$Q \geq L = \frac{1}{4\pi^2} \left\{ \int_{-\pi}^{\pi} x(t)\, dt \right\}^2$$
$$+ \left\{ \frac{1}{2\pi} \int_{-\pi}^{\pi} x(t)\, dt \right\} \left\{ \frac{1}{2\pi} \int_{-\pi}^{\pi} x(t) \cos\{2k(t - \alpha)\}\, dt \right\}$$
$$- 2\frac{1}{4\pi^2} \left\{ \int_{-\pi}^{\pi} x(t) \cos\{k(t - \alpha)\}\, dt \right\}^2.$$

Next, we use a clever little two-part math trick. For the first part, I'll write all of the above terms (each of the form of two integrals multiplied together) such that each integral in a multiplicative pair uses a different dummy variable of integration. That is,

$$4\pi^2 L = \left\{ \int_{-\pi}^{\pi} x(t)\, dt \right\} \left\{ \int_{-\pi}^{\pi} x(u)\, du \right\}$$
$$+ \left\{ \int_{-\pi}^{\pi} x(t)\, dt \right\} \left\{ \int_{-\pi}^{\pi} x(u) \cos\{2k(u - \alpha)\}\, du \right\}$$
$$- 2 \left\{ \int_{-\pi}^{\pi} x(t) \cos\{k(t - \alpha)\}\, dt \right\} \left\{ \int_{-\pi}^{\pi} x(u) \cos\{k(u - \alpha)\}\, du \right\}$$
$$= \int_{-\pi}^{\pi} \int_{-\pi}^{\pi} [x(t)x(u) + x(t)x(u) \cos\{2k(u - \alpha)\}$$
$$- 2x(t)x(u) \cos\{k(t - \alpha)\} \cos\{k(u - \alpha)\}]\, dt\, du$$
$$= \int_{-\pi}^{\pi} \int_{-\pi}^{\pi} x(t)x(u) [1 + \cos\{2k(u - \alpha)\}$$
$$- 2\cos\{k(t - \alpha)\} \cos\{k(u - \alpha)\}]\, dt\, du.$$

From the trigonometric identity $2\cos^2(\theta) = 1 + \cos(2\theta)$, we have $1 + \cos\{2k(u-\alpha)\} = 2\cos^2\{k(u-\alpha)\}$. Then,

$$4\pi^2 L = \int_{-\pi}^{\pi} \int_{-\pi}^{\pi} x(t)x(u)\left[2\cos^2\{k(u-\alpha)\}\right.$$
$$\left. -2\cos\{k(t-\alpha)\}\cos\{k(u-\alpha)\}\right] dt\, du$$

or,

$$2\pi^2 L = \int_{-\pi}^{\pi} \int_{-\pi}^{\pi} x(t)x(u)\cos\{k(u-\alpha)\}\left[\cos\{k(u-\alpha)\}\right.$$
$$\left. -\cos\{k(t-\alpha)\}\right] dt\, du.$$

Now, for the second part of our little math trick, observe that since t and u *are* simply dummy variables then if we *interchange* them we change nothing! That is,

$$2\pi^2 L = \int_{-\pi}^{\pi} \int_{-\pi}^{\pi} x(u)x(t)\cos\{k(t-\alpha)\}\left[\cos\{k(t-\alpha)\}\right.$$
$$\left. -\cos\{k(u-\alpha)\}\right] du\, dt.$$

If we now add the two expressions for $2\pi^2 L$ we have

$$4\pi^2 L = \int_{-\pi}^{\pi} \int_{-\pi}^{\pi} x(u)x(t)\left[\cos^2\{k(u-\alpha)\}\right.$$
$$\left. -2\cos\{k(u-\alpha)\}\cos\{k(t-\alpha)\} + \cos^2\{k(t-\alpha)\}\right] du\, dt$$
$$= \int_{-\pi}^{\pi} \int_{-\pi}^{\pi} x(u)x(t)\left[\cos\{k(u-\alpha)\} - \cos\{k(t-\alpha)\}\right]^2 du\, dt.$$

Since the integrand of this double integrand is *never* negative (don't forget, $x \geq 0$ was our initial assumption) then $4\pi^2 L \geq 0$, i.e., $L \geq 0$. But then $Q \geq 0$ and so

$$c_o^2 + c_0|c_{2k}| \geq 2|c_k|^2$$

as originally claimed.

Notes

1. Hugh Aitken, *The Continuous Wave: Technology and American Radio, 1990–1932*, Princeton 1985, p. 33 and p. 65.

2. Hugh Aitken, *Syntony and Spark: The Origins of Radio*, New York: Wiley 1976, 78, Note 35.

3. For more on this fantastic radio technology, see George Pickworth, "Marconi's 200 kW Transatlantic Transmitter," *Electronics World & Wireless World*, January 1994, and the same author's earlier article "The Spark That Gave Radio to the World," November 1993 (in the same journal).

4. As I mentioned back in Note 1 of Chapter 7, the Radio Act of 1912 did not specifically ban spark radio, but its language made it clear that the "spectrum splatter" of that technology would not be allowed to continue indefinitely. However, the act did send a somewhat mixed message when it also stated "When sending distress signals, the transmitter on shipboard may be tuned in such a manner as to create a *maximum of interference* [my emphasis] with maximum of radiation."

Problems

11.1. Write a computer program that calculates the distribution of ac power as a function of N and η. As a partial check on your program, what *should* it calculate as the ac power distribution table for $\eta \approx 0$, for any integer $N > 1$?

11.2. Suppose $x(t)$ is nonnegative and periodic, with period 2π. Over one period it is defined as

$$x(t) = \begin{cases} 1, & -\dfrac{\pi}{2} < t < \dfrac{\pi}{2} \\ 0, & \text{otherwise.} \end{cases}$$

Calculate the c_o and c_k ($k \neq 0$) Fourier series coefficients for $x(t)$ and verify that they satisfy the inequality $c_o^2 + c_o|c_{2k}| > 2|c_k|^2$. *Hint:* you should find, for $k = 1$, that the inequality reduces to the assertion that $\pi^2 > 8$, which is clearly true since $\pi > 3$.

11.3. Here's one last problem that can be attacked with the method of evaluating unknown constants by using the property, of certain functions, of their integrals over a period vanishing. (The trick we have used, several times, to find Fourier coefficients). Consider the equation

$$g(x) = f(x) + k \int_{-1}^{1} \cos\{\pi(x - y)\} \, g(y) \, dy,$$

an equation of a type that often occurs in advanced electrical engineering. The function $f(x)$ is given, k is a constant, and we are to find $g(x)$ which you'll notice appears both inside *and* outside of the integral. This equation is a particular example of a general class of equations called *Fredholm integral equations*, after the Swedish mathematician Erik Ivar Fredholm (1866–1927).

a. Expand the cosine function and show that

$$g(x) = f(x) + kg_1 \cos(\pi x) + kg_2 \sin(\pi x)$$

where g_1 and g_2 are the two *numbers* defined by

$$g_1 = \int_{-1}^{1} \cos(\pi y)\, g(y)\, dy, \quad g_2 = \int_{-1}^{1} \sin(\pi y)\, g(y)\, dy.$$

b. Multiply through the equation for $g(x)$ in part a by $\cos(\pi x)$, integrate term by term from -1 to 1, and show that if we define the *number* f_1 as

$$f_1 = \int_{-1}^{1} f(x) \cos(\pi x) dx$$

then $g_1 = f_1 + kg_1$. Repeat this process by multiplying through by $\sin(\pi x)$, integrating from -1 to 1, and then defining the *number* f_2 as

$$f_2 = \int_{-1}^{1} f(x) \sin(\pi x)\, dx$$

to conclude that $g_2 = f_2 + kg_2$.

c. Show how all the above leads to the solution for $g(x)$—which now appears *only* on the left-hand-side:

$$g(x) = f(x) + \frac{k}{1-k} \int_{-1}^{1} \cos\{\pi(x-y)\} f(y)\, dy, \quad k \neq 1.$$

d. Explain why, if $k = 1$, there either is *no* solution for $g(x)$ or there are *infinitely many* solutions. Hint: consider the two exhaustive possibilities of (a) at least one of f_1 and $f_2 \neq 0$; (b) $f_1 = f_2 = 0$.

Fourier Integral Theorem and the Continuous Spectrum of a Nonperiodic Time Signal

What we can say about a signal that is not periodic? Recall from Chapter 9 that I've already argued that, really, *no* signal is periodic. But I mean something stronger here. For example, what if $v(t) = 1$ for $0 \leq t \leq 1$, and is zero at all other times? Then $v(t)$ is clearly a nonperiodic pulse, and we conclude this without any philosophical fine points being invoked, as before. Clearly, we can't write a Fourier series for such a signal, as *by definition* it doesn't repeat and so there is no period. But, wait a minute— maybe there is! What if we simply think of such a signal as having an *infinitely* long period, and so it is periodic (the signal is still executing, and always will be, the "present" period). This is, of course, nothing but a devious trick but you'll soon see that it will lead.to something of great interest and value.

Recall from the two previous sections that if T is the period of $v(t)$, then

$$v(t) = \sum_{k=-\infty}^{\infty} c_k e^{jk\omega_0 t}, \quad \omega_0 = \frac{2\pi}{T}$$

$$c_k = \frac{1}{T} \int_{-T/2}^{T/2} v(t) e^{-jk\omega_0 t}\, dt.$$

We can use *any* interval of width T in the integral for c_k, but picking it to be symmetrical about the origin will give us a particularly nice result. I will now "play" with these two expressions in a rough-and-ready way in what is sometimes derisively called "engineers' math." I certainly make no claims here for precision, but what is most important to realize is that *it doesn't matter!* Once we have the mathematical result of this chapter, we can literally forget how we got it. We can treat it as a *definition*, useful because of the physical significance it has (developed in the next chapter) and, in fact, in books on this topic written by mathematicians the axiomatic approach is often taken. For engineers and physicists, of course, this should be less than satisfying. You should demand *some* motivation, and so here it is.

First, rewrite the second expression as

$$T c_k = \int_{-T/2}^{T/2} v(t) e^{-jk\omega_0 t}\, dt = V(k\omega_0),$$

where the $V(k\omega_0)$ notation simply means that, after the time integration is done, only k and ω_0 (as the product $k\omega_0$) remain as variables, i.e., the product $T c_k$ is some function of the product $k\omega_0$. Next imagine $T \to \infty$ and so $\omega_0 \to 0$. Since frequency "jumps" in steps of ω_0 in a Fourier series, we will think of the frequency steps becoming *differential* increments and so we'll replace ω_0 with $d\omega$. In addition, we'll replace $k\omega_0 = kd\omega$ with ω, a *continuous* variable (yes, this *is* sloppy, but remember, it doesn't matter). Thus, the $T c_k$ integral becomes

$$V(\omega) = \int_{-\infty}^{\infty} v(t) e^{-j\omega t}\, dt.$$

$V(\omega)$ is called the *Fourier transform* or *spectrum* of $v(t)$. You should not yet be worried about what it means, but rather just accept this as the curious result of some rather dubious symbol pushing. The ultimate justification comes in the next chapter where it is shown the Fourier transform has deep physical significance.

Now, looking back at the original series expression for $v(t)$, we can write it as

$$v(t) = \frac{1}{T} \sum_{k=-\infty}^{\infty} Tc_k e^{jk\omega_0 t} = \frac{\omega_0}{2\pi} \sum_{k=-\infty}^{\infty} Tc_k e^{jk\omega_0 t}$$

$$= \frac{1}{2\pi} \sum_{k=-\infty}^{\infty} Tc_k e^{jk\omega_0 t} \omega_0.$$

As $T \to \infty$ we replace ω_0 with $d\omega$, and Tc_k with $V(\omega)$, and imagine the summation going over into an integral:

$$v(t) = \frac{1}{2\pi} \int_{-\infty}^{\infty} V(\omega) e^{j\omega t} \, d\omega.$$

The last expression is called the *inverse* Fourier transform of $V(\omega)$.

If we change variables from ω to f (from frequency in radians per second, to frequency in hertz) then we have $\omega = 2\pi f$ and we get the nicely symmetric form of the Fourier transform *pair*,

$$V(f) = \int_{-\infty}^{\infty} v(t) e^{-j2\pi ft} \, dt,$$

$$v(t) = \int_{-\infty}^{\infty} V(f) e^{j2\pi ft} \, df.$$

These two expressions show that $v(t) \leftrightarrow V(f)$, i.e., the double-headed arrow indicates that there is a unique, one-to-one correspondence between the time function $v(t)$ and its spectrum $V(f)$, and that one can be "recovered" from the other. (Our convention will always be to write a transform pair with the time function at the left-pointing arrowhead, and the spectrum at the right-pointing arrowhead). Since integration is a linear operation, it is immediately obvious that the Fourier transform is linear, i.e., the transform of a sum is the sum of the transforms.

For those who want to see a less spirited derivation of the Fourier transform pair, see Appendix H. You should find that completely understandable after reading Chapter 14 on impulses. What we have in this book is the *one-dimensional* Fourier transform, because $v(t)$ and $V(\omega)$ are each functions of a single variable. In more advanced applications, such as in image transmission theory, the two-dimensional Fourier transform is used. For AM radio, however, all we need is the one-dimensional version.

The integral frequency representation for a nonperiodic time signal can be compared to the sum frequency representation for a periodic time signal. In these two cases we have, respectively,

$$v(t) = \frac{1}{2\pi} \int_{-\infty}^{\infty} V(j\omega)e^{j\omega t} \, d\omega$$

and

$$x(t) = \sum_{k=-\infty}^{\infty} c_k e^{jk\omega_0 t},$$

where $V(\omega)$ has been written as $V(j\omega)$, i.e., the j is written explicitly to emphasize that the Fourier transform is generally complex. For the periodic case, $x(t)$ is an infinite sum of harmonics, where the kth harmonic (at frequency $k\omega_0$) is the real quantity

$$c_{-k}e^{-jk\omega_0 t} + c_k e^{jk\omega_0 t} = c_k^* e^{-jk\omega_0 t} + c_k e^{jk\omega_0 t}$$
$$= c_k e^{jk\omega_0 t} + (c_k e^{jk\omega_0 t})^*,$$

which has (as shown in Chapter 9) the finite peak amplitude $2|c_k|$. For the nonperiodic case, $v(t)$ is an infinite sum of harmonics, too (but now the harmonics have *zero* spacing), where the harmonic at frequency ω is the real quantity

$$\frac{1}{2\pi}V(-j\omega)e^{-j\omega t}d\omega + \frac{1}{2\pi}V(j\omega)e^{j\omega t}d\omega$$
$$= \frac{d\omega}{2\pi}\left[V^*(j\omega)e^{-j\omega t} + V(j\omega)e^{j\omega t}\right]$$
$$= \frac{d\omega}{2\pi}\left[V(j\omega)e^{j\omega t} + \{V(j\omega)e^{j\omega t}\}^*\right],$$

which has the *differential* peak amplitude

$$(2d\omega/2\pi|V(j\omega)| = 2|V(j2\pi f)|df.$$

As a simple but important example of a Fourier transform pair, suppose $v(t)$ is defined as shown in the top half of Figure 12.1, i.e.,

$$v(t) = \begin{cases} 1, & t \le \left|\frac{\tau}{2}\right| \\ 0, & \text{otherwise.} \end{cases}$$

For the special case of $\tau = 1$, this $v(t)$ is often written as $\pi(t)$ and called the *unit gate* function because, for any function $\varphi(t)$, $\pi(t)\varphi(t)$ is zero except when the gate is "open" (equal to one). When the gate is open, then the product is $\varphi(t)$. Now, in the general case,

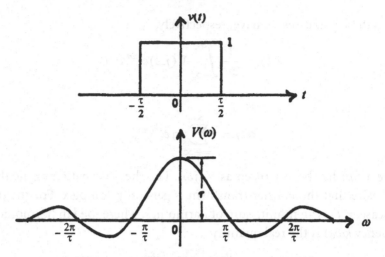

Figure 12.1   A time signal and its Fourier transform.

$$V(j\omega) = \int_{-\infty}^{\infty} v(t)e^{-j\omega t}\, dt = \int_{-\tau/2}^{\tau/2} e^{-j\omega t}\, dt = \left(\frac{e^{-j\omega t}}{-j\omega}\right)\bigg|_{-\frac{\tau}{2}}^{\frac{\tau}{2}}$$

$$= \frac{e^{-j\omega\tau/2} - e^{j\omega\tau/2}}{-j\omega} = \frac{-j2\sin\left(\dfrac{\omega\tau}{2}\right)}{-j\omega} = \tau\,\frac{\sin\left(\dfrac{\omega\tau}{2}\right)}{\left(\dfrac{\omega\tau}{2}\right)}.$$

This result is shown in the bottom half of Figure 12.1, and it illustrates the general property of *reciprocal spreading*. That is, as the signal becomes narrower in the time domain (as $\tau \to 0$), the spectrum becomes wider ($\pi/\tau \to \infty$) in the frequency domain.

Reciprocal spreading is really just a different name for the so-called time/frequency scaling theorem, which says that if $v(t) \leftrightarrow V(j\omega)$ then $v(at) \leftrightarrow (1/|a|)V(j\omega/a)$ for a any real nonzero constant. This is easily established by simply transforming $v(at)$ and making the obvious change of variable, treating the cases $a > 0$ and $a < 0$ separately. To understand how reciprocal spreading comes out of this, suppose $v(t)$ is a pulse signal existing from 0 to T. Then $v(at)$ is a pulse signal existing from 0 to $\frac{T}{a}$. Suppose $a > 1$. Then $v(at)$ exists over a *narrower* interval than does $v(t)$. But the spectrum of $v(at)$ is *spread* in the frequency domain because the factor a occurs in the denominator of the argument of $V(j\omega/a)$—the "$1/|a|$" in front of $V(j\omega/a)$ is of course just an *amplitude* scaling. Time-scaled signals commonly occur as recorded signals played back at a speed different from the recording speed. Thus, $a > 1$ corresponds to fast playback, $0 < a < 1$

corresponds to slow playback, and $a < 0$ corresponds to playing a recording *backwards* (giving a time-reversed signal).

The property of reciprocal spreading is sometimes called the *uncertainty principle* in Fourier transform theory. See, for example, Athanasious Papoulis, *The Fourier Integral and Its Applications*, McGraw-Hill 1962, pp. 62–64. This idea has found its way into fiction, for example Carl Sagan's novel *Contact* (Simon and Schuster 1985). At one point his heroine, a radio astronomer searching for radio messages from extraterrestrials, is troubled by the apparent lack of such signals. As she runs through possible explanations for the failure to detect intelligent communications from deep space, her train of thought is as follows (pp. 65–66):

> maybe the aliens are "fast talkers, manic little creatures perhaps, moving with quick and jerky motions, who transmitted a complete radio message—the equivalent of hundreds of pages of English text—in a nanosecond. Of course, if you had a very narrow band-pass to your receiver, so you were listening only to a tiny range of frequencies, you were forced to accept the long time-constant. [This is the reciprocal spreading effect.] You would never be able to detect a rapid modulation. It was a simple consequence of the Fourier Integral Theorem, and closely related to the Heisenberg Uncertainty Principle. So, for example, if you had a bandpass of a kilohertz, you couldn't make out a signal that was modulated at faster than a millisecond. It would be a kind of sonic blur. [Her radio receivers'] bandpasses were narrower than a hertz, so to be detected the [alien] transmitters must be modulating very slowly, slower than one bit of information a second."

Time scaling was used in the Second World War by German submarines in an attempt to avoid being radio-located. Messages were prerecorded at normal speed and then transmitted at greatly increased speed. The sound of a U-boat's high-speed radio signal was described as having a "warbling" tone, and it was very brief. One of the many amazing electronic inventions that came out of that war was the development of high-frequency direction finding (HF/DF) receivers (commonly called "Huff Duff"). Such receivers could determine a bearing angle on a U-boat burst signal in just a fraction of a second (see Brian Johnson, *The Secret War*, Methuen 1978, pp. 210–212). Huff Duff was one of the reasons the U boats were so successfully hunted down one by one, that, by war's end, they were virtually extinct. For a more recent military application of the time scaling of recorded messages, see Frederick Forsyth's fictional treatment of the 1991 Gulf War, *The Fist of God*, Bantam 1994. Forsyth uses a value of $a = 200$.

The last example, of the gate function, resulted in $V(j\omega)$ being real, but as mentioned before it is generally complex. But it can't be just *anything*. For example, if we impose the physically reasonable constraint on $v(t)$ that it is real (as signals in actual hardware always are!), then we can show that $|V(j\omega)|^2$ is always even. To prove this, write

$$V(j\omega) = \int_{-\infty}^{\infty} v(t)e^{-j\omega t}\, dt = \int_{-\infty}^{\infty} v(t)\cos(\omega t)\, dt - j\int_{-\infty}^{\infty} v(t)\sin(\omega t)\, dt.$$

Writing $V(j\omega) = R(\omega) + jX(\omega)$, then the real and imaginary parts of the spectrum are [since $v(t)$ is real]

$$R(\omega) = \int_{-\infty}^{\infty} v(t)\cos(\omega t)\, dt,$$

$$X(\omega) = -\int_{-\infty}^{\infty} v(t)\sin(\omega t)\, dt.$$

Since $\cos(\omega t)$ and $\sin(\omega t)$ are even and odd, respectively, we immediately have

$$R(-\omega) = \int_{-\infty}^{\infty} v(t)\cos(-\omega t)\, dt = R(\omega),$$

$$X(-\omega) = -\int_{-\infty}^{\infty} v(t)\sin(-\omega t)\, dt = -X(\omega).$$

That is, $R(\omega)$ is even and $X(\omega)$ is odd. Now, since $|V(j\omega)|^2 = R^2(\omega) + X^2(\omega)$, and since R^2 and X^2 are each even, then $|V(j\omega)|^2$ must be even. It is also easy to show that if $v(t)$ is even (odd) then $V(j\omega)$ is real (imaginary).

We can also write $V(j\omega)$ in polar form as $V(j\omega) = A(\omega)e^{j\theta(\omega)}$, where the real functions $A(\theta)$ and $\theta(\omega)$ are called the *amplitude* and the *phase* spectrums of $v(t)$, respectively. Obviously, $|V(j\omega)| = |A(\omega)|$ as $|e^{j\theta(\omega)}| = 1$. It is easy to show that $A(\omega)$ is even and that $\theta(\omega)$ is odd [express them in terms of $R(\omega)$ and $X(\omega)$ to see this]. An important property of Fourier transforms is that a *time shift* appears as a *phase* shift in the frequency domain. That is, if $v(t) \leftrightarrow V(\omega)$, then $v(t - t_0) \leftrightarrow e^{-j\omega t_0} V(\omega)$, a result easily established by transforming $v(t - t_0)$ with the obvious change of variable (you should do this). Notice, in particular, that a time *delay* ($t_0 > 0$) is associated with a *negative* phase shift.

As an example of transforming a nonsymmetrical time signal, define the so-called step function $u(t)$ as

$$u(t) = \begin{cases} 0, & t < 0 \\ 1, & t > 0. \end{cases}$$

This discontinuous function gets its name from its appearance when graphed. For the present, I'll leave $u(0)$ undefined. If we attempt to calculate the Fourier transform by direct evaluation of the integral we run into trouble, i.e.,

$$U(j\omega) = \int_{-\infty}^{\infty} u(t)e^{-j\omega t}\,dt = \int_{0}^{\infty} e^{-j\omega t}\,dt = \frac{e^{-j\omega t}}{-j\omega}\bigg|_{0}^{\infty} = ?,$$

which has no meaning at the upper limit. That is, we can't assign a value to $e^{-j\infty}$, since the real and imaginary parts of $e^{-j\omega t}$ *oscillate* forever between ± 1, and never approach any particular value as $t \to \infty$. So, let's try to be clever and ask, instead, what is the Fourier transform of $v(t) = e^{-\sigma t}u(t)$, where $\sigma > 0$? We have

$$V(j\omega) = \int_{-\infty}^{\infty} v(t)e^{-j\omega t}\,dt = \int_{0}^{\infty} e^{-(\sigma+j\omega)t}\,dt$$

$$= \frac{e^{-(\sigma+j\omega)t}}{-(\sigma+j\omega)}\bigg|_{0}^{\infty} = \frac{1}{\sigma+j\omega}.$$

This works at the upper limit of $t = \infty$. (The $e^{-\sigma t}$ is called a *convergence factor*.) It would seem to be perfectly natural, then, to conclude that since $\lim_{\sigma \to 0} v(t) = \lim_{\sigma \to 0} e^{-\sigma t}u(t) = u(t)$, then the transform of $u(t)$ would be $\lim_{\sigma \to 0} V(j\omega) = \frac{1}{j\omega}$. In fact, this is incorrect and the math itself is telling us that something is wrong because this result is purely imaginary. As stated before, such a transform is associated with an odd function of time, and $u(t)$ is *not* odd. I'll return to this calculation in Chapter 14 and show that $\frac{1}{j\omega}$ is the correct imaginary part of $U(j\omega)$, but that there is a nonzero real part, too.

> The inclusion of a convergence factor in the Fourier integral leads directly to the Laplace transform. Many time functions that do not have a Fourier transform (unless we resort to impulses, as will be discussed in Chapter 14), such as $u(t)$, have perfectly straightforward Laplace transforms. I won't pursue this any further in this book.

As an example of some aggressive engineers' math that *does* happen to work, consider the integral $\int_{0}^{\infty} e^{-x}/\sqrt{x}\,dx$, which can be shown to be equal to $\sqrt{\pi}$ [change variables to $y = \sqrt{x}$ and use a result derived in Appendix G: $\int_{0}^{\infty} e^{-y^2}\,dy = (\frac{1}{2})\sqrt{\pi}$]. If we make the change of variable $x = pt$, with p a constant, then we can write $\int_{0}^{\infty} e^{-pt}/\sqrt{t}\,dt = \sqrt{\pi/p}$. If we now pick $p = j\omega$, then this becomes

$$\int_0^\infty \frac{e^{-j\omega t}}{\sqrt{t}}\, dt = \sqrt{\frac{\pi}{j\omega}}.$$

But the integral is the Fourier transform of the time signal $\frac{u(t)}{\sqrt{t}}$, i.e., we have the transform pair

$$\frac{u(t)}{\sqrt{t}} \leftrightarrow \sqrt{\frac{\pi}{j\omega}} = \sqrt{\frac{\pi}{2\omega}}(1 - j).$$

This derivation is really a cheat because, in the change of the variable step, I implicitly assumed that p is a real positive number (notice how the integration limits transformed)—but in the very next step I used an *imaginary* value for p! So, given all this symbol pushing at its best (worst?), how do we know our transform pair is correct?

One way to answer that concern is to simply continue with reckless abandon and see if perhaps we can derive some additional results from our analysis that we *are* able to say are correct. So, what we have is the claim that

$$\int_0^\infty \frac{e^{-j\omega t}}{\sqrt{t}}\, dt = \sqrt{\frac{\pi}{j\omega}},$$

which as it stands is a purely mathematical statement, independent of our "radio" interpretation of ω as a frequency variable. It is supposed to be an identity for any value of ω. So, suppose $\omega = -1$. Then,

$$\int_0^\infty \frac{e^{jt}}{\sqrt{t}}\, dt = \sqrt{\frac{\pi}{-j}} = \sqrt{\pi j} = \sqrt{\frac{\pi}{2}} + j\sqrt{\frac{\pi}{2}}.$$

If we change variables to $t = y^2$, then this becomes the claim that

$$\int_0^\infty e^{jy^2}\, dy = \frac{1}{2}\left[\sqrt{\frac{\pi}{2}} + j\sqrt{\frac{\pi}{2}}\right].$$

Using Euler's identity on the integrand, we thus arrive at the conclusions that

$$\int_0^\infty \cos(y^2)\, dy = \int_0^\infty \sin(y^2)\, dy = \frac{1}{2}\sqrt{\frac{\pi}{2}}.$$

These two integrals (called *Fresnel* integrals) are known by other means to be, in fact, correct (see Problem G7). Thus, our transform pair, of suspicious birth, appears to be legitimate after all. The major point here is that doing such wild and crazy symbol pushing *is okay*—as long as you are always aware of how close to the edge of potential catastrophe you are walking, and that you keep alert for indications your results may no longer be making any physical sense.

As a final example, suppose $v(t)$ is the nonperiodic signal shown in Figure 12.2 [two cycles of $\sin(\omega_0 t)$], along with the absolute value of its Fourier transform (the transform, itself, is of course imaginary as $v(t)$ is odd—see Problem 12.5). While $|V(j\omega)|$ is generally nonzero at almost all frequencies, you can see it peaks at $\omega = \pm\omega_0$. In fact, if instead of *two* cycles of $\sin(\omega_0 t)$ we have $v(t)$ consist of n cycles then with a little work you can show that $|V(j\omega)|$ will still have the same general behavior (small at all frequencies except for peaks at $\omega = \omega_0$, with the amplitudes of the peaks proportional to n). That is, $|V(j\omega)|$ will tend to just two *infinite spikes* at $\omega = \pm\omega_0$ as n goes to infinity. I'll return to this example (which has an important role in AM radio) in Chapter 15.

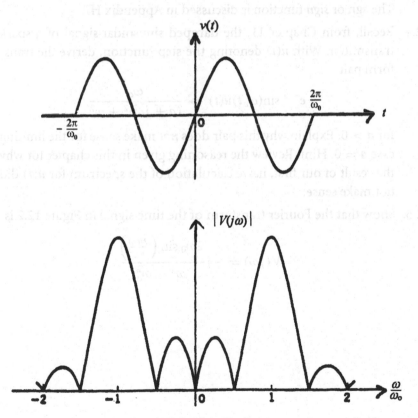

Figure 12.2   Another time signal and its Fourier transform amplitude.

Problems

12.1. Show that the Fourier transform of any odd function of time is zero at $\omega = 0$, i.e., show that $V(j\omega)|_{\omega=0} = V(0) = 0$ if $v(t)$ is odd.

12.2. If $v(t) \leftrightarrow V(j\omega)$ is a Fourier transform pair, then show that the transform of $\frac{dv}{dt}$ is $j\omega V(j\omega)$, and that the transform if $tv(t)$ is $j\,dV/d\omega$. Hint: write $v(t) = (1/2\pi) \int_{-\infty}^{\infty} V(j\omega)e^{j\omega t}\,d\omega$ and differentiate with respect to t for the first case. Using a similar approach with the integral representation for $V(j\omega)$ gives the second result. (Read Appendix G for how to differentiate an integral).

12.3. Show that the Fourier transforms of $\frac{1}{(t^2+1)}$ and $\frac{t}{(t^2+1)}$ are $\pi e^{-|\omega|}$ and $-j\pi e^{-|\omega|}\,\mathrm{sgn}(\omega)$, respectively, where

$$\mathrm{sgn}(\omega) = \begin{cases} 1, & \omega > 0; \\ 0, & \omega = 0; \\ -1, & \omega < 0. \end{cases}$$

The sgn or *sign* function is discussed in Appendix H.

12.4. Recall, from Chapter 11, the damped sinusoidal signal of a spark transmitter. With $u(t)$ denoting the step function, derive the transform pair

$$e^{-at}\sin(\omega_c t)u(t) \leftrightarrow \frac{\omega_c}{(a + j\omega)^2 + \omega_c^2}$$

for $a > 0$. Explain why this pair does *not* make sense for the limiting case $a = 0$. Hint: Review the reasoning given in this chapter for why the result of our first, naive calculation of the spectrum for $u(t)$ did not make sense.

12.5. Show that the Fourier transform of the time signal in Figure 12.2 is

$$V(j\omega) = -j\,\frac{2\omega_0 \sin\left(\frac{2\pi\omega}{\omega_0}\right)}{\omega^2 - \omega_0^2}.$$

Physical Meaning of the Fourier Transform

From the previous chapter we have the pair of integrals $v(t) \leftrightarrow V(j\omega)$ for the nonperiodic signal $v(t)$, where

$$V(j\omega) = \int_{-\infty}^{\infty} v(t)e^{-j\omega t}\, dt,$$

$$v(t) = \frac{1}{2\pi}\int_{-\infty}^{\infty} V(j\omega)e^{j\omega t}\, d\omega.$$

Pretty, yes, but what do they *mean*? To answer this, recall from Chapter 9 that we defined the energy of the periodic signal $x(t)$ (over a period) as

$$W = \int_{\text{period}} x^2(t)\, dt.$$

For the case of a nonperiodic signal (with a "period" we take as infinity), we have

$$W = \int_{-\infty}^{\infty} v^2(t)\, dt.$$

Thus,

$$W = \int_{-\infty}^{\infty} v(t)v(t)\, dt = \int_{-\infty}^{\infty} v(t)\left[\frac{1}{2\pi}\int_{-\infty}^{\infty} V(j\omega)e^{j\omega t}\, d\omega\right] dt$$

or, interchanging the order of integration (see Appendix G),

$$W = \int_{-\infty}^{\infty} \frac{1}{2\pi}V(j\omega)\left[\int_{-\infty}^{\infty} v(t)e^{j\omega t}\, dt\right] d\omega.$$

The inner integral is [for real $v(t)$] just $V^*(j\omega)$, and so

$$W = \int_{-\infty}^{\infty} \frac{1}{2\pi} V(j\omega) V^*(j\omega)\, d\omega = \int_{-\infty}^{\infty} \frac{1}{2\pi} |V(j\omega)|^2\, d\omega.$$

That is, we can calculate the energy of $v(t)$ either in the time domain *or* in the frequency domain as

$$W = \int_{-\infty}^{\infty} v^2(t)\, dt = \int_{-\infty}^{\infty} \frac{1}{2\pi} |V(j\omega)|^2\, d\omega,$$

an important result called *Rayleigh's energy theorem* (which is Parseval's theorem for nonperiodic signals).

Lord Rayleigh (John William Strutt) was one of the very big men of English science in the late nineteenth century and early twentieth century. His genius cut across practically every aspect of the physics of his time. Rayleigh (1842–1919), a member of the House of Lords by birthright, received the 1904 Nobel Prize in physics for his role in the discovery of a new element, the inert gas argon. His energy theorem can be found in his 1889 paper "On the Character of the Complete Radiation at a given Temperature," in *Scientific Papers of Lord Rayleigh*, Vol. 3, Dover, 1964. More on Rayleigh's contributions to electrical science can be found in my book *Oliver Heaviside, Sage in Solitude*, IEEE Press, 1988 (reprinted, The Johns Hopkins University Press 2001).

The second integral is interpreted as describing how the energy of $v(t)$ is distributed over frequency, i.e., $|V(j\omega)|^2$ is called the *energy spectral density* (ESD), for $-\infty < \omega < \infty$. In more detail,

$$\text{energy in the interval } \omega_1 \leq \omega \leq \omega_2 = \int_{\omega_1}^{\omega_2} \frac{1}{2\pi} |V(j\omega)|^2\, d\omega.$$

It is this statement that gives physical significance to the Fourier transform, as the Fourier spectrum tells us where the energy of a time signal is located in frequency. Knowing the frequency distribution of energy will, in particular, guide us in the next chapter in understanding how multipliers work at radio frequency.

There are several different-looking versions of the ESD that you will see in other books (but they are all really equivalent), e.g., as shown in the previous chapter, $|V(j\omega)|^2$ is even if $v(t)$ is real and so

$$W = \int_{0}^{\infty} \frac{1}{\pi} |V(j\omega)|^2\, d\omega,$$

which says the *one-sided* ESD is $(1/\pi)|V(j\omega)|^2$, for $0 \leq \omega < \infty$.

As an example of how useful Rayleigh's theorem is *right now*, even before we see how it will help us understand radio, recall the following $v(t) \leftrightarrow V(j\omega)$ Fourier transform pair from Chapter 12:

$$v(t) = \begin{cases} 1, & |t| \leq \frac{\tau}{2} \\ 0, & \text{otherwise} \end{cases}$$

$$V(j\omega) = \tau \frac{\sin\left(\frac{\omega\tau}{2}\right)}{\left(\frac{\omega\tau}{2}\right)}.$$

Rayleigh's theorem then tells us that

$$W = \int_{-\tau/2}^{\tau/2} v^2(t)\, dt = \int_{-\tau/2}^{\tau/2} dt = \int_{-\infty}^{\infty} \frac{1}{2\pi} |V(j\omega)|^2\, d\omega$$

$$= \int_{-\infty}^{\infty} \frac{\tau^2}{2\pi} \frac{\sin^2\left(\frac{\omega\tau}{2}\right)}{\left(\frac{\omega\tau}{2}\right)^2}\, d\omega.$$

Thus,

$$\tau = \frac{\tau^2}{2\pi} \int_{-\infty}^{\infty} \frac{\sin^2\left(\frac{\omega\tau}{2}\right)}{\left(\frac{\omega\tau}{2}\right)^2}\, d\omega.$$

If you change variable to $x = \frac{\omega\tau}{2}$, then this immediately becomes

$$\int_{-\infty}^{\infty} \frac{\sin^2(x)}{x^2}\, dx = \pi$$

or, equivalently, as the integrand is even,

$$\int_{0}^{\infty} \frac{\sin^2(x)}{x^2}\, dx = \frac{\pi}{2},$$

a definite integral that you will see many more times in advanced engineering and science (and which is much harder to derive by other means). You are asked to do similar calculations in Problems 13.1, 13.7, 13.8, 13.9, and 13.10.

In Problem G1 it is claimed that for any "well-behaved" function $v(t)$, its Fourier transform is such that $\lim_{\omega \to \pm\infty} V(j\omega) = 0$ (as pointed out there, this is a special case of the so-called Riemann–Lebesque lemma). Rayleigh's theorem then tells us immediately that the ESD of any well-behaved function "rolls off" to zero as we go ever higher in frequency. The proof of this is direct and elegant. We have

$$V(j\omega) = \int_{-\infty}^{\infty} v(t)e^{-j\omega t}\, dt,$$

which, if we change variables to $u = t - \frac{\pi}{\omega}$, becomes

$$V(j\omega) = \int_{-\infty}^{\infty} v\left(u + \frac{\pi}{\omega}\right) e^{-j\omega(u+\pi/\omega)} \, du$$

$$= \int_{-\infty}^{\infty} v\left(u + \frac{\pi}{\omega}\right) e^{-j\omega u} e^{-j\pi} \, du$$

$$= -\int_{-\infty}^{\infty} v\left(u + \frac{\pi}{\omega}\right) e^{-j\omega u} \, du.$$

Thus,

$$V(j\omega) = \frac{1}{2} \int_{-\infty}^{\infty} v(t) e^{-j\omega t} \, dt + \frac{1}{2}\left[-\int_{-\infty}^{\infty} v\left(u + \frac{\pi}{\omega}\right) e^{-j\omega u} \, du \right].$$

Now, since the absolute value of an integral is less than or equal to the integral of the absolute value of the integrand, and since the absolute value of a product is the product of the absolute values, and since $|e^{-j\omega t}| = 1$, then we can write

$$\left| V(j\omega) \right| \leq \frac{1}{2} \int_{-\infty}^{\infty} \left| v(t) - v\left(t + \frac{\pi}{\omega}\right) \right| \, dt.$$

From this we can conclude (assuming we remember the very definition of a derivative, and also assuming $\lim_{t \to \pm\infty} v(t) = 0$) that $\lim_{\omega \to \pm\infty} |V(j\omega)| = 0$, from which Rayleigh's theorem gives the final conclusion about the high-frequency roll-off of the ESD of $v(t)$.

Finally, recall from the end of Chapter 9 the expression for $y(t)$, the output of a linear system with the periodic signal $v(t)$ as the input:

$$y(t) = \sum_{k=-\infty}^{\infty} c_k H(jk\omega_0) e^{jk\omega_0 t}, \qquad \omega_0 = \frac{2\pi}{T}.$$

In this expression $H(j\omega)$ is the transfer function of the system. We can convert this to the case where $v(t)$ is not periodic, using the approach of the previous chapter. Thus,

$$y(t) = \frac{1}{T} \sum_{k=-\infty}^{\infty} T c_k H(jk\omega_0) e^{jk\omega_0 t}$$

$$= \frac{\omega_0}{2\pi} \sum_{k=-\infty}^{\infty} T c_k H(jk\omega_0) e^{jk\omega_0 t}$$

$$= \frac{1}{2\pi} \sum_{k=-\infty}^{\infty} T c_k H(jk\omega_0) e^{jk\omega_0 t} \omega_0.$$

Now, as $T \to \infty$ we have $\omega_0 \to d\omega$, $k\omega_0 \to \omega$ and $Tc_k \to V(j\omega)$. Thus, imagining the summation goes over into an integration,

$$y(t) = \frac{1}{2\pi} \int_{-\infty}^{\infty} V(j\omega)H(j\omega)e^{j\omega t}\, d\omega$$

is the output of a linear system [with transfer function $H(j\omega)$] with the nonperiodic $v(t)$ as its input. But, if $Y(j\omega)$ is the Fourier transform of $y(t)$, then

$$y(t) = \frac{1}{2\pi} \int_{-\infty}^{\infty} Y(j\omega)e^{j\omega t}\, d\omega,$$

and so we immediately have the important result $Y(j\omega) = V(j\omega)H(j\omega)$. That is, the spectrum of a linear system's output is simply the product of the system's transfer function and the input spectrum. You'll see in Chapter 15 how this result has important implications for radio.

Problems

13.1. Starting with the time signal

$$v(t) = \begin{cases} -1, & -a < t < 0 \\ 1, & 0 < t < a \\ 0, & \text{otherwise,} \end{cases}$$

use Rayleigh's energy theorem to show that

$$\int_0^{\infty} \left\{ \frac{\sin^2(x)}{x} \right\}^2 dx = \frac{\pi}{4}.$$

[Note that a much more concise way of writing this $v(t)$ with step functions is $v(t) = -u(t+a) + 2u(t) - u(t-a)$.] Write a MATLAB program to check this result.

13.2. Show that the Fourier transform pair derived in the previous chapter, $e^{-\sigma t}u(t) \leftrightarrow \frac{1}{(\sigma+j\omega)}$, satisfies the Rayleigh energy theorem for all $\sigma > 0$.

13.3. Show that the ESD at the output of a linear system with transfer function $H(j\omega)$ is $|H(j\omega)|^2$ times the ESD at the input.

13.4. A time function of some theoretical importance in electrical engineering is $|t|$. This function not only has infinite energy [something we've seen before in such functions as $\cos(\omega t)$ and the step], but it is also *unbounded*, i.e., it grows without limit. It shouldn't be surprising then (perhaps) that this function may present special difficulties when we try to find its Fourier transform. For now, explain why $|t| = \int_0^t \text{sgn}(x)\, dx$. (Recall Problem 12.3 for the definition of the sgn function.) Next, combine this expression with the result derived in Appendix H,

$$\left(\frac{1}{\pi}\right) \int_{-\infty}^{\infty} \frac{\sin(\omega x)}{\omega}\, d\omega = \text{sgn}(x),$$

and so derive the curious formula

$$|t| = \left(\frac{1}{\pi}\right) \int_{-\infty}^{\infty} \{1 - \cos(\omega t)\}/\omega^2\, d\omega.$$

(Hint: substitute the first statement into the second and reverse the order of integration.) Notice that the integrand is well behaved over the entire interval of integration, including at $\omega = 0$. You should check this assertion by making a power series expansion of the cosine and observing what the integrand is like near and at $\omega = 0$. The Fourier transform of $|t|$ is discussed in Problem 14.4.

13.5. Find the Fourier transform of the so-called Gaussian pulse $v(t) = e^{-at^2}$, for $|t| < \infty$ and a any positive constant. Verify that the resulting transform pair satisfies Rayleigh's energy theorem. More on this pair can be found in Problem 14.5. Hint: write $V(j\omega) = \int_{-\infty}^{\infty} e^{-at^2} e^{-j\omega t}\, dt$ and integrate by parts. Next, notice that differentiating through the transform integral sign gives

$$\frac{dV}{d\omega} = \int_{-\infty}^{\infty} -jt e^{-at^2} e^{-j\omega t}\, dt$$

which, when combined with the result of the integration by parts will allow you to show that $V(j\omega)$ satisfies the differential equation $\frac{dV}{d\omega} = -\left(\frac{\omega}{2a}\right) V$. This can then easily be integrated, subject to the constraint

$$V(0) = \int_{-\infty}^{\infty} e^{-at^2}\, dt = \sqrt{\frac{\pi}{a}},$$

a result you can verify by making the appropriate change of variable in an integral derived in Appendix G, i.e., in $\int_{-\infty}^{\infty} e^{-x^2}\, dx = \sqrt{\pi}$.

13.6. Show that a time delay of one-quarter of a period in the signal $\cos(\omega t)$ is equivalent to subtracting 90° from the phase of the positive frequency part and adding 90° to the phase of the negative

frequency part (recall from Appendix A what positive and negative frequency mean). This result will be useful later when I discuss
single-sideband radio in Chapter 20.

13.7. Find the Fourier transform of $v(t) = e^{-|t|}$, $-\infty < t < \infty$. Use the
result, with Rayleigh's theorem, to show that the energy of $v(t)$ in the
frequency interval $|\omega| < \omega_1$ is

$$\frac{2}{\pi}\left[\frac{\omega_1}{1+\omega_1^2} + \tan^{-1}(\omega_1)\right].$$

Notice that, by letting $\omega_1 = \infty$, this result says the total energy of
$v(t)$ is 1, which can be checked by evaluating the energy in the time
domain, i.e., by showing that $\int_{-\infty}^{\infty} v^2(t)\, dt = 1$. (Hint: $e^{-|t|}$ is e^{-t}
for $t > 0$, and e^t for $t < 0$.) Determine f_1 $(= \frac{\omega_1}{2\pi})$, accurate to four
decimal places, such that 99.9% of the total energy of $v(t)$ is in the
interval $|f| < f_1$. (Answer: 1.1876 Hz.) Repeat for 99.99%.

13.8. Apply Rayleigh's energy theorem to the signal

$$v(t) = \begin{cases} e^{-t}, & 0 \le t \le 1 \\ 0, & \text{otherwise} \end{cases}$$

and so derive the value of the integral $\int_0^\infty \frac{\cos(\omega)}{1+\omega^2}\, d\omega$.

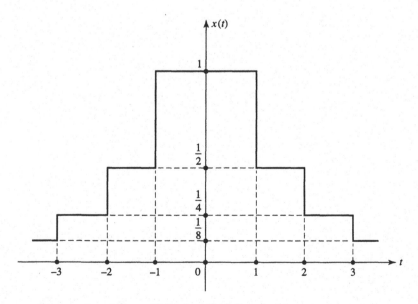

Figure 13.1   A curious time function.

13.9. Apply Rayleigh's energy theorem to the time function shown in Figure 13.1, and so evaluate the integral

$$\int_{-\infty}^{\infty} \frac{\sin^2(\omega)}{\omega^2 \left\{ \frac{5}{4} - \cos(\omega) \right\}^2} \, d\omega.$$

13.10. Apply Rayleigh's energy theorem to the time function

$$x(t) = \begin{cases} +1, & 0 < t < a \\ -1, & -a < t < 0 \\ 0, & \text{otherwise} \end{cases}$$

and so evaluate the integral

$$\int_{0}^{\infty} \frac{\sin^4(\omega)}{\omega^2} \, d\omega.$$

CHAPTER
14

Impulse "Functions" in Time and Frequency

One of the most important ideas in electrical engineering in general, and certainly in radio theory, is the concept of an *impulse*. Impulses occur in analyses whenever some physical quantity is concentrated in time, space, frequency, etc. To be simplistic, an impulse denotes something "happening all at once." To get a mathematical grip on the impulse idea (one I am copying from Dirac), imagine the time signal in Figure 14.1, a narrow pulse of height $\frac{1}{b}$ and width b, centered on $t = 0$. This pulse, which I'll call $x(t)$, is zero for all times $|t| > b/2$. For *any* nonzero value of b, $x(t)$ clearly bounds unit area. It is a perfectly ordinary, well-behaved signal.

The impulse is also commonly called the *Dirac delta*, after the great English physicist Paul Dirac (1902–1984). First trained as an electrical engineer, Dirac followed his undergraduate degree in electrical engineering (first-class honors in the 1921 class of the University of Bristol) with a Ph.D in mathematics, and a share in the 1933 physics Nobel Prize for his work in quantum mechanics. He introduced physicists to the properties of the delta function in his famous paper "The Physical Interpretation of the Quantum Mechanics," *Proceedings of the Royal Society of London A* 113, January 1, 1927, pp. 621–641, published while he was still just twenty-five years old. He spent the last twelve years of his life as Professor of Physics at

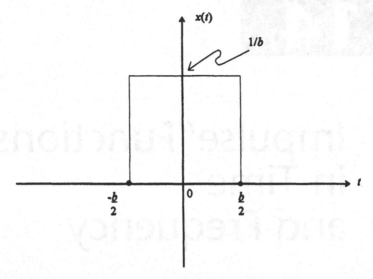

Figure 14.1 A finite width pulse with unit area.

Florida State University in Tallahassee. Impulses (also called *improper* or *singular functions*) had actually been used for decades before Dirac, most successfully by the eccentric English electrical engineer and physicist Oliver Heaviside (1850–1925). Dirac first encountered singular functions via his reading of Heaviside's books while an undergraduate (one still occasionally finds the step function called the *Heaviside step*). For much more on Heaviside, and in particular his mathematics, see my book *Oliver Heaviside, Sage in Solitude*, IEEE Press, 1988.

Imagine next that we multiply $x(t)$ by some other arbitrary continuous function $\varphi(t)$, and then integrate the product over all time, i.e., let's form the integral

$$I = \int_{-\infty}^{\infty} x(t)\varphi(t)\,dt = \int_{-b/2}^{b/2} \frac{1}{b}\varphi(t)\,dt = \frac{1}{b}\int_{-b/2}^{b/2} \varphi(t)\,dt.$$

Finally, imagine that $b \to 0$, which physically means the pulse's height becomes very big and the interval of integration (the pulse's duration) becomes very narrow. If $\varphi(t)$ is any function in the real engineering world, which means it is certainly *at least* continuous, then I'll argue that $\varphi(t)$ cannot change very much over this narrow interval (if it does, simply make the interval a billion times shorter!). That is, $\varphi(t)$ is very nearly equal to $\varphi(0)$

over the entire interval of integration, which is of width b. Thus,

$$\lim_{b \to 0} I = \lim_{b \to 0} \frac{1}{b} \int_{-b/2}^{b/2} \varphi(t)\, dt = \lim_{b \to 0} \frac{1}{b}\varphi(0)b = \varphi(0).$$

I will be quite casual about the nature of $\varphi(t)$. Generally, all I will require is that $\varphi(t)$ not only is continuous but also is differentiable (as many times as needed for the particular problem at hand) and that it behave "properly" at infinity (which is usually meant to mean $\lim_{t \to \pm\infty} \varphi(t) = 0$). Mathematicians often call such functions "good," "moderately good," or "fairly good." Dirac used the term "regular." $\varphi(t)$ is sometimes called a *testing function*.

The limit of $x(t)$ as $b \to 0$ is shown in Figure 14.2, which tries to indicate that the pulse becomes an infinitely high spike of zero width (which is *not* an ordinary function at all!). Since I can't draw an infinitely high spike, the figure simply shows an upward arrow, with "1" written next to it to indicate it is a so-called unit area or *unit strength impulse*. More formally, we write this limit as $\lim_{b \to 0} x(t) = \delta(t)$, and while it is very difficult to visualize such an amazing thing as infinite height and zero width with unit area, we know how $\delta(t)$ behaves *under the integral sign*. That is, for any continuous $\varphi(t)$ we have just seen that

$$\int_{-\infty}^{\infty} \delta(t)\varphi(t)\, dt = \varphi(0).$$

There is, in fact, nothing particularly special about $t = 0$ and we can locate the impulse at any time, say $t = t_0$, simply by writing $\delta(t - t_0)$.

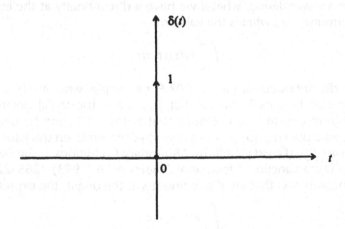

Figure 14.2   When the pulse has zero width.

Then, if $\varphi(t)$ is continuous at $t = t_0$,

$$\int_{-\infty}^{\infty} \delta(t - t_0)\varphi(t)\, dt = \varphi(t_0),$$

an important result called the *sampling property* of the impulse, one we'll use a *lot* from now on. That is, when an impulse occurs inside an integral, the value of the integral is determined by first determining the location of the impulse (by setting the argument of the impulse equal to zero) and then evaluating the rest of the integrand at that location.

Consider next the integral of the impulse, i.e., define $s(t) = \int_{-\infty}^{t} \delta(z)\, dz$, where z is, of course, just a dummy variable. What is the behavior of $s(t)$? The impulse is located at $z = 0$. Thus, if $t < 0$ the impulse is *not* inside the interval of integration and so $s(t) = 0$. If $t > 0$, however then the impulse is inside the interval and the integral is the area bounded by the impulse (which is, by the very way we constructed the impulse, unity). That is, $s(t) = 1$ if $t > 0$. We conclude then, that $s(t) = u(t)$, the unit step function introduced at the end of Chapter 12. Of course, we can position the "start," or step, of the step function anywhere we like simply by writing $u(t - t_0)$, which is 1 for $t > t_0$ and 0 to $t < t_0$. The integration of the impulse to give the step means that we can formally write $\delta(t - t_0) = \frac{d}{dt}[u(t - t_0)]$. Impulses always occur whenever a discontinuous function is differentiated, and the strength (area bounded) by the impulse is simply the value of the discontinuous (i.e., step) change.

I have been emphasizing the *continuous* nature of $\varphi(t)$, at least as far as it being continuous at the instant an impulse is applied to it. But, you may be wondering, what if we have a discontinuity at the instant of the impulse, i.e., what is the value of

$$\int_{-\infty}^{\infty} f(t)\delta(t)\, dt$$

if $f(t)$ is discontinuous at time $t = 0$? For example, what if $f(t) = u(t)$, the unit-step function? (Notice that I've made the trivial notational change from $\varphi(t)$ to $f(t)$; I've done that because I'll soon be quoting authors who use $f(t)$, not $\varphi(t)$.) As one physicist wrote on this issue only a few years ago (David J. Griffiths, "Boundary Conditions at the Derivative of a Delta Function," *Journal of Physics A* 26 (1993):2265–2267), "For functions $f(x)$ that are discontinuous at the origin, the expression

$$\int_{-\varepsilon}^{\varepsilon} \delta(x)f(x)\, dx$$

is not, in general, well-defined. However, if we stipulate that the delta function is [even, as shown in Figure 14.1—see also Problem 14.1] then

$$\int_{-\varepsilon}^{\varepsilon} \delta(x)f(x)dx = \frac{1}{2}[f(0+) + f(0-)]."$$

That is, Professor Griffiths was saying the integral is the *average* of the values of $f(x)$ on each side of the discontinuity. From now on, I'll call this the "average assertion," and so for the case of $f(t) = u(t)$ the average assertion says

$$\int_{-\varepsilon}^{\varepsilon} \delta(t)u(t)dt = \frac{1}{2}$$

since $u(0-) = 0$ and $u(0+) = 1$.

The average assertion has been used by physicists for a long time, and a historical discussion of that use can be found in M. G. Calkin and D. Kiang, "Proper Treatment of the Delta Function Potential in the One-Dimensional Dirac Equation," *American Journal of Physics* 55 (August 1987):737–739. As it turns out, however, the average assertion is simply *not true*. Here's why.

Suppose you are faced with the problem of solving the differential equation

$$\frac{df}{dt} = kf(t)\delta(t),$$

where k is any real constant. This equation is separable and we can directly solve for $f(t)$ by writing

$$\frac{df}{f} = k\delta(t)dt$$

and integrating. That is, denoting the value of $f(t)$ as $t \to -\infty$ by $f(-\infty)$, we have

$$\int_{f(-\infty)}^{f(t)} \frac{df}{f} = k\int_{-\infty}^{t} \delta(s)\,ds$$

and so

$$\ln\left\{\frac{f(t)}{f(-\infty)}\right\} = ku(t).$$

Thus,

$$f(t) = f(-\infty)e^{ku(t)}$$

and so

$$f(t) = \begin{cases} f(-\infty), & t < 0 \\ f(+\infty) = f(-\infty)e^k, & t > 0. \end{cases}$$

So, the solution to the differential equation is a function that is equal to a constant for all values of $t < 0$ (what I've called $f(-\infty)$), and to another constant for all values of $t > 0$ (what I've called $f(+\infty)$). Further, these two constants are such that

$$\frac{f(+\infty)}{f(-\infty)} = e^k.$$

This is not the only way to approach the differential equation, however. I'll now manipulate it in a way that explicitly uses the average assertion. Doing that, we'll find that we arrive at a different, i.e., *incorrect*, conclusion for the ratio $\frac{f(+\infty)}{f(-\infty)}$. So, without separating the variables in the differential equation, let's integrate across the discontinuity, i.e., for $\varepsilon > 0$,

$$\int_{-\varepsilon}^{\varepsilon} \frac{df}{dt} \, dt = k \int_{-\varepsilon}^{\varepsilon} f(t)\delta(t) \, dt.$$

Then,

$$\int_{-\varepsilon}^{\varepsilon} df = f(t)|_{-\varepsilon}^{\varepsilon} = f(\varepsilon) - f(-\varepsilon) = k \int_{-\varepsilon}^{\varepsilon} f(t)\delta(t) \, dt.$$

From the first analysis we know that $f(t)$ is a constant on each side of the discontinuity. Thus, $f(\varepsilon) = f(+\infty)$ and $f(-\varepsilon) = f(-\infty)$, and so

$$\int_{-\varepsilon}^{\varepsilon} f(t)\delta(t) \, dt = \frac{f(+\infty) - f(-\infty)}{k}.$$

But the average assertion says

$$\int_{-\varepsilon}^{\varepsilon} f(t)\delta(t) \, dt = \frac{1}{2}[f(0+) + f(0-)] = \frac{1}{2}[f(+\infty) + f(-\infty)].$$

Thus,

$$\frac{f(+\infty) - f(-\infty)}{k} = \frac{1}{2}[f(+\infty) + f(-\infty)]$$

or, after some simple algebra,

$$\frac{f(+\infty)}{f(-\infty)} = \frac{2+k}{2-k} \neq e^k$$

except for the trivial case of $k = 0$ (in which case we no longer have an impulse in the original differential equation!)

So, the average assumption has, in this particular case, lead to an incorrect conclusion and, as Professor Griffiths wrote, integrating an impulse against a discontinuous function is, indeed, "not well defined."

All of the preceding is really an "engineer's derivation," and pure mathematicians will grind their teeth and fight waves of nausea as they read my development of the impulse. In particular, when I pulled the b outside of the integral in the first equation of this chapter (leaving just $\varphi(t)$ inside), I was treating b as a constant. But then I let $b \to 0$ which implies b is *not* constant. Since the operation of integration, itself, is defined in terms of a limiting operation, what I was really doing (in the midst of much smoke and fog) was reversing the order of taking two limits. How do I know that's mathematically valid? Well, I don't!—and often it isn't! Being engineers and physicists, however, we'll not let this paralyze us into inaction. We'll simply assume it's okay and go ahead until something awful happens in the math that tells us we've pushed the symbols too hard. And, after all, *physics* Nobel laureate Dirac was professor of *mathematics* at Cambridge, and if my approach to impulses was good enough for him I see no reason to apologize.

The mathematics of impulses has, I should say (because you may be thinking my attitude toward mathematical rigor is just a bit *too* cavalier) been placed on firm foundations. This achievement is generally credited to the French mathematician Laurent Schwartz (born 1915), with the publication of his two books *Theory of Distributions* (1950, 1951). Schwartz received the 1950 Field's Medal for his work, an award often called the "Nobel Prize for mathematics." While much of the preliminary work had been done since 1936 by the Russian mathematician Sergei L. Sobolev (1908–1989), it was Schwartz who was the more concerned about the applications of his work to the problems of physics and electrical engineering (see Jesper Lützen, *The Prehistory of the Theory of Distributions*, Springer-Verlag, 1982). A number of books on these matters, suitable for undergraduate engineers and physicists, have appeared in the decades since; one I can particularly recommend (which is at a level only moderately higher than that of this book) is M. J. Lighthill, *Introduction to Fourier Analysis and Generalized Functions*, Cambridge University Press, 1959.

> Dirac of course knew that he was being nonrigorous with his use of impulses. As he wrote in his 1927 paper, "Strictly, of course, $\delta(x)$ is not a proper function of x, but can be regarded only as a limit of a certain sequence of functions. All the same one can use $\delta(x)$ as though it were a proper function for practically all the purposes of quantum mechanics without getting incorrect results. One can also use the [derivatives] of $\delta(x)$, namely $\delta'(x)$, $\delta''(x) \ldots$, which are even more discontinuous and less 'proper' than $\delta(x)$ itself."

Many years later Dirac gave credit to his youthful *engineering* training for his ability to break free of a too restrictive loyalty to the chains of rigor: "I would like to try to explain the effect of this engineering training on me. I did not make any further use of the detailed applications of this work, but it did change my whole outlook to a very large extent. Previously, I was interested only in exact equations. Well, the engineering training which I received did teach me to tolerate approximations, and I was able to see that even theories based on approximations could sometimes have a considerable amount of beauty in them.... I think that if I had not had this engineering training, I should not have had any success with the kind of work that I did later on.... I continued in my later work to use mostly the nonrigorous mathematics of the engineers, and I think that you will find that most of my later writings do involve nonrigorous mathematics.... *The pure mathematician who wants to set up all of his work with absolute accuracy is not likely to get very far in physics* [my emphasis]." (Quoted from *Paul Dirac: the man and his work*, Cambridge University Press 1998, p. 3.)

The impulse has a particularly simple Fourier transform. Thus, if we insert $\delta(t - t_0)$ into the transform integral and use the sampling property, we get

$$\int_{-\infty}^{\infty} \delta(t - t_0)e^{-j\omega t}\, dt = e^{-j\omega t_0}.$$

In particular, the unit impulse located at the origin ($t_0 = 0$) has the simplest of all Fourier transforms, namely *one*. Since $|e^{-j\omega t_0}| = 1$ for any real value of t_0, then the ESD of any impulse is a constant over all frequencies, and the impulse is therefore said to have a *flat* spectrum. This is simply the reciprocal spreading effect, mentioned in Chapter 12, taken to the extreme. This, in turn, means $\delta(t)$ is a signal with *infinite* energy, which while certainly out of the ordinary shouldn't surprise us—we already know $\delta(t)$ is an odd beast, indeed! The flat spectrum of the impulse of course violates the result of the previous section that said if $f(t)$ is "well behaved" then $\lim_{\omega \to \pm\infty} F(j\omega) = 0$. But, again $\delta(t)$ is simply *not* well behaved. The flat spectrum of the impulse is also often called *white*. This is done in analogy to white light, a uniform mixture of all visible photon frequencies. Extending the analogy with light even further, time signals that have nonflat (nonwhite) spectrums are said to be *colored* (or, alternatively, *pink*).

Notice that the conclusion $\delta(t)$ has infinite energy is obvious from one half of Rayleigh's energy theorem [$W = \int_{-\infty}^{\infty} \frac{1}{2\pi}|X(j\omega)|^2 d\omega$ blows up if $|X(j\omega)|$ is a nonzero constant over all ω], but is *not* so obvious from the other half [$W = \int_{-\infty}^{\infty} x^2(t)\, dt$]. After all, what could $\int_{-\infty}^{\infty} \delta^2(t)\, dt$ *mean*? One *might* argue that $\int_{-\infty}^{\infty} \delta(t)\delta(t)\, dt = \delta(0) = \infty$, but this really doesn't make any sense, as it is letting $\varphi(t) = \delta(t)$ and this goes far beyond what was said earlier about the mathematical nature of $\varphi(t)$.

With the Fourier transform pair $\delta(t) \leftrightarrow 1$ we can use the inverse Fourier transform to write

$$\delta(t) = \frac{1}{2\pi} \int_{-\infty}^{\infty} e^{j\omega t}\, d\omega.$$

This is an astonishing statement because the integral simply doesn't make any sense if we attempt to evaluate it, as $e^{j\omega t}$ does not approach a limit as $|\omega| \to \infty$. The only way we can make any sense at all of this, at the level of this book, is that the right-hand side is just a collection of ink squiggles that denotes the same concept (an impulse) that the ink squiggles on the left do. Any time we encounter the right-hand side squiggles, we will simply replace them with $\delta(t)$. Notice, too, that if we interchange the variables ω and t on both sides (thus retaining the truth of the statement), we arrive at

$$\delta(\omega) = \frac{1}{2\pi} \int_{-\infty}^{\infty} e^{j\omega t}\, dt,$$

an impulse in the *frequency* domain! (This trick is based on the observation that the particular ink squiggles we use in our equations are all historical accidents—the only constraint is to be consistent.) And finally, if we find the time function associated with $\delta(\omega)$ [call it $x(t)$] by putting $\delta(\omega)$ into the inverse transform integral, we get (using the sampling property)

$$x(t) = \frac{1}{2\pi} \int_{-\infty}^{\infty} \delta(\omega)e^{j\omega t}\, d\omega = \frac{1}{2\pi},$$

a *constant*. This makes physical sense, too, because a constant has all its energy at dc, i.e., at $\omega = 0$, which is precisely where $\delta(\omega)$ is located. This gives us the transform pair $1 \leftrightarrow 2\pi\delta(\omega)$. As you'll see in the next chapter, these integral representations of time and frequency impulses are no mere academic observations—such expressions do occur *often* in AM radio theory.

The above notational trick can be used to establish a general and most useful theorem, one I'll use in the next section to solve a problem that would otherwise be very difficult to do. Suppose we have the transform pair $g(t) \leftrightarrow G(j\omega)$. Then, from the inverse transform,

$$g(t) = \frac{1}{2\pi} \int_{-\infty}^{\infty} G(j\omega)e^{j\omega t}\, d\omega,$$

or, replacing t with $-t$ on both sides (which leaves the truth of the statement unaltered),

$$g(-t) = \frac{1}{2\pi} \int_{-\infty}^{\infty} G(j\omega)e^{-j\omega t} \, d\omega.$$

Next, using the symbol interchange trick, we have

$$g(-\omega) = \frac{1}{2\pi} \int_{-\infty}^{\infty} G(jt)e^{-jt\omega} \, dt$$

or,

$$2\pi g(-\omega) = \int_{-\infty}^{\infty} G(jt)e^{-j\omega t} \, dt.$$

The integral is simply the Fourier transform of the time function $G(jt)$, and so we have the transform pair $G(jt) \leftrightarrow 2\pi g(-\omega)$. This result is often called the Fourier transform's property of *duality* and an application of it appears in the final example of this chapter.

Before going on any further with this symbol pushing, let me show you how the idea of the impulse can be given some physical plausibility. Consider the real periodic signal $x(t)$ in Figure 14.3, consisting of an infinitely long sequence (or *train*) of unit impulses with period $T = 1$. If we expand $x(t)$ into a Fourier series, recall from Chapter 9 that

$$x(t) = \sum_{k=-\infty}^{\infty} c_k e^{jk\omega_0 t}, \quad \omega_0 = \frac{2\pi}{T}$$

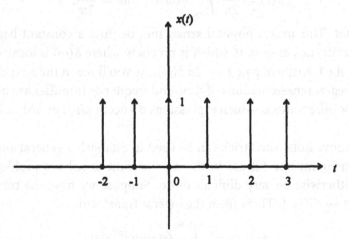

Figure 14.3   The periodic impulse train $x(t) = \sum_{n=-\infty}^{\infty} \delta(t - n)$.

where the coefficients are given by

$$c_k = \frac{1}{T} \int_{\text{period}} x(t)e^{-jk\omega_0 t}\, dt.$$

The interval of integration (of length T) can be anywhere, but let's pick it to be $-T/2 < t < T/2$ which will make it easy to see exactly what is happening. If we use the perhaps more obvious interval of 0 to 1 then we'd have *two* impulses in the integration interval, one at each end. Or would it be two "half" impulses, whatever *that* might be? The fact is, it isn't clear just *what* we should think for such a choice, so that's why we don't make that choice. Thus, since only the single impulse at $t = 0$ is in our integration interval, and as $T = 1$,

$$c_k = \frac{1}{T} \int_{-T/2}^{T/2} \delta(t)e^{-jk\omega_0 t}\, dt = \frac{1}{T} = 1.$$

Then, $x(t)$ has the Fourier series (with $\omega_0 = \frac{2\pi}{T} = \frac{2\pi}{1} = 2\pi$)

$$x(t) = \sum_{k=-\infty}^{\infty} e^{jk2\pi t} = \sum_{k=-\infty}^{\infty} \left\{ \cos(k2\pi t) + j\sin(k2\pi t) \right\}.$$

If you write the sums out term by term, then you should be able to see that

$$\sum_{k=-\infty}^{\infty} \sin(k2\pi t) = 0,$$

which is satisfying since we would be surprised to see an imaginary part to a real $x(t)$!, and that

$$\sum_{k=-\infty}^{\infty} \cos(k2\pi t) = 1 + 2\sum_{k=1}^{\infty} \cos(k2\pi t).$$

These results follow directly from the oddness and the evenness of the sine and the cosine, respectively. Thus, the formal mathematics of impulses is telling us that

$$x(t) = \sum_{n=-\infty}^{\infty} \delta(t - n) = 1 + 2\sum_{k=1}^{\infty} \cos(k2\pi t).$$

Is this true?

The right-hand side of this claim is easy to generate on a computer for a finite number of terms, and Figures 14.4, 14.5, and 14.6 show the re-

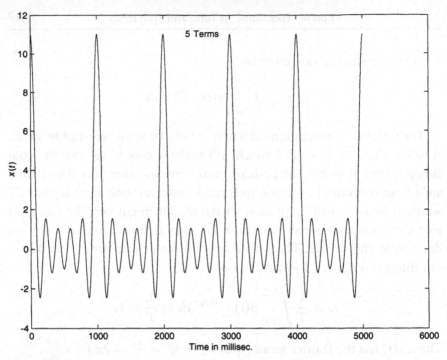

Figure 14.4   Fourier series (truncated) approximation of impulse sequence.

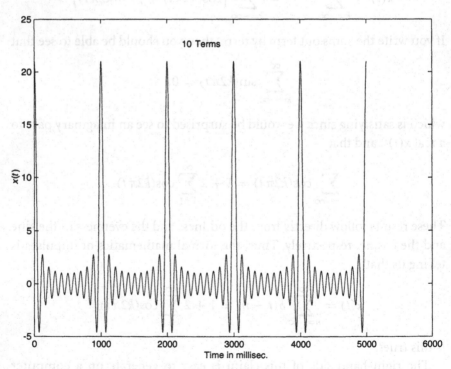

Figure 14.5   Fourier series (truncated) approximation of impulse sequence.

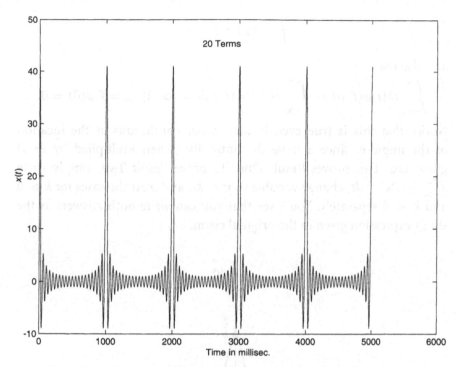

Figure 14.6   Fourier series (truncated) approximation of impulse sequence.

sults when we use the first five, ten, and twenty terms of the sum. And, by gosh, these plots *do* look like periodic impulses are indeed building up, with the harmonically related cosine waves constructively adding at $t = \ldots, 0, 1, 2, \ldots$, but destructively interfering at all other times. These plots are very suggestive that, as we add in more and more terms, we will see the "impulse building" effect become even more pronounced. This isn't a proof, of course, but it is quite compelling (as well as amazing, I think), and it gives reason to believe there *is* a method behind the apparent madness of my symbol pushing.

Continuing with this formal manipulation of symbols, let me next demonstrate two very important results that we will use in later work.

Result One: $t\delta(t) = 0$.

Result Two: $\delta(kt) = (1/|k|))\delta(t)$, for k any real nonzero constant.

To understand what these statements mean, remember that impulses really only have operational meaning when inside an integral (a point particularly emphasized by Dirac in his uses of impulses in physics). Thus, to say $t\delta(t) = 0$ means that, placed *inside* an integral, $t\delta(t)$ and 0 produce the same result. So, for some arbitrary $\varphi(t)$, we see first that, trivially,

$$\int_{-\infty}^{\infty} 0\,\varphi(t)\,dt = 0,$$

and also that

$$\int_{-\infty}^{\infty} t\delta(t)\varphi(t)\,dt = \int_{-\infty}^{\infty} \delta(t)\,\{t\varphi(t)\}\,dt = t\varphi(t)|_{t=0} = 0\,\varphi(0) = 0.$$

Notice that this is true even if $\varphi(t)$ is *not* continuous at the location of the impulse, since a *finite* discontinuity (when multiplied by zero) gives zero. This proves Result One. To prove Result Two, simply write $\int_{-\infty}^{\infty} \delta(kt)\varphi(t)\,dt$, change variable to $u = kt$, and treat the cases for $k > 0$ and $k < 0$ separately. You'll see that you can write both answers as the single expression given in the original claim.

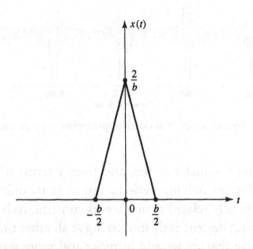

Figure 14.7   A triangular approximation to the unit impulse.

Another way to "make" an impulse out of a perfectly respectable time function is shown in Figure 14.7. Unlike the signal in Figure 14.1, I'm now using a triangular shape (that still bounds unit area). Again, as $b \to 0$ we see $x(t)$ collapsing to a signal with vanishing duration and enormous amplitude—and yet always bounding unit area. In other words, as $b \to 0$ we have $x(t) \to \delta(t)$.

Why is the new signal shape of interest? Because we can *differentiate* it. We can't differentiate the $x(t)$ in Figure 14.1 without getting what appear to be two impulses, at $x = \pm\frac{1}{2}b$, which is not attractive since what we are trying to accomplish with all this is an *explanation* of impulses, i.e., differentiating Figure 14.1 leads us down a circular path.

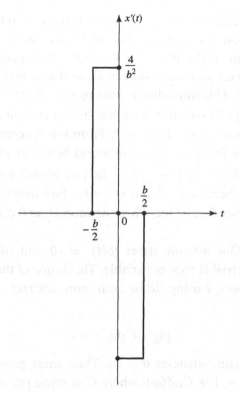

Figure 14.8 The derivative of the triangular approximation.

The derivative of the $x(t)$ in Figure 14.7, however, is the always well-behaved signal of Figure 14.8.

We see from Figure 14.8 that $x'(t)$ approaches two vanishingly brief spikes of opposite polarity and unbounded amplitude as $b \to 0$. This is the derivative of the impulse, i.e., $\lim_{b \to 0} x'(t) = \delta(t)$—see also Problem 14.2. Unlike the impulse, itself, however, this so-called *unit doublet* function has a total net area of zero, while the area under each spike is *unbounded* (the magnitude of the area under each spike is $\frac{2}{b}$). Somewhat poetically we might say that each spike of the doublet is individually a mighty signal and yet together they are but a wisp of vacuum!

Now, let's be less poetic and more businesslike, and ask how $\delta'(t)$ behaves under the integral sign. If $\varphi(t)$ is a *differentiable* function (more restrictive than demanding mere continuity) then this is easy to answer. Simply integrate by parts and find that

$$\int_{-\infty}^{\infty} \delta'(t)\varphi(t)\, dt = \varphi(t)\delta(t)\Big|_{-\infty}^{\infty} - \int_{-\infty}^{\infty} \delta(t)\frac{d\varphi}{dt}\, dt = -\varphi'(0).$$

And now, finally, we are at last in a position to return to the calculation from near the end of Chapter 12: the determination of the Fourier transform of the unit step function, $u(t)$. Recall the result of Problem 12.2, where you were asked to show that if $v(t) \leftrightarrow V(j\omega)$ then $\frac{dv}{dt} \leftrightarrow j\omega V(j\omega)$. This immediately tells us that if $u(t) \leftrightarrow U(j\omega)$ then $\frac{du}{dt} = \delta(t) \leftrightarrow j\omega U(j\omega)$. But since we know the Fourier transform of $\delta(t)$ is 1, then we have $j\omega U(j\omega) = 1$. From this it seems we *again* have $U(j\omega) = \frac{1}{j\omega}$, the incorrect result we found before as we let $\sigma \to 0$ in the transform pair $e^{-\sigma t} u(t) \leftrightarrow \frac{1}{(\sigma + j\omega)}$. *But*, we haven't been quite general enough with this, because $j\omega U(j\omega) = 1 + 0$ which may seem silly to write until you remember Result One, where we found that 0 can also be written as $\omega \delta(\omega)$.

Now, Result One actually states $t\delta(t) = 0$, but of course writing $\omega \delta(\omega) = 0$ is a trivial change of variable. This is one of the great strengths of the notation we are using. To be even more abstract (silly?), we could even write

$$\text{“pig” } \delta(\text{“pig”}) = 0$$

in the "pig" domain (whatever that is). Thus, most generally, we should write $j\omega U(j\omega) = 1 + C\omega\delta(\omega)$, where C is some yet to be determined constant. (I have been able to trace this approach back to Dirac, who discusses it in his famous book *The Principles of Quantum Mechanics*, first published in 1930). So, let's write $U(j\omega) = 1/j\omega - jC\delta(\omega)$ and calculate the value of C.

From the inverse Fourier integral itself we have

$$u(t) = \frac{1}{2\pi} \int_{-\infty}^{\infty} U(j\omega) e^{j\omega t} \, d\omega.$$

Thus,

$$u(0) = \frac{1}{2\pi} \int_{-\infty}^{\infty} \left[\frac{1}{j\omega} - jC\delta(\omega) \right] d\omega$$

$$= \frac{1}{j2\pi} \int_{-\infty}^{\infty} \frac{d\omega}{\omega} - j\frac{C}{2\pi} \int_{-\infty}^{\infty} \delta(\omega) \, d\omega.$$

We'll take the first of the last two integrals as zero (because $\frac{1}{\omega}$ is odd), and the second integral is one (by definition!) Thus, $u(0) = -j\frac{C}{2\pi}$ or, $C = j2\pi u(0)$.

So, at last, we can no longer avoid the question of the value of $u(0)$. This might, in fact, seem to be an ambiguous question. After all, all we have said in defining $u(t)$ is that it is zero for $t < 0$, and one for $t > 0$. What could *force* $u(t)$ to be any *particular* value right at $t = 0$? Maybe, in fact, it could be *anything* we want? But *that* can't be right because then we could say $u(0) = 0$ (and so $C = 0$, which would give $U(j\omega) = \frac{1}{j\omega}$, the incorrect result that keeps coming up). As I'll now show, $u(0)$ *cannot* be just anything and still be consistent with Fourier theory.

Starting with a pair $e^{-\sigma t}u(t) \leftrightarrow \frac{1}{\sigma+j\omega}$, we can use the inverse transform integral to write

$$e^{-\sigma t}u(t) = \frac{1}{2\pi} \int_{-\infty}^{\infty} \frac{e^{j\omega t}}{\sigma + j\omega}\, d\omega.$$

Writing $e^{-\sigma t}u(t)$ is simply a mathematically convenient way of forcing the time function to be zero for $t < 0$ and $e^{-\sigma t}$ for $t > 0$. At $t = 0$ it is $u(0)$ (independent of σ), which is of course precisely the value we are trying to determine. Setting $t = 0$, we get

$$u(0) = \frac{1}{2\pi} \int_{-\infty}^{\infty} \frac{d\omega}{\sigma + j\omega} = \frac{1}{2\pi} \int_{-\infty}^{\infty} \frac{\sigma - j\omega}{\sigma^2 + \omega^2}\, d\omega.$$

The imaginary part of the integrand is an odd function, and so its contribution to the integral is zero [which is good, since the mysterious $u(0)$ must at least be real!]. Since the real part of the integrand is even, then we can write

$$u(0) = \frac{1}{\pi} \int_{0}^{\infty} \frac{\sigma}{\sigma^2 + \omega^2}\, d\omega$$

which is easily evaluated to give $u(0) = \frac{1}{2}$. Thus $C = j2\pi u(0) = j\pi$, and so we have the pair

$$u(t) \leftrightarrow U(j\omega) = \frac{1}{j\omega} + \pi\delta(\omega).$$

Notice that $u(t)$ is an infinite energy signal.

The derivation of $U(j\omega)$ in this chapter, following Dirac's approach, is not the way it is done in most electrical engineering texts. I like it, however, because it explicitly shows how the answer depends on the value of $u(0)$. See Appendix I for the usual engineering approach to finding the transform of a step (in that appendix it is a step in the frequency domain, but that is a trivial difference). In contrast to $\delta(t)$, the infinite energy content of $u(t)$ is obvious in the *time* domain from

$$W = \int_{-\infty}^{\infty} u^2(t)\, dt = \int_0^{\infty} dt = \infty.$$

Finally, we can apply the duality theorem to this pair to derive another pair with an exotic nature (and with enormous importance in single-sideband radio theory). Duality tells us that $\pi\delta(t) + \frac{1}{jt} \leftrightarrow 2\pi u(-\omega)$ or, $(\frac{1}{2})\delta(t) - j\frac{1}{(2\pi t)} \leftrightarrow u(-\omega)$. Then, the time/frequency scaling theorem from Chapter 12 [i.e., $v(at) \leftrightarrow (\frac{1}{|a|})V(\frac{j\omega}{a})$], with $a = -1$, says $(\frac{1}{2})\delta(-t) + j\frac{1}{(2\pi t)} \leftrightarrow u(\omega)$ or, as the impulse is even, we have the pair $(\frac{1}{2})\delta(t) + j\frac{1}{(2\pi t)} \leftrightarrow u(\omega)$. Observe carefully that $u(\omega)$ is *not* $U(j\omega)$! $u(\omega)$, a step in the frequency domain, is the transform of the rather complicated time signal on the left-hand side of the last pair (which you'll see again in Chapter 20), while $U(j\omega)$ is the transform of the step in the *time* domain.

With the completion of this chapter you've seen a fair number of Fourier transform calculations and, no doubt, have been struck by the technical ingenuity that even rather benign-looking time functions can require. So, now is the time for me to tell you something wonderful; MATLAB can do *symbolic* manipulations as well as the usual numerical ones and, in particular, it can do Fourier transforms. MATLAB makes it as easy as simply typing the time function to be transformed. For example, in Problem 12.3 you were asked to show that

$$\frac{1}{t^2 + 1} \leftrightarrow \pi e^{-|\omega|}$$

is a transform pair. With >> denoting the MATLAB Command Window prompt, here's how you ask MATLAB to do all the work.

You type the following two lines, where the first one defines w (representing ω) and t as *symbols* (*not* as variable names of specific numerical quantities), and the second line asks for the Fourier transform of the symbolic expression within the quote marks:

```
>> syms t w
>> fourier(sym('1/(t^2 + 1)'))
```

After a few seconds, MATLAB returns its answer as

```
ans =
exp(w)*pi*Heaviside(-w)+exp(-w)*pi*Heaviside(w)
```

where Heaviside(w) $= u(\omega)$ is MATLAB's way of denoting the step function. This looks complicated, but a little thought should convince you that it is a perfectly valid way to write $\pi e^{-|\omega|}$.

If impulses occur in a Fourier transform then MATLAB will include uses of Dirac in its answer, where Dirac $(w) = \delta(\omega)$. For example, consider the problem we solved at the end of this chapter, that of finding the Fourier transform of $u(t)$. Using MATLAB, we type

```
>> syms t w
>> fourier(sym('Heaviside(t)'))
```

and MATLAB replies with

```
ans =
pi*Dirac(w)-i/w
```

which does, indeed, agree with our theoretical result (MATLAB defaults to using $i = \sqrt{-1}$ rather than j).

MATLAB can do inverse Fourier transforms, too, and we can immediately apply the `ifourier` command to the answer we just got as follows:

```
>>ifourier(ans,w,t)
```

which produces

```
ans =
1/2+1/2*Heaviside(t)-1/2*Heaviside(-t)
```

which simply says that

$$\frac{1}{2} + \frac{1}{2}u(t) - \frac{1}{2}u(-t) = u(t).$$

This is clearly true.

As a final example of the inverse transform, consider the step function in the frequency domain. We can get the time function that is the other half of the pair by typing

```
>> syms w t
>> ifourier(sym('Heaviside(w)'),t)
```

which produces

```
ans =
1/2*(pi*Dirac(t)*t+i)/t/pi
```

which is MATLAB's somewhat long winded (but who can really complain on that point!) way of writing the correct result of

$$\frac{1}{2}\delta(t) + i\frac{1}{2\pi t}.$$

Now, here's a little puzzle for you. If we ask MATLAB for the Fourier transform of $tu(t)$, then here's what happens:

```
>> syms t w
>> fourier(sym('t*Heaviside(t)'))

ans =
i*(pi*Dirac(1,w)+i/w^2)
```

What, do you suppose, does MATLAB mean by "Dirac (1,w)?" (See Problem 14.8).

Problems

14.1. Show that $\delta(t)$ is even, i.e., that $\delta(-t) = \delta(t)$. Show also that $\delta(t^2 - a^2) = (1/2|a|)[\delta(t - a) + \delta(t + a)]$ for $a \neq 0$. Hint: write $t^2 - a^2 = (t - a)(t + a)$ and use Result Two.

14.2. If we write $\delta'(t)$ to denote the derivative of an impulse, then show that $t\delta'(t) = -\delta(t)$. Hint: for $\varphi(t)$ an arbitrary *differentiable* function write $\int_{-\infty}^{\infty} t\delta'(t)\varphi(t)\,dt$ and integrate by parts.

14.3. Recall the sgn or sign function defined in Problem 12.3. Derive the transform pair sgn$(t) \leftrightarrow 2/j\omega$. This purely imaginary transform raises no concerns [as did my naive calculation of $U(j\omega)$ in Chapter 12, the transform of the step] because sgn(t) is, indeed, an odd function of time (in the mathematical sense). Hint: write sgn$(t) = 2u(t) - 1$ and use the known transforms for the time functions $u(t)$ and 1.

14.4. Recall the theorem (from Problem 12.2) $\frac{dv}{dt} \leftrightarrow j\omega V(j\omega)$. Using the pair derived in the previous problem, sgn$(t) \leftrightarrow 2/j\omega$, and the observation sgn$(t) = (\frac{d}{dt})|t|$, derive the pair $|t| \leftrightarrow \frac{-2}{\omega^2}$. Now, consider the following and ask yourself if this pair actually makes sense. From the Fourier transform integral we have, at $\omega = 0$, $V(0) = \int_{-\infty}^{\infty} v(t)\,dt = +\infty$ if $v(t) = |t|$. The quantity $-2/\omega^2$ does indeed blow up at $\omega = 0$, *but the sign is wrong* (it blows up to *minus* infinity). So, let's try a different approach to calculating $V(j\omega)$ for $v(t) = |t|$. Recall the result from Problem 13.4,

$$|t| = \frac{1}{\pi} \int_{-\infty}^{\infty} \frac{1 - \cos(\omega t)}{\omega^2}\,d\omega,$$

and substitute it directly into the transform integral. Then,

$$V(j\omega) = \int_{-\infty}^{\infty} \left\{ \frac{1}{\pi} \int_{-\infty}^{\infty} \frac{1 - \cos(ut)}{u^2} \, du \right\} e^{-j\omega t} \, dt$$

where I've changed the dummy variable in the inner integral from ω to u, to avoid confusing it with the *independent* variable ω in the outer integral (I'm about to reverse the order of integration). In fact, *you* do that and write

$$V(j\omega) = \int_{-\infty}^{\infty} \frac{1}{\pi u^2} \left\{ \int_{-\infty}^{\infty} \{1 - \cos(ut)\} e^{-j\omega t} \, dt \right\} du.$$

With what you now know from this chapter you should be able to show that the inner integral is *three impulses*, i.e.,

$$\int_{-\infty}^{\infty} \{1 - \cos(ut)\} e^{-j\omega t} \, dt = 2\pi \delta(\omega) - \pi \delta(\omega + u) - \pi \delta(\omega - u).$$

Inserting this into the double integral for $V(j\omega)$ immediately gives

$$V(j\omega) = -\frac{2}{\omega^2} + 2\delta(\omega) \int_{-\infty}^{\infty} \frac{du}{u^2}.$$

So, it appears that the result $-2/\omega^2$ is okay, as long as $\omega \neq 0$, i.e.,

$$|t| \leftrightarrow -\frac{2}{\omega^2}, \quad \omega \neq 0.$$

Right *at* $\omega = 0$, however, the transform has an impulse with *infinite positive* strength (the integral $\int_{-\infty}^{\infty} du/u^2$ clearly diverges). Finally, notice that while $|t|$ has infinite total energy *and* is unbounded, there is just a *finite* amount of energy in any frequency interval (even of infinite width) that does not include $\omega = 0$. Indeed, use Rayleigh's energy theorem to show that the energy of $|t|$ in the broken interval $|\omega| > \omega_1$ is $\frac{4}{(3\pi \omega_1^3)}$, which is finite no matter how small ω_1 may be. The infinite energy of $|t|$ is, therefore, packed into the infinitesimally tiny frequency interval around $\omega = 0$ (contrast this to the infinite energy of $\delta(t)$ which is smeared out evenly over all frequencies).

14.5. If you did Problem 13.5 then you found the Fourier transform pair $e^{-at^2} \leftrightarrow \sqrt{\pi/a} e^{-\omega^2/4a}$, $a > 0$. Notice, in particular, the reciprocal spreading effect. That is, as $a \to 0$ the time signal spreads ever wider, approaching a dc value of one for all t, while the transform collapses into a spike centered on $\omega = 0$ with an ever increasing amplitude. Now, we know that the transform of a dc time signal is an impulse in frequency, i.e., recall the pair $1 \leftrightarrow 2\pi\delta(\omega)$ derived in the text. Letting $a \to 0$ in the Gaussian pulse gives $1 \leftrightarrow \lim_{a\to 0}\sqrt{\pi/a}e^{-\omega^2/4a}$, which implies $\delta(\omega) = \lim_{a\to 0}(\frac{1}{2\pi})\sqrt{\pi/a}e^{-\omega^2/4a}$. If this is in fact so, then since $\int_{-\infty}^{\infty}\delta(\omega)\,d\omega = 1$ we must have $\int_{-\infty}^{\infty}\lim_{a\to 0}\frac{1}{(2\sqrt{\pi a})}e^{-\omega^2/4a}\,d\omega = 1$. Show that for *any* value of $a > 0$ the value of this last integral is indeed one. This result gives us an everywhere continuous approximation to the impulse (an approximation so smooth that it can be endlessly differentiated everywhere), as opposed to the discontinuous pulse approximation I used in the text.

14.6. One of the strangest of the Dirac impulse identities (discussed in his 1927 paper), one that almost always leaves students with a stunned, glazed stare, is

$$\int_{\infty}^{\infty}\delta(u - x)\delta(x - v)\,dx = \delta(u - v).$$

When asked to prove it, most of the students I have had over the years reply, in effect: "*Prove* it? I don't even know what it is suppose to mean!" Here's how to do it. Write $\varphi(u)$ as a testing function. Then, first evaluate $\int_{-\infty}^{\infty}\varphi(u)\left\{\int_{-\infty}^{\infty}\delta(u - x)\delta(x - v)\,dx\right\}du$ and show that it is equal to $\varphi(v)$. (Hint: reverse the order of integration.) Then, evaluate $\int_{-\infty}^{\infty}\varphi(u)\delta(u - v)\,du$ and show that this, too, is $\varphi(v)$. Hence, since they behave in the same way *under an integral sign* when multiplied with a testing function, then we say $\int_{-\infty}^{\infty}\delta(u - x)\delta(x - v)\,dx = \delta(u - v)$. [You could start with $\varphi(v)$ as the testing function just as well.] Is it all clear now? If it makes you feel any better, even though I can push the symbols around and get the formal answer it is *still* mysterious to me, too. As the historian of mathematics Jesper Lützen has put it in a masterful understatement, "All of this shows Dirac as a skillful manipulator of the δ-function." Indeed.

14.7. Dirac invented the new symbol $\delta(t)$ to represent a spike of infinite amplitude and zero duration. The question here is, is it possible to represent a time signal that is a spike of zero duration and *finite* amplitude without having to invent any new symbols? The answer is yes, and as an example the following integral is a zero-duration spike of amplitude 2 located at $t = 1$:

$$f(t) = \int_{-\pi}^{\pi}\frac{\{t + \cos(\alpha)\}\{1 + t\cos(\alpha)\}}{\{1 + t^2 + 2t\cos(\alpha)\}^{3/2}}\,d\alpha.$$

```
%fig14p9.m created by PJNahin for "Science of Radio" 6/17/98
alpha=linspace(-pi,pi,500);
trig=cos(alpha);
for index=1:199
    t=index/100;
    if t==1
        integrand=0.5*sqrt((1+trig)/2);
    else
        numerator=(t+trig).*(1+t*trig);
        denominator=(1+t*t+2*t*trig).^1.5;
        integrand=numerator./denominator;
    end
    area=trapz(alpha,integrand);
    result(index)=area;
end
time=10:10:1990;
plot(time,result)
axis([0 2000 -1 3])
xlabel('time (milliseconds)')
ylabel('f(t)')
title('FIGURE 14.9. A Finite Amplitude, Almost-Zero-Duration Spike')
figure(1)
```

You are to fill in the details for a proof of this, following the outline below, but before doing that take a look at the MATLAB program fig14p9.m. It plots $f(t)$ over the interval $0 < t < 2$ in steps of $\Delta t = 0.01$. This plot, in Figure 14.9, should be enough, I think, to convince you the claim has at least a chance of being correct. Question: do you see the reason for the interior "if-else" check for $t = 1$? It's there because even though the entire integrand is well-behaved for all t, the denominator is zero at both of the integration limits when $t = 1$. Because of a cancellation effect with the numerator this actually is not a theoretical problem, but MATLAB doesn't know that and a *practical* problem develops (try running the program without the check and observe what MATLAB does).

a. Use the identity $t = t[\sin^2(\alpha) + \cos^2(\alpha)]$ to show that

$$f(t) = \int_{-\pi}^{\pi} \left\{ \frac{t \sin^2(\alpha)}{[1 + t^2 + 2t\cos(\alpha)]^{3/2}} + \frac{\cos(\alpha)}{[1 + t^2 + 2t\cos(\alpha)]^{1/2}} \right\} d\alpha.$$

b. Define the function

$$g(t, \alpha) = \frac{\sin(\alpha)}{[1 + t^2 + 2t\cos(\alpha)]^{1/2}}$$

and show that

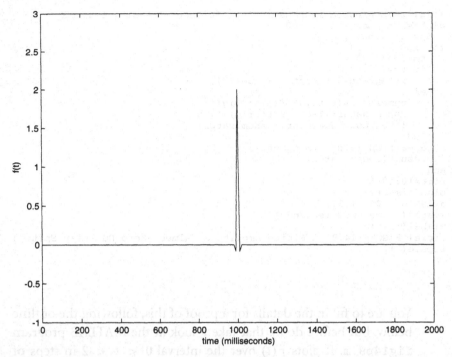

Figure 14.9   A finite amplitude, almost-zero-duration spike.

$$f(t) = \int_{-\pi}^{\pi} \frac{\partial g}{\partial \alpha}\, d\alpha = g(t, \pi) - g(t, -\pi).$$

c. If $t \neq 1$, then show that $g(t, \pi) = g(t, -\pi) = 0$ and so $f(t) = 0$ for $t \neq 1$.

d. Set $t = 1$ in $[1 + t^2 + 2t \cos(a)]^{1/2}$, show it reduces to $2 \cos\left(\frac{1}{2}\alpha\right)$, and so $g(1, \alpha) = \sin\left(\frac{1}{2}\alpha\right)$ if $\alpha \neq \pm\pi$. Use this to conclude that $\lim_{\alpha \to \pm\pi} g(1, a) = \pm 1$.

e. Use the results of parts b and d to show that $f(1) = g(1, \pi) - g(1, -\pi) = 1 - (-1) = 2$.

14.8. What is the Fourier transform of $tu(t)$? Recall that at the end of this chapter we found that MATLAB says the answer is

```
i*(pi*Dirac(1,w) + i/w^2) = iπDirac(1,w) - 1/w².
```

$$\text{i*(pi*Dirac(1,w) + i/w\^2)} = i\pi\text{Dirac(1,w)} - \frac{1}{w^2}.$$

To see what Dirac(1,w) means, recall the general transform formula $t\upsilon(t) \leftrightarrow j\frac{dV}{d\omega}$, and set $\upsilon(t) = u(t)$ and so $V(j\omega) = \pi\delta(\omega) - i\frac{1}{\omega}$. Can you now see what Dirac(2,w) means? Hint: write $t^2 u(t) = t[tu(t)]$, set $\upsilon(t) = tu(t)$, and use the general transform formula again.

Let me conclude this somewhat wild, symbol-pushing chapter with a little philosophical preaching. Rigorous mathematicians are, of course, appalled at the sort of devil-may-care manipulations I have taken you through in this chapter. And, of course, they are intellectually correct in demanding caution. Mathematicians hate "hand-waving" arguments, even ones that are "convincing," but when asked how to "do it right," they all too often drag out so much mathematical artillery most electrical engineers and physicists are soon sorry that they asked. My personal position on this is that electrical engineers and physicists shouldn't be afraid to plunge in *ahead* of the rigor. If you make a blunder, you will soon know—perhaps a circuit you've designed will melt! Much more likely, however, is that you will simply end up with an obviously unphysical result. That's just the math telling you that you twisted the rubber band too tight, and somewhere in your analysis something "snapped." That's the proper time to go back and be rigorous. When Dirac was once told by a former student that he (the student) was uneasy with some of his own nonrigorous proofs, his mentor replied "I am not interested in proofs, but only in what nature does." Dirac knew that if the proofs are wrong, nature will let you know. So, for a first attempt at analysis, dare to be bold.

Convolution Theorems, Frequency Shifts, and Causal Time Signals

Linear systems are very important in electrical engineering, but radio would simply not be possible if they were all there is. As discussed in Chapter 6, AM radio depends on the ability to shift baseband frequencies (dc to several kilohertz for human speech) up to radio frequencies (i.e., rf) on the order of half a megahertz and more. A baseband spectrum *by definition* is zero for all frequencies above some maximum frequency (and this maximum frequency is certainly *far* below rf). More generally, a signal with a transform that is nonzero only over a finite interval or band of frequencies is said to be *bandlimited*. A baseband signal $x(t)$, with $|X(j\omega)| = 0$ for $|\omega| > \omega_0$, as shown in Figure 15.1, is then a special case of bandlimited signal. Since this is a generic figure, I have used a triangle as metaphor for spectrum, i.e., the actual shape of $|X(j\omega)|$ depends on the specific details of $x(t)$. However, since $x(t)$ is taken to be real, we at least know that $|X(j\omega)|$ is even and that is how I have drawn it.

As shown at the end of Chapter 13, the relationship between the input and the output spectra of a linear system, $X(j\omega)$ and $Y(j\omega)$, respectively, is $Y(j\omega) = H(j\omega)X(j\omega)$, where $H(j\omega)$ is the transfer function of the system. Therefore, because of the

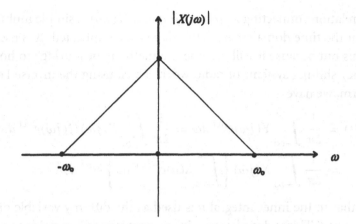

Figure 15.1   The generic spectrum for a baseband signal.

bandlimited nature of $X(j\omega)$, the output spectrum must satisfy the same constraint, $|Y(j\omega)| = 0$ for $|\omega| > \omega_0$ no matter what $H(j\omega)$ may be. If we want to shift baseband energy up to rf, then the conclusion is that a linear system just can't do it.

Linear systems can still be useful to us, however, as a prelude to studying how certain nonlinear systems can perform the necessary frequency shift. We begin with an elementary question. We know $X(j\omega)$ is the transform of the input signal $x(t)$, and that $Y(j\omega)$ is the transform of the output signal $y(t)$, but what time signal is $H(j\omega)$ associated with? We can write it as $h(t)$, of course, but what *is* $h(t)$? This is actually quite easy to answer once you recall the pair $\delta(t) \leftrightarrow 1$. Thus, if $x(t) = \delta(t)$, then $Y(j\omega) = H(j\omega)$ since $X(j\omega) = 1$, and so $y(t) = h(t)$. That is, $h(t)$ is the system output if the input is an impulse [and so $h(t)$ is, logically enough, called the *impulse response*]. This is the time domain interpretation of the transfer function, a concept originally defined in the frequency domain using impedance and voltage divider ideas (see Appendix C).

If you want to experimentally measure $h(t)$ for an actual system, matters aren't so straightforward as simply applying $\delta(t)$ to the input and observing the response. It simply isn't possible to actually generate $\delta(t)$—remember it has infinite energy—and even if you could it might damage the system. In a mechanical system, for example, an approximation to $\delta(t)$ would be the brief application of a huge force, e.g., a massive bash with a hammer. See Problem 15.1 for how one can realistically measure the $h(t)$ of a real system.

The relation connecting $X(j\omega)$, $Y(j\omega)$ and $H(j\omega)$ is simple multiplication, but the time domain connection is more complicated. It is useful to work this out because it will provide the mathematical bridge to how the frequency shifting systems of radio work. Thus, using the inverse Fourier transform, we have

$$y(t) = \frac{1}{2\pi} \int_{-\infty}^{\infty} Y(j\omega)e^{j\omega t}\, d\omega = \frac{1}{2\pi} \int_{-\infty}^{\infty} X(j\omega)H(j\omega)e^{j\omega t}\, d\omega$$

$$= \frac{1}{2\pi} \int_{-\infty}^{\infty} X(j\omega) \left\{ \int_{-\infty}^{\infty} h(u)e^{-j\omega u}\, du \right\} e^{j\omega t}\, d\omega.$$

Notice that in the inner integral u is used as the dummy variable of integration, not t. This is done because in the next step we are going to reverse the order of the two integrations, and we don't want to confuse the dummy variable in the inner integral with the independent variable t in the outer integral. So, continuing,

$$y(t) = \frac{1}{2\pi} \int_{-\infty}^{\infty} h(u) \left\{ \int_{-\infty}^{\infty} X(j\omega)e^{j\omega(t-u)}\, d\omega \right\} du$$

$$= \int_{-\infty}^{\infty} h(u) \left\{ \frac{1}{2\pi} \int_{-\infty}^{\infty} X(j\omega)e^{j\omega(t-u)}\, d\omega \right\} du$$

or, recognizing the inner integral as the inverse Fourier transform of $x(t - u)$,

$$y(t) = \int_{-\infty}^{\infty} h(u)x(t - u)\, du \overset{\Delta}{=} h(t)^*x(t).$$

That is, the $*$ is *defined* to be the symbol we'll use to represent the above integral operation called *time convolution*. (My other use of $*$ to denote the conjugate of a complex quantity will always be clear by context). As Problem 15.2 asks you to show, convolution is *associative*, i.e., $y(t) = h(t)^*x(t) = x(t)^*h(t)$, and so we can also write

$$y(t) = \int_{-\infty}^{\infty} x(u)h(t - u)\, du.$$

In any case, we have the pair $h(t)^*x(t) \leftrightarrow H(j\omega)X(j\omega)$. And since $y(t) = h(t)$ when $x(t) = \delta(t)$, we have the useful relation $h(t)^*\delta(t) = h(t)$. We can generalize this for linear systems which also happen to be time invariant (which means a time shift in the input results in an equal time shift in the output—see Appendix B). Then, the input $\delta(t - t_0)$ must result in the output $h(t - t_0)$ or, $h(t)^*\delta(t - t_0) = h(t - t_0)$.

Direct evaluation of a convolution integral in the time domain is generally a complicated business, and the following example (and the next shaded box) are the only direct evaluations I'll do in the main text of this book (see Appendix I for the one other time it's done in this book). It is almost always easier, you see, to calculate $Y(j\omega)$ from $H(j\omega)X(j\omega)$ and then to find $y(t)$ by applying the inverse transform to $Y(j\omega)$.

For an example of the inner workings of the time-convolution integral, recall the unit gate function mentioned in Chapter 12, i.e.,

$$\pi(t) = \begin{cases} 1 & \text{for } |t| < 1/2 \\ 0 & \text{otherwise.} \end{cases}$$

I will now calculate the convolution of $\pi(t)$ with itself, which we write as

$$\pi(t)^*\pi(t) = \int_{-\infty}^{\infty} \pi(u)\pi(t-u)\,du.$$

$\pi(-u)$ is $\pi(u)$ reflected through (or folded around) the vertical axis, and $\pi(t-u)$ is then $\pi(-u)$ shifted to the left by t. The first two sketches in Figure 15.2 show these two functions (because of its inherent symmetry $\pi(-u) = \pi(u)$, but this evenness is a peculiarity of the special problem here). Now, to find the convolution integral for any particular value of t we simply multiply the two plots together and compute the area bounded by the result. (I have, to be honest, picked $\pi(t)^*\pi(t)$ as my example of self-convolution simply because this process is then easy to do!) When t is very negative, there is no overlap of $\pi(u)$ with $\pi(t-u)$ and so their product is zero everywhere. As t increases, however, there comes a time when the two plots do begin to overlap; this occurs when $\frac{1}{2} + t = -\frac{1}{2}$ (when $t = -1$). As t increases from $t = -1$ the overlap increases and the area bounded by the product curve increases *linearly* until, at $t = 0$, there is perfect alignment of $\pi(u)$ and $\pi(-u)$. Thus, at $t = 0$ the integral is maximum and has value equal to 1. Then, as t increases beyond $t = 0$, the overlap linearly *decreases* and so the area bounded by the product curve decreases. The overlap reaches zero when $-\frac{1}{2} + t = \frac{1}{2}$ (when $t = 1$). The final result for $\pi(t)^*\pi(t)$ is thus the triangle in the bottom sketch of Figure 15.2.

We can use the convolution integral as a new way to determine $u(0)$, the calculation done in the previous chapter with a Fourier argument. Let's define the two time signals $x(t) = u(-t)$ and $h(t) = e^{-t}u(t)$, i.e., x and h are functions that turn *off* and *on*, respectively, at $t = 0$. Then, using s as the dummy variable of integration (to avoid confusion with

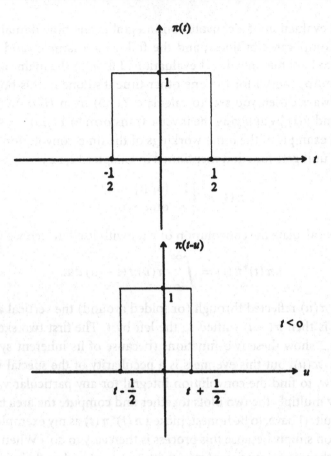

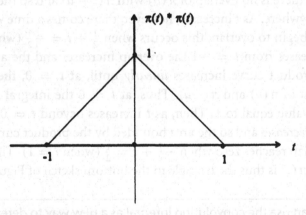

Figure 15.2   The unit gate function (top), its shifted reflection (middle), and the convolution of the unit gate with itself (bottom).

the earlier part of this chapter, as u now denotes the unit step function), we have

$$y(t) = x(t)*h(t) = \int_{-\infty}^{\infty} x(s)h(t-s)\,ds = \int_{-\infty}^{\infty} u(-s)e^{-(t-s)}u(t-s)\,ds$$

$$= e^{-t}\int_{-\infty}^{\infty} e^{s}u(-s)u(t-s)\,ds.$$

Figure 15.3 shows sketches of $u(-s)$, and of $u(t-s)$ for the two cases of $t < 0$ and $t > 0$. From these sketches it is immediately obvious that

$$u(-s)u(t-s) = u(-s) \text{ if } t > 0$$
$$u(-s)u(t-s) = u(t-s) \text{ if } t < 0.$$

So, we can now write

$$y(t) = e^{-t}\int_{-\infty}^{\infty} e^{s}u(-s)\,ds = e^{-t}\int_{-\infty}^{0} e^{s}\,ds = e^{-t}\left(e^{s}|_{-\infty}^{0} = e^{-t},\quad t > 0\right.$$

And,

$$y(t) = e^{-t}\int_{-\infty}^{\infty} e^{s}u(t-s)\,ds = e^{-t}\int_{-\infty}^{t} e^{s}\,ds$$
$$= e^{-t}\left(e^{s}|_{-\infty}^{t} = e^{-t}(e^{t}) = 1,\quad t < 0.\right.$$

We can write these two partial results in a single expression as

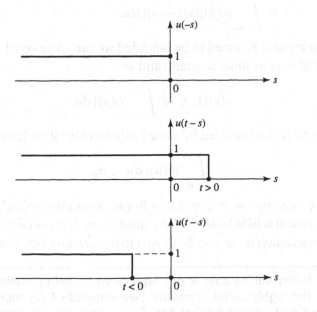

Figure 15.3   Shifted step functions in the time domain.

$$y(t) = u(-t) + e^{-t}u(t).$$

For $t = 0$, in particular, we have $y(0) = 2u(0)$.

   Now, let's go back to the start of this analysis and calculate $y(0)$, directly. That is, let's return to

$$y(t) = e^{-t} \int_{-\infty}^{\infty} e^s u(-s)u(t-s)\, ds$$

and set $t = 0$ to arrive at

$$y(0) = \int_{-\infty}^{\infty} e^s u(-s)u(-s)\, ds = \int_{-\infty}^{0} e^s\, ds = \left(e^s |_{-\infty}^{0} = 1. \right.$$

Thus, $2u(0) = 1$ and so, again, we have $u(0) = \frac{1}{2}$.

The time convolution integral allows us a very nice way of specifying what is called the *stability* of a linear system. A system is said to be stable if, for any bounded input (the input never becomes infinite), the output is also bounded. This is sometimes called BIBO stability ("bounded input, bounded output"). Thus, remembering the area interpretation of the integral, we have

$$|y(t)| = \left| \int_{-\infty}^{\infty} h(u)x(t-u)\, du \right| \le \int_{-\infty}^{\infty} |h(u)x(t-u)|\, du$$
$$= \int_{-\infty}^{\infty} |h(u)||x(t-u)|\, du.$$

Since the input is assumed to be bounded we can write $|x(t)| < M$ for all t, where M is some *finite* constant, and so

$$|y(t)| \le M \int_{-\infty}^{\infty} |h(u)|\, du.$$

Thus, $|y(t)|$ is also bounded by some finite constant if we have

$$\int_{-\infty}^{\infty} |h(u)|\, du < \infty.$$

That is, *if* the impulse response of a linear system is *absolutely integrable* then the system is BIBO stable. This condition on $h(t)$ is said to be *sufficient* to ensure stability; it can also be shown to be *necessary* (see Problem 15.3).

Not being BIBO-stable is *not* equivalent to being useless. For example, the highly useful *integrator* (see Appendix F for more on integration circuits) is not BIBO-stable. This is physically obvious since the output, $y(t)$, is given by

$$y(t) = \int_0^t x(s)\, ds$$

with $x(t)$ as the input, and if $x(t)$ is any nonzero constant (which is certainly bounded) then $y(t)$ increases without bound towards either $\pm\infty$ (depending on the sign of $x(t)$). Mathematically, we have the impulse response of the integrator as $y(t) = h(t)$ when $x(t) = \delta(t)$, and so

$$h(t) = \int_0^t \delta(s)\, ds = u(t).$$

But the step function is not absolutely integrable, since $\int_0^t u(s)\, ds = t$, which diverges as $t \to \infty$.

To move beyond linear systems, let's take a hint from the result in Chapter 6 that showed how to shift a low-frequency tone signal up to a higher so-called *carrier* frequency simply by multiplying the tone signal by a sinusoid at carrier frequency. It seems likely, then, that we could learn a lot by applying the Fourier transform to *multiplicative* systems. For example, if we have a message signal at baseband, $m(t)$, we can ask what is the spectrum of $m(t)e^{j\omega_c t}$? The answer is

$$\int_{-\infty}^{\infty} m(t)e^{j\omega_c t} e^{-j\omega t}\, dt = \int_{-\infty}^{\infty} m(t)e^{-j(\omega-\omega_c)t}\, dt = M\{j(\omega - \omega_c)\}.$$

This is a shift of $M(j\omega)$ *up* in frequency by ω_c. In an actual system, of course, we can't simply multiply by just $e^{j\omega_c t}$ (how do you generate a *complex* signal?!), but we *can* multiply by $\cos(\omega_c t)$. From Euler's identity we have the result of this multiplication as $(\frac{1}{2})m(t)e^{j\omega_c t} + (\frac{1}{2})m(t)e^{-j\omega_c t}$ which has the spectrum $(\frac{1}{2})M\{j(\omega - \omega c)\} + (\frac{1}{2})M\{j(\omega + \omega_c)\}$. This is the baseband spectrum of $m(t)$ shifted in both spectral directions, as shown in Figure 15.4. This result is often called the *modulation* or *heterodyne theorem* (recall Fessenden's circuit from Chapter 7). There are two points to notice about this figure. First, since $M(j\omega)$ is generally complex, I have sketched only magnitudes (as in Figure 15.1). Second, I have assumed $\omega_c > \omega_0$, which prevents overlap of the up-shifted spectrum of $m(t)$ (the $M\{j(\omega - \omega_c)\}$ term) with the down-shifted spectrum (toward lower frequencies) of $m(t)$. (For AM radio this condition is easily satisfied, as discussed in Chapter 6. Failure to prevent spectral overlap leads to what is called spectrum *aliasing*, an important concern that is of great interest in more advanced discussions I'll take up later in Chapter 18.)

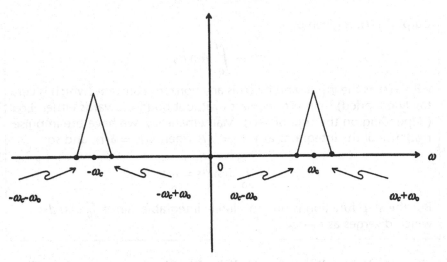

Figure 15.4   The spectrum of a heterodyned baseband signal.

As a special case of the up-down spectral shift, let's continue with the elementary message signal consisting of a single tone, which I wrote before as $m(t) = A_m \cos(\omega_m t)$. The spectrum of this $m(t)$ is

$$M(j\omega) = \int_{-\infty}^{\infty} m(t)e^{-j\omega t}\, dt = A_m \int_{-\infty}^{\infty} \cos(\omega_m t)e^{-j\omega t}\, dt$$

$$= A_m \left[\frac{1}{2}\int_{-\infty}^{\infty} e^{-j(\omega-\omega_m)t}\, dt + \frac{1}{2}\int_{-\infty}^{\infty} e^{-j(\omega+\omega_m)t}\, dt \right].$$

But, as shown in Chapter 14,

$$\delta(\omega) = \frac{1}{2\pi}\int_{-\infty}^{\infty} e^{j\omega t}\, dt$$

and so

$$\int_{-\infty}^{\infty} e^{-j(\omega-\omega_m)t}\, dt = 2\pi\delta\{-(\omega-\omega_m)\},$$

$$\int_{-\infty}^{\infty} e^{-j(\omega+\omega_m)t}\, dt = 2\pi\delta\{-(\omega+\omega_m)\}.$$

Since the impulse function is even (recall Problem 14.1), we can then write the transform pair $\cos(\omega_m t) \leftrightarrow \pi\delta(\omega+\omega_m) + \pi\delta(\omega-\omega_m)$. That is, the spectrum of $\cos(\omega_m t)$ consists of just two impulses, at $\omega = \pm\omega_m$ (recall the related example worked out in Chapter 12 that approximated these impulses). Thus,

$$M(j\omega) = A_m\pi[\delta(\omega+\omega_m) + \delta(\omega-\omega_m)]$$

and so the *shifted* spectrum has *four* impulses:

$$\frac{A_m \pi}{2} \underbrace{[\delta(\omega - \omega_c + \omega_m) + \delta(\omega - \omega_c - \omega_m)]}_{\text{two impulses at } \omega = \omega_c \pm \omega_m}$$

$$+ \frac{A_m \pi}{2} \underbrace{[\delta(\omega + \omega_c + \omega_m) + \delta(\omega + \omega_c - \omega_m)]}_{\text{two impulses at } \omega = -\omega_c \pm \omega_m}.$$

The multiplication of $m(t)$ by $\cos(\omega_c t)$ has accomplished our desired shift of the baseband spectrum of $m(t)$ up to rf (if ω_c is at rf). So, all we need to do next is discover how to multiply two time signals together at the transmitter. And at the receiver we have to discover how to take the intercepted signal *at rf* and shift its spectrum back down to baseband so human ears can hear it. As you'll soon see, the spectral downshift at the receiver is accomplished by *another* multiplication, and so learning how to do electronic multiplication is absolutely crucial to AM radio at both ends of the communication path. And that is what we'll do in the next chapter.

It will be enormously useful to generalize our previous results on multiplying two time signals together. Instead of multiplying $m(t)$ by $\cos(\omega_c t)$ as we did before, let's multiply $m(t)$ by *any* time function. That is, suppose we have two time signals, $m(t)$ and $g(t)$ with spectrums $M(j\omega)$ and $G(j\omega)$. To find the spectrum of the product $m(t)g(t)$, we evaluate

$$\int_{-\infty}^{\infty} m(t)g(t)e^{-j\omega t}\,dt = \int_{-\infty}^{\infty} m(t) \left\{ \frac{1}{2\pi} \int_{-\infty}^{\infty} G(ju)e^{jut}\,du \right\} e^{-j\omega t}\,dt,$$

where $g(t)$ has been replaced with its equivalent inverse transform. Because I am going to reverse the order of integration in the next step, the dummy variable of integration in the inner integral has been written as u, not ω, to avoid confusion with the ω in the outer integral which is *not* the dummy variable there—t is. Thus, doing the reversal, we continue by writing the spectrum of $m(t)g(t)$ as

$$\int_{-\infty}^{\infty} \frac{1}{2\pi} G(j\omega) \left\{ \int_{-\infty}^{\infty} m(t)e^{-j(\omega-u)t}\,dt \right\} du$$

$$= \frac{1}{2\pi} \int_{-\infty}^{\infty} G(ju)M\{j(\omega - u)\}\,du.$$

Comparing the form of the last integral with that of the time convolution integral, we see that it is simply a *frequency* convolution integral, i.e., we have the transform pair

$$m(t)g(t) \leftrightarrow \frac{1}{2\pi}G(j\omega)^*M(j\omega),$$

which is nicely symmetric with our earlier time convolution result, i.e., with the pair

$$m(t)^*g(t) \leftrightarrow M(j\omega)G(j\omega).$$

As yet another example of the use of time convolution, recall the earlier calculation of $\pi(t)^*\pi(t)$. There we found this is equal to the triangular signal shown in the bottom sketch of Figure 15.2. Since we have the pair (from Chapter 12) $\pi(t) \leftrightarrow \prod(j\omega) = \sin(\omega/2)/(\omega/2)$, then we immediately have the transform of $\pi(t)^*\pi(t)$ as $\prod^2(j\omega) = \sin^2(\omega/2)/(\omega/2)^2$.

The frequency convolution pair contains Rayleigh's energy theorem as a special case, which is an interesting connection to make as it suggests all of our aggressive engineers' mathematics may actually be self-consistent. Thus, writing the frequency convolution pair out in detail,

$$\int_{-\infty}^{\infty} m(t)g(t)e^{-j\omega t}dt = \frac{1}{2\pi}\int_{-\infty}^{\infty}G(j(\omega-u))M(ju)\,du.$$

In this statement ω is the independent variable, and so the statement is true for any value of ω. In particular, for $\omega = 0$ we have

$$\int_{-\infty}^{\infty} m(t)g(t)\,dt = \frac{1}{2\pi}\int_{-\infty}^{\infty}G(-ju)M(ju)\,du$$

or, because $G(-j\omega) = G^*(j\omega)$, then we have (notice that I've changed the dummy variable in the right integral from u to ω, just to match the use of t as the dummy variable in the left integral)

$$\int_{-\infty}^{\infty} m(t)g(t)\,dt = \frac{1}{2\pi}\int_{-\infty}^{\infty}M(j\omega)G^*(j\omega)\,d\omega.$$

Now, suppose $m(t) = g(t)$. Then $M(j\omega) = G(j\omega)$ and so

$$\int_{-\infty}^{\infty} m^2(t)\,dt = \frac{1}{2\pi}\int_{-\infty}^{\infty}M(j\omega)M^*(j\omega)\,d\omega = \frac{1}{2\pi}\int_{-\infty}^{\infty}|M(j\omega)|^2\,d\omega,$$

which is just Rayleigh's energy theorem.

There is one last topic we should consider, concerning systems in general, about which Fourier analysis can tell us much. This is the *engineering* question of the possibility of constructing the circuits we study, a topic that all electrical engineers and physicists should find of more than mere academic interest. What we demand of any system we wish to construct is that it be *causal*, that it obey the constraint of cause and effect. Put simply, there must be no output signal before there is an applied input signal. This may sound so trivially obvious that it seems hardly worth mentioning, but in fact some circuits that look quite benign *on paper* are *not* causal. Try as you might, they are simply impossible to build according to electrical engineering as it is presently known, and to save yourself from an endless quest it is good to know how to tell if a theoretical system design could actually be constructed. To see how this works, let's work through the details of a specific example. Later, I'll be a little more general.

> The requirement that the world always obeys causality has traditionally been thought of as a "law of physics" as basic as the fundamental conservation laws of energy and electrical charge. Most electrical engineers and physicists find the possibility of causality violation to be simply horrifying, and in this book we'll be similarly shocked at such odd doings. The universe may actually be stranger than many engineers and scientists think, however, as the recent work by quite serious physicists on the possibility of time travel hints. Time travel to the past inherently violates causality, and yet there is nothing in physics as we presently understand it that forbids time travel. For much more on this, see my book *Time Machines*, Springer-Verlag 1999.

An important theoretical circuit in AM radio is the ideal *unity-gain bandpass filter*, which allows energy located in an interval of frequencies to pass through, while stopping energy located outside of that interval. A tuneable bandpass filter, for example, occurs in the front end of an AM receiver, and its job is to select out, from all the signals intercepted by the antenna, the one station signal you want to hear. An idealized plot of the *magnitude* of the transfer function of such a filter is shown in Figure 15.5. This plot is said to be idealized because of the vertical skirts. (The term *skirt* comes from the resemblance the plot of $|H(j\omega)|$ has to nineteenth-century hoop skirts!) Actual filters exhibit a less than vertical *roll-off* of the skirts. The *bandwidth* of this filter is $2\Delta\omega$, and the frequency interval over which the filter passes energy is called the *passband*.

The bandwidth of a system is defined as follows. Let $|H(j\omega)|$ be maximum at $\omega = \omega_0$. Then, there are two other frequencies $\omega_1 < \omega_0 < \omega_2$ such that $|H(j\omega_1)| = |H(j\omega_2)| = (1/\sqrt{2})|H(j\omega_0)|$. The bandwidth is defined to be $\omega_2 - \omega_1$. The $\frac{1}{\sqrt{2}}$ factor is completely arbitrary in this definition (other than being positive and less than one). All that really matters is that we *all* use the *same* factor.

Knowledge of $|H(j\omega)|$ is not enough to completely describe the filter, of course, as it doesn't include phase information. To determine what we

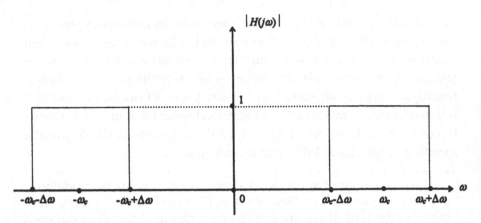

Figure 15.5   Transfer function (magnitude only) of the ideal bandpass filter.

should use for phase we impose the additional ideal constraint on the filter of *zero phase distortion*. Phase distortion is said to occur in a system if the energies at different frequencies take different times to transit the system from input to output. Physically, zero phase distortion means the input signal shape will be unaltered (although its *amplitude* may change) by its passage through the filter *if all* the energy of the signal is in the passband (see Problem 15.4).

Consider, then, the particular frequency ω, where ω is in the passband of our ideal filter. Further, suppose all energy propagating through the filter experiences the same time delay of t_0. The input signal $e^{j\omega t}$ will then produce the output signal $e^{j\omega(t-t_0)} = e^{-j\omega t_0}e^{j\omega t}$. But, by definition of the transfer function (see Appendix C), the input signal $e^{j\omega t}$ will produce the output signal $H(j\omega)e^{j\omega t}$. Thus, for the ideal unity-gain bandpass filter we have $H(j\omega) = e^{-j\omega t_0}$ where ω is any frequency in the passband [$H(j\omega) = 0$, by definition of the ideal bandpass filter, when ω is outside the passband].

That is, an ideal zero phase distorting filter has a negative phase shift that varies linearly with frequency (as shown in Figure 15.6, where $\theta(\omega) = -\omega t_0$ for $\omega_c - \Delta\omega < \omega < \omega_c + \Delta\omega$). Unfortunately, the ideal bandpass filter described by Figures 15.5 and 15.6 is impossible to build because, as I'll show next, its impulse response $h(t)$ is not zero when $t < 0$. That is, the filter would (if it could be built) respond to the input signal $\delta(t)$ (which occurs at $t = 0$) *before* $t = 0$. Such behavior is called *anticipatory*, or noncausal, and it is obviously nonphysical. Thus,

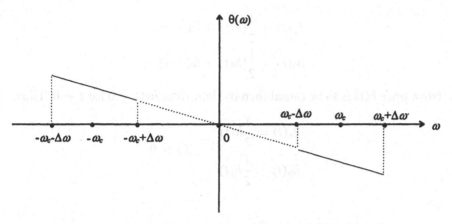

Figure 15.6   Transfer function (phase only) of the ideal bandpass filter.

$$
\begin{aligned}
h(t) &= \frac{1}{2\pi} \int_{-\infty}^{\infty} H(j\omega)e^{j\omega t}\, d\omega \\
&= \frac{1}{2\pi} \left[\int_{-\omega_c-\Delta\omega}^{-\omega_c+\Delta\omega} e^{-j\omega t_0} e^{j\omega t}\, d\omega + \int_{\omega_c-\Delta\omega}^{\omega_c+\Delta\omega} e^{-j\omega t_0} e^{j\omega t}\, d\omega \right] \\
&= \frac{1}{\pi} \frac{\sin\left[\omega_c t_0 \left(1 + \frac{\Delta\omega}{\omega_c}\right)\left(\frac{t}{t_0} - 1\right)\right] - \sin\left[\omega_c t_0 \left(1 - \frac{\Delta\omega}{\omega_c}\right)\left(\frac{t}{t_0} - 1\right)\right]}{t - t_0} \\
&= \frac{2}{\pi} \frac{\cos\left[\omega_c t_0 \left(\frac{t}{t_0} - 1\right)\right] \sin\left[\frac{\Delta\omega}{\omega_c}\left(\frac{t}{t_0} - 1\right)\right]}{t - t_0},
\end{aligned}
$$

which is clearly nonzero for $t < 0$.

We can use frequency convolution to study what imposing causality on $h(t)$ says about the structure of $H(j\omega)$, which we'll write as $H(j\omega) = R(\omega) + jX(\omega)$. I'll begin by writing $h(t)$ as the sum of even and odd functions of time, i.e., as $h(t) = h_e(t) + h_0(t)$. By this notation I mean (see Appendix A)

$$
\begin{aligned}
h_e(-t) &= h_e(t), \\
h_0(-t) &= -h_0(t).
\end{aligned}
$$

That we can actually write $h(t)$ in such a way is most directly shown by simply demonstrating what $h_e(t)$ and $h_0(t)$ are. Thus, $h(-t) = h_e(-t) + h_0(-t) = h_e(t) - h_0(t)$, and so if we add and subtract $h(-t)$ and $h(t)$ we get

$$h_e(t) = \frac{1}{2}[h(t) + h(-t)],$$

$$h_0(t) = \frac{1}{2}[h(t) - h(-t)].$$

Now, since $h(t)$ is to be causal then by definition $h(t) = 0$ for $t < 0$. Thus,

$$h_e(t) = \frac{1}{2}h(t)$$

$$\qquad\qquad\qquad \text{if } t > 0$$

$$h_0(t) = \frac{1}{2}h(t)$$

and

$$h_e(t) = \frac{1}{2}h(-t)$$

$$\qquad\qquad\qquad \text{if } t < 0.$$

$$h_0(t) = -\frac{1}{2}h(-t)$$

That is,

$$h_e(t) = h_0(t), \qquad t > 0$$

$$h_e(t) = -h_0(t), \quad t < 0.$$

These last two statements can be written more compactly, without having to explicitly give conditions on t, as

$$h_e(t) = h_0(t)\,\text{sgn}(t).$$

In the same way we can also write

$$h_0(t) = h_e(t)\,\text{sgn}(t).$$

Since $h(t) = h_e(t) + h_0(t)$, we can write $H(j\omega) = H_e(j\omega) + H_0(j\omega)$. Since $h_e(t)$ is even, then $H_e(j\omega)$ is purely real, and since $h_0(t)$ is odd then $H_0(j\omega)$ is purely imaginary, and thus

$$H_e(j\omega) = R(\omega),$$

$$H_0(j\omega) = jX(\omega).$$

From the frequency-convolution theorem, and from the transform pair $\text{sgn}(t) \leftrightarrow \frac{2}{j\omega}$ (see Problem 14.3), we have

$$R(\omega) = \frac{1}{2\pi}H_0(j\omega) * \frac{2}{j\omega} = \frac{1}{2\pi}jX(\omega) * \frac{2}{j\omega}$$

and

$$jX(\omega) = \frac{1}{2\pi}H_e(j\omega) * \frac{2}{j\omega} = \frac{1}{2\pi}R(\omega) * \frac{2}{j\omega}.$$

Or,

$$R(\omega) = \frac{1}{\pi}X(\omega) * \frac{1}{\omega} = \frac{1}{\pi}\int_{-\infty}^{\infty}\frac{X(u)}{\omega - u}\,du$$

$$X(\omega) = -\frac{1}{\pi}R(\omega) * \frac{1}{\omega} = -\frac{1}{\pi}\int_{-\infty}^{\infty}\frac{R(u)}{\omega - u}\,du.$$

Demanding that $h(t)$ be causal, then, imposes the above interdependencies on the real and imaginary parts of $H(j\omega)$. The integrals that connect $R(\omega)$ and $X(\omega)$ are called *Hilbert transforms*, and that transform is discussed in more detail in Appendix I. The point here is that if $h(t)$ is causal, then $H(j\omega)$ has constraints on it beyond that of simply requiring $|H(j\omega)|^2$ to be even [true for all real $h(t)$, causal or not].

These constraints might be called *local*, in that they show how the values of $R(\omega)$ and $X(\omega)$ are determined, for *every* ω, in terms of the integrated (or *global*) behavior of $X(\omega)$ and $R(\omega)$, respectively. We can also derive global constraints on $R(\omega)$ and $X(\omega)$ for a causal signal as follows. As shown in the text, $h(t) = h_e(t) + h_0(t)$, and $h_e(t) \leftrightarrow R(\omega)$, $h_0(t) \leftrightarrow jX(\omega)$. From Rayleigh's energy theorem, then,

$$\int_{-\infty}^{\infty} h_e^2(t)\,dt = \frac{1}{2\pi}\int_{-\infty}^{\infty}R^2(\omega)\,d\omega.$$

$$\int_{-\infty}^{\infty} h_0^2(t)\,dt = \frac{1}{2\pi}\int_{-\infty}^{\infty}X^2(\omega)\,d\omega.$$

For $h(t)$ causal, I showed $h_e(t) = h_0(t)\operatorname{sgn}(t)$, which says $h_e^2(t) = h_0^2(t)$. Thus, the two time integrals are equal and so, therefore, are the two frequency integrals. That is, for a causal signal we have the constraint $\int_{-\infty}^{\infty} R^2(\omega)\,d\omega = \int_{-\infty}^{\infty} X^2(\omega)\,d\omega$, which shows how the integrated (i.e., global) behavior of $R(\omega)$ depends on the integrated behavior of $X(\omega)$, and vice versa.

During a 1933 collaboration the American mathematician and electrical engineer Norbert Wiener (1894–1964) and the English mathematician Raymond Paley (1907–1933) discovered a necessary and sufficient condition for $A(\omega) = |H(j\omega)|$ to be the *amplitude* response of a causal filter:

$$\int_{-\infty}^{\infty}\frac{|\ln A(\omega)|}{1 + \omega^2}\,d\omega < \infty.$$

For the perfect bandpass filter, with $A(\omega) = 0$ almost everywhere, it is

> obvious that the Paley–Wiener integral diverges and so such an ampli-
> tude response is not possible in a causal filter.

As the final example of this chapter, I'll now do a problem that ties together several of the results and theorems that have been developed. So,
consider the signal $v(t) = [\frac{\sin(t)}{t}]u(t)$, a causal, damped sinusoid different
from the *exponentially* damped oscillations generated by the early spark
transmitters. If we write $v(t)$ as the sum of even and odd functions, then
in particular the even function is associated with the real part, $R(\omega)$, of
the transform of $v(t)$. This even function is $\frac{\sin(t)}{2t}$—be sure you can show
this—and to find its spectrum $R(\omega)$ we can use the duality theorem of
Chapter 14.

We start by recalling the unit gate function $\pi(t)$, and the pair $\pi(t) \leftrightarrow$
$\sin(\frac{\omega}{2})/(\frac{\omega}{2})$, from Chapter 12. From duality, then, we immediately have the
pair

$$\frac{\sin(-t/2)}{(-t/2)} = \frac{\sin(\frac{t}{2})}{\frac{t}{2}} \leftrightarrow 2\pi\, \pi(\omega).$$

Notice carefully the dual use of the same symbol "π"—once for the number and once as the symbol for the unit gate function [in the frequency
domain, i.e., $\pi(\omega) = 1$ for $|\omega| < \frac{1}{2}$ and is zero otherwise]. Next, using the
time/frequency scaling theorem from Chapter 12 (with $a = 2$), we have
the pair

$$\frac{\sin(t)}{t} \leftrightarrow \pi\, \pi\left(\frac{\omega}{2}\right)$$

and so

$$\frac{\sin(t)}{2t} \leftrightarrow \frac{\pi}{2}\pi\left(\frac{\omega}{2}\right) = R(\omega).$$

That is, $R(\omega) = \pi/2$ for $|\omega| < 1$, and is zero otherwise.

Because $v(t)$ is causal, we can now find $X(\omega)$ by taking the Hilbert transform of $R(\omega)$, i.e.,

$$X(\omega) = -\frac{1}{\pi}\int_{-\infty}^{\infty}\frac{R(u)}{\omega - u}\,du = -\frac{1}{2}\int_{-1}^{1}\frac{du}{\omega - u}.$$

Doing the integral gives

$$X(\omega) = \frac{1}{2}\ln\left|\frac{\omega - 1}{\omega + 1}\right|,$$

and so we have the rather exotic pair

$$\frac{\sin(t)}{t}u(t) \leftrightarrow \frac{\pi}{2}\pi\left(\frac{\omega}{2}\right) + j\frac{1}{2}\ln\left|\frac{\omega-1}{\omega+1}\right|.$$

"Doing" the above integral is easier said than done, however; it has a subtle problem. If you simply go ahead and make the obvious change of variable and integrate, you'll get $(\frac{1}{2})\ln\{(\omega-1)/(\omega+1)\}$ as the answer. This makes sense for $|\omega| > 1$, but for $|\omega| < 1$ it doesn't because then the log function has a negative argument. The correct answer, given above, has absolute-value signs around the argument which eliminates the problem, but where do they come from? The answer is that to properly evaluate the integral you must notice that the integrand is discontinuous at $u = \omega$, which doesn't cause any problem if $|\omega| > 1$ because then the discontinuity is outside the interval of integration. But if $|\omega| < 1$ the integrand *blows up* at the discontinuity and that *does* cause a problem! So, what's the cure? The answer is given in Appendix I, where a similar integration is done in numbing detail, but see if you can figure this out for yourself before turning to the back of the book.

Problems

15.1. As shown in the previous chapter, the Fourier transform of $u(t)$ is $U(j\omega) = \pi\delta(\omega) + \frac{1}{j\omega}$. Thus, if we let the input of a linear system [with causal impulse response $h(t)$ and transfer function $H(j\omega)$] be $u(t)$, then the transform of the system output is $Y(j\omega) = U(j\omega)H(j\omega)$. Use the inverse Fourier transform to show that the system response to a step input [easy to generate, because while it has infinite energy, it is spread over infinite time, unlike the energy of $\delta(t)$] is

$$y(t) = \frac{1}{2}H(0) + \frac{1}{2\pi}\int_{-\infty}^{\infty}\frac{1}{j\omega}H(j\omega)e^{j\omega t}\,d\omega.$$

From this conclude that $dy/dt = (\frac{1}{2}\pi)\int_{-\infty}^{\infty}H(j\omega)e^{j\omega t}\,d\omega = h(t)$, i.e., the impulse response is the derivative of the step response. (To build an excellent differentiator in real electronic hardware is, as discussed in Appendix F, a routine undergraduate laboratory exercise today, easily wired up in just minutes with perhaps two dollars worth of parts.) To squeeze every last drop out of this that we can, notice that (with g as a dummy variable of integration)

$$y(t) = h(t)^* u(t) = \int_{-\infty}^{\infty} h(g)u(t-g)\,dg = \int_0^t h(g)\,dg.$$

Combine this observation with the inverse Fourier transform of $Y(j\omega)$ to derive the pair

$$\int_0^t h(g)dg \leftrightarrow \pi H(0)\delta(\omega) + \frac{1}{j\omega}H(j\omega).$$

15.2. Show that convolution is associative, i.e., that $f^*g = g^*f$. Hint: write the left-hand side out in detail, make the obvious change of variable, and show it becomes the right-hand side.

15.3. Suppose that the impulse response of a linear system, $h(t)$, is *not* absolutely integrable. That is, suppose $\int_{-\infty}^{\infty} |h(u)|\,du = \infty$. Let the input to this system be the particular signal $x(-t) = h(t)/|h(t)|$. Notice that $|x(t)| = 1$ when $h(t) \neq 0$, and $x(t) = 0$ when $h(t) = 0$, as illustrated in Figure 15.7 for a causal system (but this is just

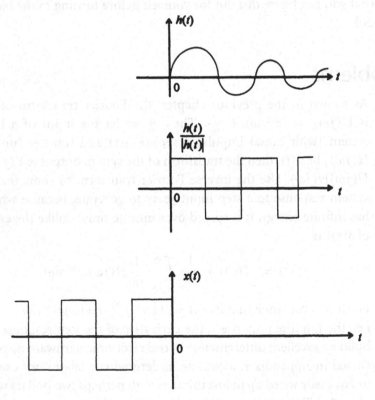

Figure 15.7 A bounded input signal constructed from the impulse response of a causal linear system.

an example—we are *not* assuming causality here). Thus, $x(t)$ is a bounded input. Observe carefully, too, that although $h(t)$ is not absolutely integrable it is still possible for it to be quite benign; always finite and, indeed, even such that $\lim_{t \to \infty} h(t) = 0$. Now, use the time convolution integral to show that the output at time $t = 0$ is unbounded, i.e., that $y(0) = \infty$. This shows that absolute integrability of $h(t)$ is a *necessary* condition for BIBO stability.

15.4. Suppose a linear system's only influences on its input are to introduce a constant time delay and a constant amplitude scaling, i.e., for *any* input $x(t)$ the output is $y(t) = Ax(t - t_0)$ where $|A| < \infty$ and $t_0 \neq 0$. Show that such a system must have *infinite* bandwidth starting at dc. Explain why such a system is a physical impossibility. (Note: the answer to the second part is *not* causality violation, as we've insured the system *is* causal with the constant time *delay*.) Hints: Show that $|H(j\omega)|$ is a constant. Also, since the matter used to build the system will unavoidably form the parts of various capacitances, think about what all capacitors do at sufficiently high frequencies.

15.5. Use the pair derived in the text for convolution of the unit gate function with itself $[\pi(t){}^*\pi(t) \leftrightarrow \prod^2(j\omega)]$, and Rayleigh's energy theorem, to show that

$$\int_0^\infty \left\{ \frac{\sin(x)}{x} \right\}^4 dx = \frac{\pi}{3}.$$

Write a MATLAB program to check this result.

15.6. Use the global causality constraints and the transform derived for $[\frac{\sin(t)}{t}]u(t)$ to show that

$$\int_0^\infty \ln^2 \left\{ \left| \frac{x-1}{x+1} \right| \right\} dx = \pi^2.$$

Write a MATLAB program to check this result.

15.7. Show that if the input $x(t)$ to a BIBO-stable system has finite energy then the output $y(t)$ will also have finite energy. That is, given that

$$\int_{-\infty}^\infty x^2(t)\, dt < \infty, \qquad \int_{-\infty}^\infty |h(t)|\, dt < \infty$$

show it then follows that

$$\int_{-\infty}^\infty y^2(t)\, dt < \infty.$$

Hint: write $H(j\omega) = \int_{-\infty}^{\infty} h(t)e^{-j\omega t}\,dt$ and observe what BIBO-stability implies about $|H(j\omega)|$. Then, simply recall Rayleigh's energy theorem and that $Y(j\omega) = X(j\omega)H(j\omega)$. Assume all the transforms exist.

Nonlinear Circuits for Multiplication

Multiplying by Squaring and Filtering

For radio transmitters to work, we need to shift baseband energy up to rf. Our previous work has shown you how to do that— simply multiply the baseband signal $m(t)$ by $\cos(\omega_c t)$. The spectrum of $m(t)$, which describes how the energy of $m(t)$ is distributed in frequency (Rayleigh's energy theorem), then shifts both up and down by ω_c, the so-called *carrier frequency*. Going next in the opposite direction, for radio receivers to work we obviously need to shift the up-shifted spectrum of $m(t)$ back down to baseband where we can hear it. Again, multiplication is the way to do it (I'll discuss this process in more detail in Section 4). Accurate analog multipliers that work at radio frequencies are expensive, difficult-to-build devices, however, and some great ingenuity has gone into developing *indirect* ways of performing multiplication (and, hence, spectrum shifting). You will be able to understand how these clever circuits work because you have (haven't you?) worked your way through the mathematics of the Fourier transform in the previous section. The Fourier transform is the key to unlocking the physics of what is happening in these circuits.

To begin, let's drop our sights a bit and, instead of directly building a multiplier, let's study the behavior of the summer/ squarer/filter shown in Figure 16.1. The output of the squarer is $m^2(t) + 2m(t)\cos(\omega_c t) + \cos^2(\omega_c t)$, which includes the desired product term $m(t)\cos(\omega_c t)$. It also includes, seemingly to

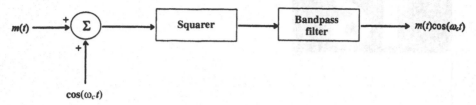

Figure 16.1   A circuit for multiplying *without* multiplying!

our misfortune, two other terms. The astonishing fact is, however, that it is possible to arrange matters so that the spectrum of the product term is distinct from the spectra of the other terms. Thus, we can apply the output of the squarer to a suitably designed bandpass filter which will pass only the energy of the product term; the *total* circuitry in Figure 16.1 (including the filter) is therefore a multiplier.

This solution for how to build a multiplier is not quite complete, of course, because it leaves us with the obvious problem of how to build a *squarer*. Intuitively, however, we might expect this to be a simpler problem, as squaring is a special, less complicated process than multiplying. After all, a squarer has just *one* input while a multiplier has two, i.e., a squarer can be built from a single multiplier (simply apply the same input signal to both multiplier inputs). In the next chapter I'll discuss the summing/squaring operation in some detail, but for now let's verify that the filtered output of the circuit in Figure 16.1 is, theoretically, the result of a pure multiplication.

As mentioned in Problem 15.1 it is duck soup to build an electronic differentiator (or integrator). Oddly, however, the less "sophisticated" operation of multiplication is *much* more difficult to implement in hardware. Just because a process is elementary doesn't mean it's trivial! A multiplier can, however, be built from *two* squarers (and some summers). Can you see how to do this? [Hint: consider the identity $(x + y)^2 - (x - y)^2 = 4xy$.] Multipliers based on this idea can be built that work very well up to 100 kHz and above, but that is still far below even the low end of the commercial AM radio band.

Consider each of the three terms in the squarer output, in turn. First, the easiest, is $2m(t)\cos(\omega_c t)$. By the heterodyne theorem of Chapter 15, the spectrum of this term is just the baseband spectrum of $m(t)$, centered on $\omega = 0$, shifted both up and down to be centered on $\omega = \pm\omega_c$. Next, the $\cos^2(\omega_c t)$ term can be expanded with a trigonometric identity to give the equivalent expression $(\frac{1}{2}) + (\frac{1}{2}) \cos(2\omega_c t)$. From Chapters 14 and 15 we know the spectrum of this is $\pi \delta(\omega) + (\frac{1}{2})\pi[\delta(\omega - 2\omega_c) + \delta(\omega + 2\omega_c)]$, i.e., three impulses located at $\omega = 0, \pm 2\omega_c$. And finally, we have the squared term $m^2(t)$. From the frequency convolution theorem of Chapter 15 we know the spectrum of this is

$$M(j\omega) * M(j\omega) = \int_{-\infty}^{\infty} M(ju)M[j(\omega - u)]\, du.$$

Using the fact that $m(t)$ is bandlimited, i.e., that $M(j\omega) = 0$ for $|\omega| > \omega_m$, where ω_m is the maximum frequency in $m(t)$, then it should be clear that $M(j\omega) * M(j\omega)$ is also bandlimited (specifically, the spectrum of $m^2(t)$ is zero for $|\omega| > 2\omega_m$). If this isn't clear, then go back to Chapter 15 and review how we argued our way through the convolution of the unit-gate function $\pi(t)$ with itself. That argument is precisely analogous (even though the domains are different) for $M(j\omega)$ convolved with itself. Notice carefully that the precise details of $M(j\omega) * M(j\omega)$ are not important, only that, whatever they are, $m^2(t)$ has no energy outside the interval $|\omega| < 2\omega_m$ simply because $m(t)$ has no energy outside the interval $|\omega| < \omega_m$.

Figure 16.2 shows the spectrum of the squarer output (the sum of the spectra for the above three terms) drawn for the case where the spectrum of the heterodyned baseband term does not overlap the spectra of the other terms. That will be the case if $2\omega_m < \omega_c - \omega_m$, i.e., if $\omega_c > 3\omega_m$. This condition is easily satisfied in AM radio, where the smallest ω_c is 2π (540 kHz) and $\omega_m = 2\pi$ (5 kHz). Thus, if the output of the squarer is then pro-

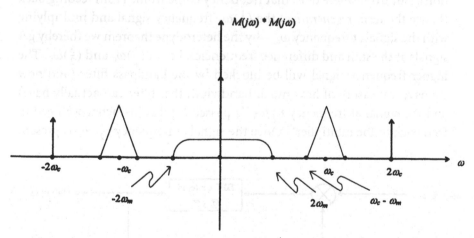

Figure 16.2   The output spectrum of the squarer in the circuit of Figure 16.1 when multiplying a baseband signal with $\cos(\omega_c t)$ for the case of $\omega_c > 3\omega_m$.

cessed by a bandpass filter centered on ω_c (and with a bandwidth of $2\omega_m$), then the output signal of the filter has the spectrum associated *only* with the $m(t)\cos(\omega_c t)$ term (to within an amplitude scaling factor). We have achieved the desired multiplication.

The 5-kHz value for ω_m is imposed by the FCC (Federal Communications Commission) on all holders of commercial AM radio licenses. That is, the baseband signal *by law*, cannot have energy above 5 kHz (low-pass filtering ensures this), and so the bandwidth of a radiated AM signal is limited to 10 kHz. For this reason alone it is not possible for AM radio to broadcast high-fidelity signals (e.g., hi-fi music). The much wider bandwidth allowed in FM radio, however, can easily achieve hi-fi quality transmission. The reason for the narrow bandwidth in AM is simply due to the scarcity of frequency in the AM broadcast band (only slightly more than 1 MHz). FM station assignments, by contrast, are spread over a much wider interval (88 to 108 MHz) and so there is proportionally more bandwidth available for each station.

A fascinating application of multipliers and of the heterodyne (or modulation) theorem is the so-called *regenerative frequency divider*. It is shown in block diagram form in Figure 16.3, where I've assumed a multiplier is available. The circuit has $\cos(\omega_c t)$ as its input, and the claim is that a sinusoidal output at one-half that frequency will result. To see that this is so, reason as follows. The bandpass filter is centered at $\omega = (\frac{1}{2})\omega_c$ and so only a signal at that frequency *could* appear (I'll assume the filter has a very narrow bandwidth at first, but you'll soon see that isn't a necessary assumption). But from where does that frequency come from? From feeding back (hence the term *regenerative*) the $(\frac{1}{2})\omega_c$ frequency signal and multiplying with the signal at frequency ω_c—by the heterodyne theorem we thereby get signals at the sum and difference frequencies, i.e., at $(\frac{1}{2})\omega_c$ and $(\frac{3}{2})\omega_c$. The higher frequency signal will be blocked by the bandpass filter (and now we have a measure of how much bandwidth that filter can actually have) and the signal at frequency $(\frac{1}{2})\omega_c$ is passed (and is just what we need to feed back to the multiplier!) Once the signal at frequency $(\frac{1}{2})\omega_c$ is present

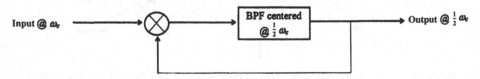

Figure 16.3   Regenerative frequency divider (÷2).

at the output, everything is okay. "But how does that frequency show up at the output *at the very start*?" you might ask. Refer back to the discussion in Chapter 8 on how an oscillator circuit gets started (the answers to both questions are the same). This ingenious circuit borders, I think, on being the electronic equivalent of "wish fulfillment"! I'll mention it again, briefly, in the next section when I discuss the problem of demodulating double-sideband AM without the presence of a strong carrier signal.

Now, all of the preceding is fine *if* we can build a multiplier from a squarer. How do we build the squarer? That's the next chapter!

Problems

16.1. The sketch in Figure 16.4 shows a *speech scrambler*, a private, portable device used to provide a moderate level of privacy over public telephone circuits. This gadget (which clamps on to the mouthpiece/receiver of a telephone) is sufficiently complex to keep the "innocent" from listening in on a conversation, but of course the CIA, FBI, and most likely even the local police department would find it easy to neutralize. Analyze the operation of this system by drawing the spectrums of the signals $x_1(t)$, $x_2(t)$, $x_3(t)$, and $y(t)$ (the scrambler output). The highpass filter is ideal, with a vertical

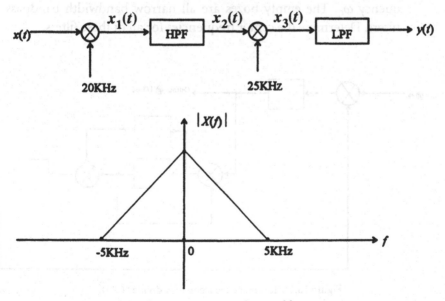

Figure 16.4 A speech scrambler.

skirt at its low-frequency cutoff of 20 kHz, and the low-pass filter is similarly ideal, with a vertical skirt at its high-frequency cutoff of 20 kHz. The sketch of $|X(f)|$ shows the input spectrum, i.e., $x(t)$ is a baseband signal, bandlimited to 5 kHz. The multipliers are also perfect. This type of scrambler is quite old, dating back to just after the First World War. It was first used commercially on the 25-mile radio–telephone circuit connecting Los Angeles and Santa Catalina Island. By the Second World War, speech scramblers had reached truly heroic levels of technical sophistication—see David Kahn's "Cryptology and the Origins of Spread Spectrum," *IEEE Spectrum*, September 1984, pp. 70–80, and the shaded box at the end of Chapter 20.

16.2. Continuing with the previous problem, a *descrambler* is obviously needed at the other end of the telephone line. An attractive feature of the system in Figure 16.4 is that it is its *own descrambler*. Verify that this is so by applying the scrambled spectrum to the scrambler input and show that the output signal has the original spectrum. Also discuss the operation of this device if the 20 kHz and/or the 25 kHz signals to the multipliers "drift" in frequency.

16.3. The regenerative frequency divider circuit can be extended to achieve division by any integer. For example, the circuit shown in Figure 16.5 will produce an output of frequency $(\frac{1}{3})\omega_c$ if the input is at frequency ω_c. The empty boxes are all narrow bandwidth bandpass filters. Determine the center frequencies for all of these filters.

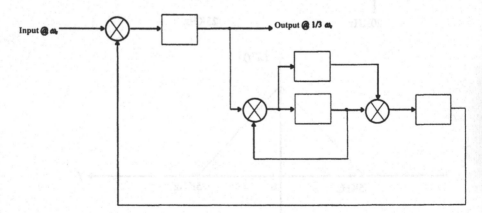

Figure 16.5   Regenerative frequency divider (÷3).

Squaring and Multiplying with Matched Nonlinearities

Consider the circuit of Figure 17.1, which uses *matched* two-terminal nonlinear components. What that means is that if we write the voltage-current relationship for each component as a power series expansion

$$i = \alpha_0 + \alpha_1 v + \alpha_2 v^2 + \alpha_3 v^3 + \cdots$$

then the α_n coefficients are the same for $i = i_1$ ($v = v_1$) and for $i = i_2$ ($v = v_2$). If we next suppose that the value (R) of the two resistors is such that the voltage drops $i_1 R$ and $i_2 R$ are small compared to the other voltages in their respective loops, then

$$v_1 = \cos(\omega_c t) + m(t),$$
$$v_2 = \cos(\omega_c t) - m(t).$$

We can avoid this approximation and make the resulting equations for v_1 and v_2 exact if we use three-terminal nonlinear components instead, such as vacuum triode tubes or field effect transistors. Then, v_1 and v_2 would be either the grid-to-cathode (see Chapter 8) or the gate-to-source (see any book on solid-state electronics) potential differences, and i_1 and i_2 would be either the plate or drain currents. Such three-terminal devices effectively isolate the controlling voltage variables from the dependent currents, something two-terminal devices can't do.

We have $e(t) = i_1 R - i_2 R = R(i_1 - i_2)$, and so

$$\frac{e(t)}{R} = \sum_{n=1}^{\infty} \alpha_n (v_1^n - v_2^n).$$

If this expression is evaluated for the first three values of n, then it can be shown (you should verify this) that

$$\frac{e(t)}{R} = (2\alpha_1 + 3\alpha_3)m(t) + 2\alpha_3 m^3(t) + 3\alpha_3 m(t)\cos(2\omega_c t)$$
$$+ 4\alpha_2 m(t)\cos(\omega_c t).$$

This is most easily done by using the binomial theorem (see Appendix A), i.e., let $x = \cos(\omega_c t)$ and $y = m(t)$, and then the general term in $e(t)/R$ is

$$\alpha_n(v_1^n - v_2^n) = \alpha_n \sum_{k=0}^{n} \binom{n}{k} x^k \{ y^{n-k} - (-y)^{n-k} \}.$$

The first two terms on the right of the expression for $e(t)/R$ are bandlimited at $|\omega| < \omega_m$ and $|\omega| < 3\omega_m$, respectively. The third term's spectrum is the spectrum of $m(t)$ shifted up (and down) to be centered around $\omega = \pm 2\omega_c$. Only the last term is the desired product term, with a spectrum centered around $\omega = \pm\omega_c$. That is, if we can assume the matched nonlinearities don't extend beyond the third power then the *filtered* output of the circuit in Figure 17.1 is the result of a multiplication. This circuit,

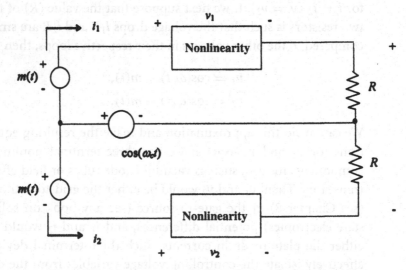

Figure 17.1   A prototype multiplier using matched nonlinearities.

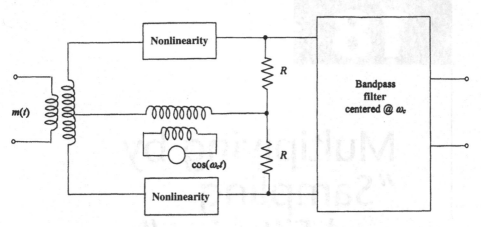

Figure 17.2 A balanced modulator (multiplier) with filter.

including a bandpass filter centered at ω_c, is called a *balanced modulator* and is shown in more detail in Figure 17.2.

We get the required squaring via the $\alpha_2 v^2$ term in the nonlinear voltage-current relationship, and the configuration of the circuitry in which the two matched nonlinearities are embedded has either canceled or rendered moot (via the frequency shifting of energy to outside of the filter's passband) the effects of the constant, linear, and cubic terms. It is only when we include the $\alpha_4 v^4$ term that deviations from a perfect multiplier are encountered (see Problem 17.1).

Problems

17.1. Show that if the matched nonlinearities in Figure 17.1 contain a fourth power term ($\alpha_4 v^4$) then that term will place energy in the passband of the filter centered at $\omega = \omega_c$. That is, the balanced modulator will deviate from being a perfect multiplier if it contains a quartic nonlinearity.

17.2. Discuss the behavior of the circuit of Figure 17.2 if $m(t)$ and $\cos(\omega_c t)$ trade places.

Multiplying by "Sampling and Filtering"

Fourier theory allows us to understand how an approach, completely different from that of the balanced modulator, can also achieve the frequency shift of the spectrum of a baseband signal up to rf. For you to understand this radical approach, I need to discuss the process of *sampling*. A sampler is simply a circuit that at regular intervals (called the *sampling period*, T) briefly transmits $m(t)$. A sampler can, as a beginning, be thought of as a mechanically rotating switch, as shown in the top half of Figure 18.1. We can mathematically model this sampler by writing the sampler output as $m_s(t) = m(t)s(t)$, where $s(t)$ is the wave shape shown in the bottom half of Figure 18.1. From this you can see sampling is, in fact, a multiplication process.

Since $s(t)$ is a periodic function we can write it as a Fourier series and so, with $\omega_s = 2\pi f_s$ as the sampling frequency,

$$m_s(t) = m(t) \sum_{n=-\infty}^{\infty} c_n e^{jn\omega_s t}, \quad \omega_s = \frac{2\pi}{T}.$$

The spectrum of $m_s(t)$ is thus given by

$$M_s(j\omega) = \int_{-\infty}^{\infty} \left\{ m(t) \sum_{n=-\infty}^{\infty} c_n e^{jn\omega_s t} \right\} e^{-j\omega t} \, dt$$

$$= \sum_{n=-\infty}^{\infty} c_n \int_{-\infty}^{\infty} m(t) e^{-j(\omega - n\omega_s)t} \, dt.$$

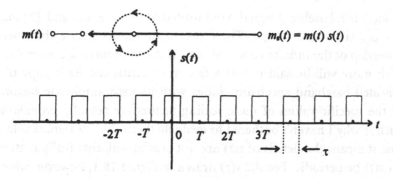

Figure 18.1 A mechanical sampler. The switch rotates once each T seconds, and completes the path once each rotation for a duration of τ seconds.

Recognizing the last integral is $M\{j(\omega - n\omega_s)\}$, we have

$$M_s(j\omega) = \sum_{n=-\infty}^{\infty} c_n M\{j(\omega - n\omega_s)\}.$$

This deceptively simple-appearing result says the spectrum of the sampler output signal is just the spectrum of the input signal repeated, endlessly, up and down the frequency axis at intervals of ω_s. Figure 18.2 illustrates what such a spectrum looks like under the two assumptions that

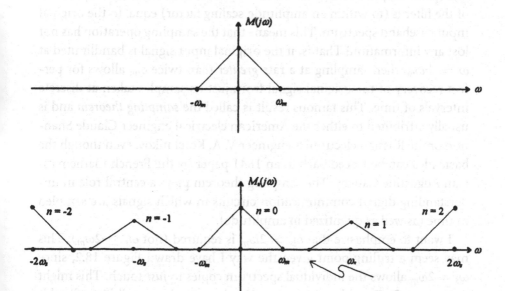

Figure 18.2 The spectrums of a baseband signal $m(t)$ and its sampled version $m_s(t)$.

(1) $m(t)$ is a baseband signal bandlimited at $\omega = \pm\omega_m$, and (2) $\omega_m <$ $\omega_s - \omega_m$ (i.e., $\omega_s > 2\omega_m$). These two assumptions ensure that there is no overlap of the infinite copies of the spectrum of $m(t)$, a concern about which more will be said in just a few more sentences. Each copy of the replicated baseband spectrum comes with its own amplitude factor, c_n, but the specific values of these scaling factors are actually *unimportant*, which is why I haven't bothered to evaluate them. This is remarkable, because it means the details of $s(t)$ are not crucial—all that really matters is that $s(t)$ be periodic. For the $s(t)$ drawn in Figure 18.1, however, which is shown as an even function, it is simple to show (you should do this) that all the c_n are real.

The process of sampling has, in particular, shifted the spectrum up ($n = 1$) and down ($n = -1$) in frequency by ω_s. By bandpass filtering the sampler output (with the passband centered at $\omega = \omega_s$), the sampler output is (to within a multiplicative scaling factor) equal to the product $m(t) \cos(\omega_c t)$. If we identify the sampling frequency ω_s with the AM carrier frequency ω_c, then we have achieved the desired spectrum shift of $m(t)$ up to rf. Since the sampler can be said to physically "chop up" the input signal $m(t)$ in the time domain, the sampler/filter combination is often called a *chopper modulator*.

Notice, in passing, that if we *low*-pass filter the sampler output to select (pass) only the energy of the $n = 0$ term, then the output spectrum of the filter is (to within an amplitude scaling factor) equal to the original input baseband spectrum. This means that the sampling operation has not lost any information! That is, if the original input signal is bandlimited at $\omega = \pm\omega_m$, then sampling at a rate *greater* than twice ω_m allows for perfect recovery of the original signal from just its samples taken at discrete intervals of time. This famous result is called the *sampling theorem* and is usually attributed to either the American electrical engineer Claude Shannon or the Russian electronics engineer V. A. Kotel'nikov, even though the basic idea can be traced back to an 1841 paper by the French mathematician Augustine Cauchy. The sampling theorem plays a central role in understanding digital communication circuits in which signals are sampled in time (as well as quantized in amplitude).

I want to emphasize that $\omega_s > 2\omega_m$ is required (not $\omega_s \geq 2\omega_m$). This may seem a trifling point given the way I have drawn Figure 18.2, since $\omega_s = 2\omega_m$ allows the individual spectrum copies to *just* touch. This might not seem to be a problem (except for the theoretical impossibility of building a real filter with a non-vertical skirt that could select just one copy of the repeated baseband spectrum), but what if the baseband spectrum has *impulses* at $\omega = \pm\omega_m$? That would occur if $m(t)$ has a sinusoidal component at $\omega = \omega_m$. Then $\omega_s = 2\omega_m$ would have the impulses in adjacent copies of the baseband spectrum fall on top of each other, causing *significant* effects. To avoid any possibility of spectrum overlap, we therefore have to insist on $\omega_s > 2\omega_m$.

If, on the other hand, $\omega_s < 2\omega_m$ there will then be overlap of adjacent copies of the baseband spectrum giving what is called an *aliased* sampled spectrum. The word *alias* is used because if one tries to recover the original signal by low-pass filtering, the filter output will contain energy from an overlapped baseband spectrum copy, i.e., energy will appear in the filter's passband at one frequency that is really energy from a different frequency. Energy originally associated with one frequency (its "name" so to speak) is thus passing under a different name (an *alias*!). An interesting example of spectrum aliasing occurs every time one watches a western movie. Such films invariably have a scene in which a wagon with spoke wheels moves across the screen—it invariably happens as well that the wheels will appear to either *not* be rotating, or even to be rotating *backwards*! This illusion is an optical aliasing effect, resulting from the fact that the image on the screen is a sampled version of reality (the motion-picture industry standard is 24 frames/sec). To see how this occurs, suppose a wheel is 3 ft in diameter and has 10 spokes. The wheel will appear to be stationary (not rotating) if, from one frame to the next ($\frac{1}{24}$ sec), each spoke rotates into the next spoke's position. This occurs if one-tenth of a rotation requires $\frac{1}{24}$ sec, i.e., if one complete revolution takes $\frac{5}{12}$ sec. Now, a wheel 3 ft in diameter moves 3π ft in one revolution, and so the wheel is moving forward at a speed of $36\pi/5$ feet/sec $= 15.4$ mph. This is the slowest speed at which the wagon can move with the appearance of nonrotating wheels. Integer multiples of this speed will have the same effect. Notice, too, that if the wagon speed is slightly *less* than this speed, then the wheels will appear to be rotating backwards. These odd effects are the result of an aliased (undersampled) spectrum from which a low-pass filter (your eyes and brain) *cannot* recover the original signal. Indeed, a spoked wheel rotating one spoke position into the next, in $\frac{1}{24}$ sec, is a periodic phenomenon with a single 24-Hz component. To see the wheel rotating properly on a movie screen would therefore require a film sampling rate greater than twice that frequency, i.e., more than 48 frames per sec.

A mechanically rotating switch is okay for *thinking* about sampling, but that clearly isn't going to be a practical way to actually implement sampling

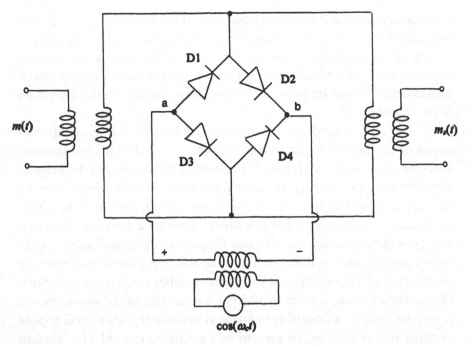

Figure 18.3   A switched diode sampler for multiplying at rf.

at AM radio frequencies (540 kHz and higher)! We need an *electronic circuit* to implement sampling at rf, and Figure 18.3 shows one ingenious solution. Here's how it works. To keep things simple, suppose the four diodes are perfect (shorted if forward biased, and opened if reverse biased). Then, on the positive half-cycles of $\cos(\omega_c t)$ (the polarity shown in the figure) the series diodes D1 and D2, and the series diodes D3 and D4, are forward biased. This brings points a and b together (electrically) and so $m_s(t) = 0$. On the negative half-cycles the series diode pairs are reversed biased and so the diodes don't conduct. Points a and b are thus electrically isolated and so $m_s(t) = m(t)$.

Thus, $m_s(t)$ is indeed a sampled version of $m(t)$, with the sampling occurring at a rate of $f_s = f_c$ samples per sec. Since f_c is (for AM broadcast radio) at least 540 kHz, and as $f_m = 5$ kHz, then the requirement $f_c > 2f_m$ is easily satisfied. Notice, too, that the duration of each sample [the τ for $s(t)$ in Figure 18.1] is one-half the period of $\cos(\omega_c t)$, i.e., $\frac{1}{(2f_c)}$ sec. Since f_c is at least 540 kHz, then the sampling duration is less than one microsecond which is *far* less than the period of the highest frequency in $m(t)$ (5 kHz corresponds to a period of 200 microseconds), i.e., $m(t)$ changes negligibly during the duration of each sample and so the sampling can be considered to be effectively instantaneous.

Applying $m_s(t)$, the output of the circuit in Figure 18.3, to a bandpass filter centered on $\omega = \omega_c$, produces the desired signal $m(t)\cos(\omega_c t)$. Because of the appearance of its schematic this chopper modulator circuit is also often called a *diode ring modulator*. Notice that this is a *time-varying* circuit, as some of its components are switched in and out of use, depending on the polarity of the carrier frequency generator.

Problem

18.1. Suppose $x(t)$ is a signal with all of its energy located in a bandwidth of $2\omega_m$, centered on ω_c. That is, the highest frequency in $x(t)$ is $\omega_c + \omega_m$. A naive application of the sampling theorem would seem to imply that, to avoid aliasing, $x(t)$ must be sampled at a frequency greater than $2(\omega_c + \omega_m)$. Explain how heterodyning and filtering can be applied to allow sampling at a frequency that need only be greater than $2\omega_m$.

Mathematics of "Unmultiplying"

During the First World War (the "Great War") the United States government came to fully appreciate the importance of radio as an instrument of national security, and so desired to bring American radio under *American* control. This realization doomed American Marconi (the subsidiary of a *British* company). At the same time, however, Congress was unwilling to allow radio to come under total government control (the U.S. Navy, in particular, lobbied hard for just such control); the lack of public support for the way the government had seized and operated such vital public monopolies as the telephone system and the railroads during the war convinced many that government radio would suffer a similar unhappy fate. So, as a compromise, Congress forced American Marconi to sell its American assets and then created the Radio Corporation of America (RCA) in 1919. RCA became the central holder of nearly all of the important radio patents that had previously been distributed among various competing, private corporations such as General Electric, American Telephone and Telegraph, and Westinghouse. Mutually beneficial cross-licensing production arrangements were made between RCA and those corporations, however, and they would all prosper in the coming "radio boom" years.

Synchronous Demodulation and its Problems

The previous section has shown us how, given a baseband signal $m(t)$, we can shift its spectrum up to rf by multiplying $m(t)$ by $\cos(\omega_c t)$. To form the AM signal that can be detected by the envelope detector circuit discussed in Chapter 6, we must next add to this product a constant carrier term (see Figure 6.1). This gives what is called a "double-sideband, large carrier," or DSB-LC, AM signal. I discussed in Chapter 6 why a *strong* carrier term is necessary for the proper operation of the envelope detector— the output of the detector is proportional to $m(t)$ if a sufficiently strong carrier is present, but if there is no carrier or even if it is present but too weak, then the detector produces a signal proportional to $|m(t)|$ (see Figure 6.2) which is unintelligible. However, I also discussed how the insertion of a carrier at the transmitter wastes energy. A natural question, then, is to ask if there isn't *some* way we can extract (detect) $m(t)$ from $m(t) \cos(\omega_c t)$ without having to assume the transmitter has added in a carrier term? Such a signal is called "double-sideband, suppressed carrier," or DSB-SC, and the answer is *yes*. The following discussion shows how and also why, nevertheless, commercial AM radio does *not* use DSB-SC.

It is an easy and educational laboratory experiment to show that $|m(t)|$ is unintelligible. $|m(t)|$ is a full-wave rectification (see any book on electronics for the elementary four diode circuit of such a rectifier) of $m(t)$. So, simply take the wires leading to the loudspeaker of a radio and insert the rectifier. When you hear the result, you'll agree it's unintelligible!

To recover $m(t)$ from a DSB-SC signal is, on paper, actually a trivial problem. To *demodulate* the DSB-SC signal $m(t) \cos(\omega_c t)$ we'll just use the modulation or heterodyne theorem again, the same theorem we used to shift the spectrum of $m(t)$ up to rf in the first place. Thus, if we multiply $m(t) \cos(\omega_c t)$ by $\cos(\omega_c t)$ we'll shift the spectrum of the DSB-SC signal both *up* (to be centered around $\pm 2\omega_c$) and *down* (to be centered around $\omega = 0$). But a spectrum centered around $\omega = 0$ is just the original baseband spectrum and is just what we want. The energy at and around $\pm 2\omega_c$ is, on the other hand, at a very high frequency, over at least a megahertz for AM radio, and can be safely ignored (if applied to the input terminals of a loudspeaker, for example, the speaker would reproduce the baseband spectrum—which is at *audio* frequencies—but, because of mechanical inertia, would be completely unresponsive to energy at rf, i.e., the speaker is a mechanical low-pass filter).

The practical problem with this approach is that it assumes that the *receiver*, which may be separated from the transmitter by tens, hundreds, even thousands of miles, has the signal $\cos(\omega_c t)$ available. This is a problem because for this approach to work it is necessary for the receiver's carrier frequency sinusoid to be *precisely* at frequency ω_c *and* to be *phase coherent* with the transmitter's sinusoid, i.e., the receiver must *locally* generate $\cos(\omega_c t)$ and not $\cos\{(\omega_c + \Delta\omega)t + \theta\}$. That is, the receiver's local oscillator must be synchronized with the transmitter's carrier, with $\Delta\omega$ and θ both equal to zero, and so this approach is called *synchronous demodulation*. To see why synchronization is required, I'll calculate what happens if there are phase or frequency mismatches between transmitter and receiver.

Consider first a phase mismatch, only. Receiving $m(t) \cos(\omega_c t)$ the receiver then multiplies by $\cos(\omega_c t + \theta)$ to give

$$m(t) \cos(\omega_c t) \cos(\omega_c t + \theta) = \frac{1}{2} m(t) \left[\cos(\theta) + \cos(2\omega_c t) \right],$$

from which a low-pass filter can recover a term proportional to $m(t) \cos(\theta)$. Phase mismatch appears therefore as the amplitude attenuation factor $\cos(\theta)$, which is not serious until (if) θ approaches 90°. At $\theta = 90°$, however, the filter output signal vanishes! Even a phase mismatch is big as $\theta = 90°$ represents precise timing, i.e., in the middle of the AM radio band, at one megahertz, each period of 1 μsec represents 360° and so a 90° mismatch means the transmitter and the receiver are only 250 *nano*seconds out of time alignment. Well, you might counter, except for the $\theta = 90°$ case we can simply negate the amplitude attenuation by increasing the receiver's amplification. The problem with that is that θ is generally an unknown *function of time*, because the received signal has made its way through space via a time-varying path (e.g., consider the case of a receiver on an airplane that isn't flying in a circle around the transmitter, or radio signals scattered back down to earth from an altitude-fluctuating ionosphere).

As bad as phase mismatch is for synchronous demodulation, frequency mismatch is arguably (far) worse. To see this, suppose after receiving $m(t)\cos(\omega_c t)$ the receiver multiplies by $\cos\{(\omega_c + \Delta\omega)\}$ to give

$$m(t)\cos(\omega_c t)\cos\{(\omega_c + \Delta\omega)t\} = \frac{1}{2}m(t)\left[\cos\{(\Delta\omega)t\} + \cos\{(2\omega_c + \Delta\omega)t\}\right],$$

from which a low-pass filter removes a term proportional to $m(t)\cos\{\Delta\omega)t\}$. That is, a frequency mismatch appears as a *time-varying* amplitude factor, and this factor is significant for even very small mismatches. For example, if $f_c = 1$ MHz and $\Delta f = 1$ Hz (an error of just one part in a *million*!) then the amplitude of the low-pass filter output will vary from zero to maximum and back to zero *twice a second*. This would be a catastrophic effect, rendering human speech unintelligible (as you can simulate for yourself simply by rolling the volume control on a radio back and forth twice a second).

The conclusion from all this is that to successfully demodulate DSB-SC we need some way for the receiver to *accurately* reconstruct the carrier. And there *is* a way, as shown in Figure 19.1. The received signal $m(t)\cos(\omega_c t)$ is

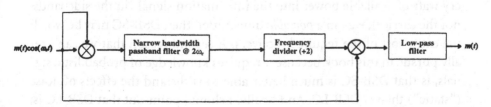

Figure 19.1 A receiver for double-sideband, suppressed carrier (DSB-SC) radio signals.

immediately squared, giving $(\frac{1}{2})m^2(t) + (\frac{1}{2})m^2(t)\cos(2\omega_c t)$. The first term is a baseband signal (limited to $|\omega| < 2\omega_m$ where ω_m is the maximum frequency in $m(t)$), and the second term is that same baseband signal shifted up to rf with its spectrum centered on $\omega = \pm 2\omega_c$. The crucial observation at this point is that this *ensures* us there will be energy at $\omega = 2\omega_c$, because $m^2(t)$ is *certain* to have a positive dc value over its duration (because a squared quantity is never negative). For more on this, see Problem 19.1.

Thus, a (*very*) narrow bandwidth bandpass filter centered at $\omega = 2\omega_c$ will have an output at twice the carrier frequency. The *narrower* the bandwidth of this filter the better. It is beyond the level of this book, but an exotic circuit called a *phase locked loop* (PLL) can be then used for the task of reconstructing the carrier.[1] Such a circuit can lock onto both the frequency and the phase of the transmitter's carrier, and then *track* it if either one (or both) of the carrier frequency and phase change with time (which they both will surely do in real life). We can drop the filter output frequency down to carrier frequency using the regenerative frequency divider circuit discussed in Chapter 16. This reconstructed carrier signal is then used to heterodyne the original $m(t)\cos(\omega_c t)$ signal, and in particular to reproduce the baseband signal $m(t)$. There is an obvious problem with this receiver, however, as the circuit of Figure 19.1 cannot be said to be "simple." The use of DSB-SC is not attractive for commercial AM broadcast radio, with its need for *tens of millions* (or even more) of *cheap* receivers to make listening attractive to the potential customers of the advertisers whose checkbooks make the whole business possible. This, quite simply, is the technical reason for why synchronous demodulation isn't the way broadcast AM radio historically evolved. It's okay if the single transmitter is expensive, but each of the many receivers had better be low-cost if you want people to buy them.

At this point, with only its deficiencies discussed, you might well be wondering if synchronous demodulation is *ever* used? The answer is yes, under certain circumstances. I've already discussed the energy inefficiency of DSB-LC in Chapter 6, and in situations where it is important to put every watt of available power into the information signal (in the sidebands, not the carrier), e.g., in a *portable* transmitter, then DSB-SC may be worth the extra cost of the required receivers. A second reason that we can't really pursue in this book because it requires knowledge of probabilistic signals, is that DSB-SC is much better able to withstand the effects of noise ("static") than is DSB-LC. And finally, I should point out that DSB-LC is, really, demodulated in an AM radio with a *synchronous* detector in a bit of disguise! This is so because the process of envelope detection requires the use of a diode that conducts only during one-half of each cycle of the input signal—and the timing of this condition is controlled by the carrier signal (which is *not* locally generated as it would be in a true synchronous receiver).

Synchronous communication systems do have one extraordinary property that should be mentioned before we leave them. Consider the transmitter and receiver circuits shown in Figure 19.2, where it is assumed that the receiver has a perfect replica of the carrier available. As I'll show next, these circuits allow us to do the seemingly impossible—to transmit two entirely different baseband signals, $m_1(t)$ and $m_2(t)$, at the same carrier frequency *at the same time* without mutual interference! The key is to use two carrier signals at the same frequency but 90° out of phase, which is why this approach is called *quadrature amplitude multiplex* (QAM). From the figure we see that the transmitted signal is

$$r(t) = m_1(t) \cos(\omega_c t) + m_2(t) \sin(\omega_c t).$$

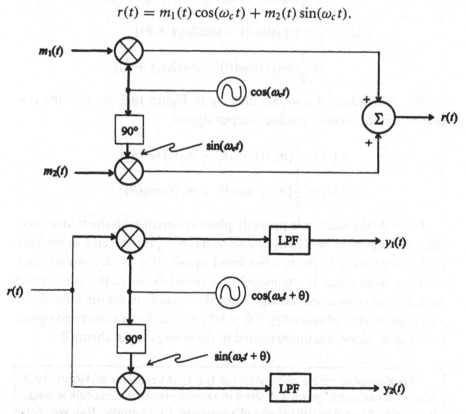

Figure 19.2   The transmitter (top) and receiver (bottom) for quadrature amplitude multiplexed DSB-SC AM radio, with a phase mismatch of θ at the receiver.

Now, following our earlier approach to DSB-SC, let's assume there is a phase mismatch of θ at the receiver. Thus, the signals immediately after the multipliers in the upper and lower receiver channels, $s_1(t)$ and $s_2(t)$, respectively, are

$$s_1(t) = r(t)\cos(\omega_c t + \theta) = m_1(t)\cos(\omega_c t)\cos(\omega_c t + \theta)$$
$$+ m_2(t)\sin(\omega_c t)\cos(\omega_c t + \theta),$$
$$s_2(t) = r(t)\sin(\omega_c t + \theta) = m_1(t)\cos(\omega_c t)\sin(\omega_c t + \theta)$$
$$+ m_2(t)\sin(\omega_c t)\sin(\omega_c t + \theta).$$

Expanding these two expressions using the obvious trigonometric identities, we get

$$s_1(t) = \frac{1}{2}m_1(t)\{\cos(\theta) + \cos(2\omega_c t + \theta)\}$$
$$+ \frac{1}{2}m_2(t)\{-\sin(\theta) + \sin(2\omega_c t + \theta)\},$$
$$s_2(t) = \frac{1}{2}m_1(t)\{\sin(\theta) + \sin(2\omega_c t + \theta)\}$$
$$+ \frac{1}{2}m_2(t)\{\cos(\theta) - \cos(2\omega_c t + \theta)\}$$

or, after the indicated low-pass filtering in Figure 19.2, we find the two channels in the receiver produce output signals

$$y_1(t) = \frac{1}{2}[m_1(t)\cos(\theta) - m_2(t)\sin(\theta)],$$
$$y_2(t) = \frac{1}{2}[m_1(t)\sin(\theta) + m_2(t)\cos(\theta)].$$

If $\theta = 0$ (the receiver is perfectly phase coherent with the transmitter) then we find $y_1(t) = (\frac{1}{2})m_1(t)$ and $y_2(t) = (\frac{1}{2})m_2(t)$, and so we have perfect separation of the two baseband signals. If $\theta \neq 0$, however, then each baseband signal is attenuated by a $\cos(\theta)$ factor in its own channel as well as suffering from "leakage" (called *cross-talk*) of the other baseband signal [proportional to $\sin(\theta)$]. If $\theta = 90°$, in fact, the two baseband signals each appear alone and unattenuated in the *wrong* output channel!

The allowable phase mismatch for the QAM receiver of Figure 19.2 is easy to calculate, given a desired maximum level of cross-talk leakage from one channel to the other. Let's suppose, for example, that we wish the cross-talk of $m_2(t)$ into $m_1(t)$'s channel to be no more than -50 dB.

That is, we demand that

$$20 \log_{10} \left[\frac{m_2 \sin(\theta)}{m_1 \cos(\theta)} \right] = 50.$$

Then, if we further assume that m_1 and m_2 are of equal magnitude, we have

$$\frac{\sin(\theta)}{\cos(\theta)} = \tan(\theta) = 10^{-2.5} = 0.0032.$$

Since $\tan(\theta) \approx \theta$ when θ is "small," as it clearly is here, then $\theta = 0.0032$ radians $= 0.18°$ is the maximum allowable phase mismatch. This example illustrates the near-perfect phase coherence needed between a QAM transmitter and receiver.

QAM allows twice as many signals to be broadcast in the same frequency band as does DSB with a distinct carrier frequency for each signal, but spectrum was cheap in the early days of radio and the complexity of achieving synchronization at a QAM receiver is much more of a negative than spectrum conversation is a plus. Today quadrature amplitude multiplexing is, however, used in color television. The receiver's local versions of both phases of the carrier are kept synchronized with the transmitter's via the periodic insertion of a short burst of carrier signal (called the *color burst*) in the transmitter signal, $r(t)$.

Note

1. The classic DSB-SC receiver circuit using a PLL is called a *Costas demodulator*, first discussed in an important paper by J. P. Costas, "Synchronous Communications," *Proceedings of the Institute of Radio Engineers* 44, December 1956, pp. 1713–1718.

Problems

19.1. Why go to all the trouble of squaring $m(t) \cos(\omega_c t)$ just to get our hands on $2\omega_c$ (so we can then divide $2\omega_c$ in half!)? Why not just include in the receiver a very narrow bandwidth bandpass filter centered on $\omega = \omega_c$, which is where the spectrum of the received signal is centered? Hint: is there any energy *at* $\omega = \omega_c$, in the received DSB-SC signal? Remember that the energy at $\omega = \omega_c$ *if any*, is the dc energy in $m(t)$—and then ask yourself what *is* the dc (or average) value of music or human speech over its duration?

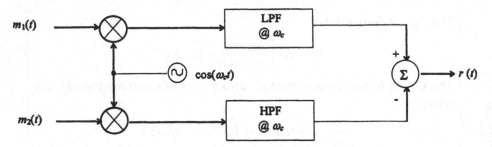

Figure 19.3   Another circuit for sending two baseband signals at the same time on the same carrier.

19.2. In the *non*quadrature transmitter circuit shown in Figure 19.3 the two filters are ideal, with vertical skirts at $\omega = \omega_c$. Both $m_1(t)$ and $m_2(t)$ are baseband signals. Without writing any equations, but simply by sketching spectrums, show that $r(t)$ transmits both signals without mutual interference on the same carrier. Can you also see, directly from $|R(j\omega)|$, how to make a synchronous demodulator that recovers $m_1(t)$ and $m_2(t)$ from $r(t)$? Hint: your receiver circuit should contain *four* ideal filters (one high-pass and three low-pass).

Analytic Signals and Single-Sideband Radio

In this chapter I'll discuss single-sideband (SSB) AM which, while dramatically different from QAM, possesses that modulation scheme's property of conserving bandwidth. SSB is historically quite old (as radio goes), with the first patent application filed in 1915 (U.S. Patent 1,449,382 granted in 1923) by the American electrical engineer John R. Carson. The importance of that invention can be measured by his obituary notice in the *New York Times* (he died young, at age 54 in 1940), which specifically cited it, alone, of all of the many accomplishments of Carson's productive career.

After thinking about the symmetry of the spectrum of a real signal, Carson reasoned that not only did the carrier in ordinary AM not contain any information, but that the spectrum itself was redundant; information in the positive frequency half is duplicated in the negative frequency half, and so only one of the two halves (*either* one) need be transmitted. Carson's thinking along this line was motivated by his employment with the American Telephone and Telegraph Company, during times when most commercially transmitted information was sent over copper wires. The available frequency bandwidth on such wires is greatly constrained compared to today's fiber-optic cables, and any way to "compress" more message-carrying capability into the available bandwidth was eagerly sought.

In 1918 the first commercial use of SSB occurred in wired telephony with a line between Baltimore and Pittsburgh, and four years later the time was ripe to widen the concept's use in establishing a transoceanic radio-telephone link between America and Europe. Experiments beginning in 1922 led to a successful demonstration of trans-Atlantic SSB radio by the Western Electric engineer Raymond A. Heising (1888–1965) in January 1923. This was soon followed by the establishment of commercial SSB radio service in 1927, between New York City and London. SSB was partic-

Figure 20.1   On the obviously cold day of April 23, 1921 Albert Einstein was the guest of the then new Radio Corporation of America (RCA), and of the General Electric Company (GE). Einstein is shown standing next to GE's electrical wizard Charles Proteus Steinmetz (with a cigar stuck in his mouth). All are in front of the RCA high-speed (50 words/min) transoceanic telegraphic transmitter facility in New Brunswick, NJ. After Einstein won the Nobel Prize in physics later that year GE began to use a retouched version of this photograph in its efforts to mythologize Steinmetz through association with the great physicist. The doctored photo (all but Einstein and Steinmetz were removed) can be found in an article in *Technology and Culture* (January 1989, p. 64), in which the author writes that in the original image "Einstein and Steinmetz [are] separated by an unidentified subject." In fact, that stern-looking fellow is John Carson, who was there as AT&T's expert in transoceanic radio engineering. (The pudgy man, fourth from the left, leaning into the camera and clearly eager not to be overlooked, is David Sarnoff. Sarnoff, then RCA's General Manager, will appear in the epilogue as a significant player in the development of commercial radio.)

ularly attractive at that time for two reasons. First, all available transmitter power could be put into doing useful work in sending nonredundant information, with none of that power wasted on the informationless but power-hungry carrier needed in ordinary AM for envelope detection. Second, in those days of relatively primitive antenna design, it was easier to make low-frequency antennas resonant over the narrow bandwidth of SSB (relative to the wider bandwidths of DSB). The higher the frequency of the SSB transmission, the less important is the second concern, of course, because the fixed bandwidth of *either* SSB or DSB becomes relatively narrower compared to the transmission frequencies. The early days of SSB radio, however, used frequencies far below today's commercial AM radio band (the London–New York SSB radio link transmitted its sideband with a bandwidth of 2.7 kHz, which could be placed anywhere in the interval extending from 41 to 71 kHz). Now, with all that said, how does SSB work?

We begin the analysis, as always, with $m(t)$ a real baseband signal [with a symmetrical spectrum, of course, $M(j\omega)$]. When we multiply $m(t)$ by $\cos(\omega_c t)$ we generate a DSB signal *with no carrier term* (which is our first goal, already accomplished), as shown in our previous work. That is, the carrier is *suppressed*. The spectrum of $m(t)$, and of $m(t)\cos(\omega_c t)$, are shown in the first two parts of Figure 20.2. The spectrum of $m(t)\cos(\omega_c t)$ is simply the spectrum of $m(t)$ shifted both up and down in frequency by ω_c. If you now concentrate your attention on the positive frequencies of the DSB-SC spectrum (the middle sketch of the figure), you'll see that the positive frequency half of $M(j\omega)$ forms the upper sideband of $m(t)\cos(\omega_c t)$, and the negative frequency half of $M(j\omega)$ forms the lower sideband. Similarly, looking next at the negative frequencies of the DSB-SC spectrum, you'll see that the negative frequency half of $M(j\omega)$ forms the upper sideband of $m(t)\cos(\omega_c t)$ and the positive frequency half of $M(j\omega)$ forms the lower sideband.

Now, as stated before, since $m(t)$ is real, the information content of $m(t)$ is duplicated in each half of $M(j\omega)$, i.e., in each sideband. So, why use power to transmit both sidebands? Why not, instead, send the signal whose spectrum looks like the bottom sketch of the figure (the upper sideband)? If $m(t)$ has a bandwidth of 5 kHz, for example, then the upper *and* lower sidebands in the DSB spectrum have a total bandwidth of 10 kHz—but the SSB spectrum has a bandwidth of only 5 kHz, which is an attractive reduction in the use of spectrum.

The most obvious way to generate the signal with the SSB spectrum shown in Figure 20.2 is to simply run the DSB signal through a high-pass

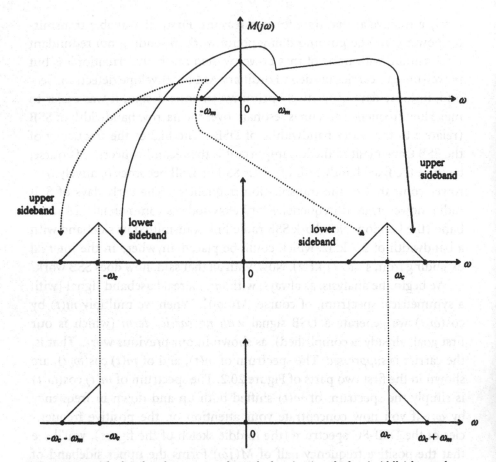

Figure 20.2   The baseband spectrum of $m(t)$ before (top) and after (middle) heterodyning. The spectrum of the upper sideband SSB signal is shown in the bottom sketch.

filter which passes energy at frequencies $|\omega| > \omega_c$. This works *on paper*, but it requires the filter to "cut off" with a vertical skirt at $\omega = \omega_c$, an impossibility (recall Chapter 15 where it was shown that a bandpass filter with vertical skirts is impossible). A "real-life" filter would either have to let a bit of the rejected lower sideband "leak through," or else the filter would have to cut off a bit of the desired upper sideband. Similar problems occur if we attempt to transmit only the lower sideband by rejecting the upper sideband with a perfect low-pass filter that passes energy only at frequencies $|\omega| < \omega_c$. Still, good engineering can to a large extent overcome these problems and, in fact, the original London–New York SSB circuitry used sideband rejection filters.

More elegant than simply filtering a DSB signal, however, is to design a transmitter circuit that *directly* generates the signal whose spectrum looks like the bottom sketch in Figure 20.2. The inventor of this form of SSB (called the "phase shift" method) was Carson's colleague Ralph Vinton Lyon Hartley at AT&T, who received U.S. Patent 1,666,206 in 1928 (filed in 1925). Born in 1888 and educated as a physicist, Hartley made many important contributions to electronics and information theory during a career plagued by illness at the Bell Telephone Laboratories (created by AT&T in 1925). He died in 1970. In the analysis that follows you'll see how all the elegant Fourier theory developed in Section Two shows how Hartley neatly solved the problem of how to directly generate an SSB signal.

We start with the pair $m(t) \leftrightarrow M(j\omega)$. Let's now define a new signal $z_+(t) \leftrightarrow Z_+(j\omega)$, where $Z_+(j\omega) = M(j\omega)u(\omega)$. That is, since $u(\omega)$ is the unit step in the frequency domain, then $z_+(t)$ is the time signal that has a spectrum that is zero for negative frequencies and equal to the positive frequency half of $M(j\omega)$. Clearly $Z_+(j\omega)$ is not a symmetrical spectrum, and so we know $z_+(t)$ is not a real time function. Don't worry about this—we are not going to try to actually generate $z_+(t)$, which would be futile since it's complex. The top sketch in Figure 20.3 shows $Z_+(j\omega)$. Now, as shown at the end of Chapter 14, $(\frac{1}{2})\delta(t) + j\frac{1}{(2\pi t)} \leftrightarrow u(\omega)$, and so from the time convolution theorem we have

$$z_+(t) = m(t) * \left\{\frac{1}{2}\delta(t) + j\frac{1}{2\pi t}\right\} = \frac{1}{2}m(t) * \delta(t) + j\frac{1}{2\pi}m(t) * \frac{1}{t},$$

or

$$z_+(t) = \frac{1}{2}\left[m(t) + j\frac{1}{\pi}\int_{-\infty}^{\infty}\frac{m(u)}{t-u}\,du\right].$$

This really quite odd-looking time function is called an *analytic signal*.[1] Indeed, *any* time signal that has a single-sided spectrum is said to be analytic. Recall from Chapter 15 that the integral that is the imaginary part of $z_+(t)$ is the Hilbert transform of $m(t)$, which I'll write as $\overline{m}(t)$. Thus, $z_+(t) = (\frac{1}{2})[m(t) + j\overline{m}(t)]$. Since $m(t)$ is a baseband signal, then $z_+(t)$ is also a baseband signal [look again at $Z_+(j\omega)$ in Figure 20.3]. To shift the spectrum of $z_+(t)$ up to rf, we'll simply multiply by $e^{j\omega_c t}$. Doing this, we get

$$z_+(t)e^{j\omega_c t} = \frac{1}{2}[m(t) + j\overline{m}(t)][\cos(\omega_c t) + j\sin(\omega_c t)]$$

$$= \frac{1}{2}[m(t)\cos(\omega_c t) - \overline{m}(t)\sin(\omega_c t)]$$

$$+ j\frac{1}{2}[\overline{m}(t)\cos(\omega_c t) + m(t)\sin(\omega_c t)].$$

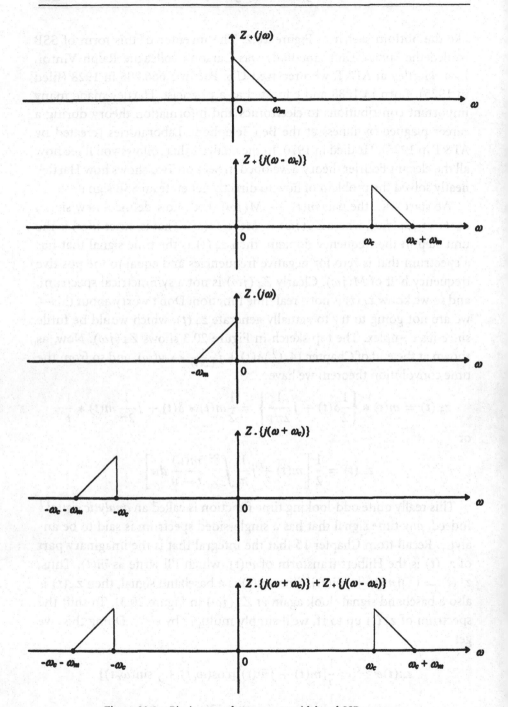

Figure 20.3   Piecing together an upper sideband SSB spectrum.

This last expression is the complex time signal that has a spectrum that is the *positive* frequency part of an upper SSB signal, as shown in the second sketch of Figure 20.3.

To get a real time signal we can physically generate, we of course need a *symmetrical* spectrum. So, we repeat the whole business, but this time we put together the time signal that gives us the *negative* frequency part of the SSB signal. So, let's write $z_-(t) \leftrightarrow Z_-(j\omega)$ where $Z_-(j\omega) = M(j\omega)u(-\omega)$, as shown in the third sketch of Figure 20.3. That is, $z_-(t)$ is the time signal that has a spectrum that is zero for positive frequencies and equal to the negative frequency half of $M(j\omega)$. Since we have $(\frac{1}{2})\delta(t) - j\frac{1}{(2\pi t)} \leftrightarrow u(-\omega)$ then our new analytic signal is

$$z_-(t) = m(t) * \left\{ \frac{1}{2}\delta(t) - j\frac{1}{2\pi t} \right\} = \frac{1}{2}[m(t) - j\overline{m}(t)].$$

Multiplying this by $e^{-j\omega_c t}$ to shift the spectrum of $z_-(t)$ *down* the frequency axis by ω_c, we have

$$z_-(t)e^{-j\omega_c t} = \frac{1}{2}[m(t)\cos(\omega_c t) - \overline{m}(t)\sin(\omega_c t)]$$
$$- j\frac{1}{2}[\overline{m}(t)\cos(\omega_c t) + m(t)\sin(\omega_c t)].$$

Now, if we combine the spectrums (as in the bottom sketch of Figure 20.3) of the two complex time signals $z_+(t)e^{j\omega_c t}$ and $z_-(t)e^{-j\omega_c t}$ we have our SSB spectrum. And best of all, the sum of these two complex time signals is real and so we can generate it. Thus, our desired SSB signal is

$$r(t) = z_+(t)e^{j\omega_c t} + z_-(t)e^{-j\omega_c t} = m(t)\cos(\omega_c t) - \overline{m}(t)\sin(\omega_c t).$$

This signal gives us upper sideband SSB. If you repeat this entire analysis for lower sideband SSB (you should do this), you'll find the required time signal is $m(t)\cos(\omega_c t) + \overline{m}(t)\sin(\omega_c t)$. This is the same as for the upper sideband except for the $+$ sign. Thus, by adding a simple switch to the final summer in the Hilbert signal path (of the circuit shown in Figure 20.4) we can generate upper or lower sideband SSB with the flip of a switch. (For the historically minded, the pioneer London–New York SSB radio link used the *lower* sideband.)

Historically, this is *not* the way SSB was discussed in the early technical literature. In Raymond Heising's classic SSB paper, for example ("Production of Single Sideband for Trans-Atlantic Radio Telephony," *Proceedings of the IRE*, June 1925, pp. 291–312) you'll find only

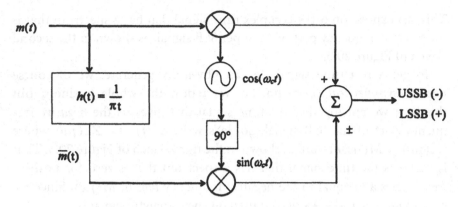

Figure 20.4   An SSB radio transmitter (phase-shift method).

trigonometric identities. The passage of decades did little to change that. In December 1956 the Institute of Radio Engineers devoted its entire *Proceedings* to the single topic of SSB. *Nowhere* in that entire issue do analytic signals appear. This was 10 years *after* Dennis Gabor's pioneering paper on analytic signals (see Appendix I), and so we can make two interesting observations from this: (1) the radio pioneers Carson, Heising and Hartley were *very* clever people who didn't need the fancy math to invent SSB (but the fancy math makes it a *lot* easier to understand), and (2) it took a long time for even the more academic electrical engineers to see how analytic signals fit into the scheme of things.

The central problem with the SSB transmitter we have just "designed" is the Hilbert transform box [with impulse response $h(t) = \frac{1}{\pi t}$, $|t| < \infty$]. Since $h(t) \neq 0$ to $t < 0$, this box is not causal, i.e., we can't build it! Matters aren't completely grim, however. As shown in Appendix I, the frequency domain description of $h(t)$ is remarkable—it is an all-pass 90° phase shifter. That is, a Hilbert box does not affect the amplitude of any of the sinusoidal components of its input signal, but rather merely phase shifts them, i.e., all positive frequency components are shifted by −90° and all negative frequency components are shifted by +90°. (If you skipped over Problem 13.6, now is the time to go back and take a look at it.)

As already stated, the Hilbert box is impossible to build, but it *is* possible to build a box that has *two* sets of output terminals that, over a *finite* bandwidth [the bandwidth of the input baseband signal, $m(t)$], provide signals with the same amplitude but with a *nearly* constant 90° *difference* in phase. Thus, instead of having a fixed 0° shift in the upper channel of

Figure 20.4 and a fixed $90°$ shift in the lower channel, we can have a shift of α (which can vary with ω) in the upper channel and a shift of $\alpha + 90°$ in the lower channel, and still have our SSB. Such realizable boxes, using only resistors and capacitors, have been described in the technical literature, and in fact are quite easy to construct (although their theoretical treatment is far from trivial, significantly beyond the level of this book).[2] To see how this works, focus your attention on just one particular frequency component in $m(t)$, e.g., the component at frequency ω_p, which I'll write as $\cos(\omega_p t)$. Boosted up to rf by multiplication with $\cos(\omega_c t)$, this baseband component would produce sideband tones at frequencies $\omega_c + \omega_p$ (upper sidetone) and $\omega_c - \omega_p$ (lower sidetone). Now, after passing through the all-pass phase shifters, the inputs to the upper and lower channel multipliers are (upper) $\cos(\omega_p t - \alpha)$ and (lower) $\cos(\omega_p t - \alpha - \frac{\pi}{2})$. Thus, the multiplier outputs are (after some *careful* trigonometric manipulations):

In the upper channel

$$\cos(\omega_p t - \alpha)\cos(\omega_c t) = \frac{1}{2}\cos\left[\{(\omega_c + \omega_p)t\}\cos(\alpha)\right.$$
$$+ \sin\{(\omega_c + \omega_p)t\}\sin(\alpha)$$
$$+ \cos\{(\omega_c - \omega_p)t\}\cos(\alpha)$$
$$\left. - \sin\{(\omega_c - \omega_p)t\}\sin(\alpha)\right];$$

In the lower channel

$$\cos\left(\omega_p t - \alpha - \frac{\pi}{2}\right)\sin(\omega_c t) = \frac{1}{2}\left[-\sin\{(\omega_c + \omega_p)t\}\sin(\alpha)\right.$$
$$- \cos\{(\omega_c + \omega_p)t\}\cos(\alpha)$$
$$- \sin\{(\omega_c - \omega_p)t\}\sin(\alpha)$$
$$\left. + \cos\{(\omega_c - \omega_p)t\}\cos(\alpha)\right].$$

If we now add the upper and lower channel multiplier outputs (as in Figure 20.4) we get $\cos\{(\omega_c - \omega_p)t\}\cos(\alpha) - \sin\{(\omega_c - \omega_p)t\}\sin(\alpha) = \cos\{(\omega_c - \omega_p)t + \alpha\}$ which is, indeed, a signal at *only* the lower sidetone frequency. And if we subtract the lower channel sign from the upper channel signal, we get $\cos\{(\omega_c + \omega_p)t\}\cos(\alpha) + \sin\{(\omega_c + \omega_p)t\}\sin(\alpha) = \cos\{(\omega_c + \omega_p)t - \alpha\}$ which is, indeed, a signal at *only* the upper sidetone frequency. But what, you may be wondering, about that α which appears as a phase shift in the SSB output? After all, if α is a function of frequency, then each frequency component of $m(t)$ will experience a different phase shift. Won't that scramble $m(t)$ up at the receiver? "Yes," theoretically, but "no" practically because, as explained near the end of this chapter, for $m(t)$ signals meant to be *heard* (speech), the human ear is fairly insensitive to phase. This is not true for the eye, however, and so this method of generating SSB would be a bad choice for transmitting video signals meant to be *seen*.

The detection of an SSB signal by a receiver is identical to that of a DSB signal. We simply multiply the received signal by $\cos(\omega_c t)$ and then low-pass filter. To show this, as well as to determine the effects of both phase and frequency mismatches, I'll now calculate what this multiplication process gives when we use $\cos\{(\omega_c + \Delta\omega)t + \theta\}$, where both $\Delta\omega$ and θ are zero for perfect synchronous demodulation. Thus,

$$
\begin{aligned}
r(t)\cos\{(\omega_c + \Delta\omega)t + \theta\} &= [m(t)\cos(\omega_c t) \\
&\quad \pm \overline{m}(t)\sin(\omega_c t)]\cos\{(\omega_c + \Delta\omega)t + \theta\} \\
&= \frac{1}{2}m(t)\,[\cos\{(\Delta\omega)t + \theta\} \\
&\quad + \cos\{(2\omega_c + \Delta\omega)t + \theta\}] \\
&\quad \pm \frac{1}{2}\overline{m}(t)\,[\sin\{(\Delta\omega)t + \theta\} \\
&\quad + \sin\{(2\omega_c + \Delta\omega)t + \theta\}]\,.
\end{aligned}
$$

After low-pass filtering, the detected signal is thus

$$
\frac{1}{2}m(t)\cos\{(\Delta\omega)t + \theta\} \pm \frac{1}{2}\overline{m}(t)\sin\{(\Delta\omega)t + \theta\}\,.
$$

Notice that if $\Delta\omega$ and θ are both zero, then the detected signal is just $(\frac{1}{2})m(t)$, which means the original baseband signal has been exactly recovered, as claimed.

Now, suppose that we have a perfect match in frequency at the receiver ($\Delta\omega = 0$) but not in phase. Then the detected signal is $(\frac{1}{2})m(t)\cos(\theta) \pm (\frac{1}{2})\overline{m}(t)\sin(\theta)$. The first term is just what we got in the previous chapter for a phase mismatch in demodulating DSB-SC, but now with SSB we also have a second term involving the Hilbert transform of $m(t)$. What does that mean? Actually, for normal speech transmission it is *good* news. This is so because since $\overline{m}(t)$ is simply $m(t)$ with all of its various frequency components shifted by 90°, and since the human ear is experimentally found to be fairly *insensitive* to phase, then $\overline{m}(t)$ generally sounds the same as $m(t)$. Thus, when θ is approaching 90° and the $\cos(\theta)$ factor is attenuating $m(t)$ (just as in DSB-SC), the opposite effect is occurring for the $\overline{m}(t)$ term and so the *total* detected signal never vanishes for any θ.

For $m(t)$ message signals that change rapidly, however (e.g., pulses), then $\overline{m}(t)$ can have arbitrarily large variations [see Appendix I for the detailed calculation of $\overline{m}(t)$ when $m(t)$ is a pulse] and such variations cannot be reproduced in any real receiver. For this reason, SSB is not a good choice for transmitting pulselike signals. In addition, unlike the ear, the eye is an excellent phase detector and it can distinguish (with great sensitivity) $m(t)$ from $\overline{m}(t)$ when both are presented on a video screen in the form of time-varying luminous patterns. Thus, phase mismatch in the demodulation of an SSB television signal would be easily *seen*.

Now, with a frequency mismatch at the receiver during the detection of SSB, an equally curious effect occurs that is also completely unlike the result for DSB-SC. In the SSB case, the detected signal after low-pass filtering is (with $\theta = 0$) $(\frac{1}{2})m(t)\cos\{(\Delta\omega)t\} \pm (\frac{1}{2})\overline{m}(t)\sin\{(\Delta\omega)t\}$. Again, the first term is what we got for DSB-SC, and by itself represents a time-varying amplitude effect which can be catastrophic even for a very small $\Delta\omega$. But now we also have a second term involving the Hilbert transform of $m(t)$. And so, again, we ask what that means? To answer this question, let's focus our attention on a particular frequency component of $m(t)$; call it $\cos(\omega_p t)$.

As shown in Appendix I, the Hilbert transform of $\cos(\omega_p t)$ is just $\sin(\omega_p t)$—which is of course simply $\cos(\omega_p t)$ shifted by 90°—and so the detected signal at frequency ω_p is

$$\frac{1}{2}\cos(\omega_p t)\cos\{(\Delta\omega)t\} \pm \frac{1}{2}\sin(\omega_p t)\sin\{(\Delta\omega)t\}$$

$$= \frac{1}{2}\left[\cos\{(\omega_p + \Delta\omega)t\} + \cos\{(\omega_p - \Delta\omega)t\}\right]$$

$$\pm \frac{1}{2}\left[\cos\{(\omega_p - \Delta\omega)t\} - \cos\{(\omega_p + \Delta\omega)t\}\right].$$

Therefore, depending on which sideband we are using (in Figure 20.4's $\pm$ notation, $-$ goes with the upper sideband and $+$ goes with the lower sideband), we have the detected signal as

$$\cos\{(\omega_p + \Delta\omega)t\} \text{ for upper sideband SSB}$$

and

$$\cos\{(\omega_p - \Delta\omega)t\} \text{ for lower sideband SSB}.$$

Thus, in either case we see that the presence of the Hilbert transform term has resulted in shifting ω_p by $\Delta\omega$. Since ω_p is arbitrary, then *every* frequency component of $m(t)$ is shifted by the *same* amount, equal to the frequency mismatch. And, as with phase mismatch, this effect is *not* catastrophic (as is frequency mismatch in demodulating DSB-SC). Since every frequency is shifted by the same absolute value, then the harmonic relationships that exist among the components of the original $m(t)$ are destroyed, but unless $\Delta\omega$ is fairly large this does *not* destroy the intelligibility of the receiver output. Indeed, if Δf can be held to within ± 30 Hz or so, speech reproduction is typically good. (This effect of frequency mismatch in demodulating SSB is *not* the same as simply changing the pitch of speech. A pitch change is the effect one would get by, for example, playing a tape recording either too fast or too slow. In either case, each recorded frequency component is shifted *proportionally*, which leaves the harmonic relationships among the various components unaltered.) For Δf greater than about ± 30 Hz, the result for SSB is to make people sound a bit like Donald Duck (but even then intelligibility is not necessarily completely lost). To aid the receiver in achieving synchronism, a small amount of carrier can also be transmitted, which the receiver could extract with the use of a phase-locked loop.

Indeed, suppose that a single-sideband signal is transmitted not just with a small carrier but, in fact, with an arbitrarily large carrier present. Such a signal is called SSB-LC, and while it no longer has the energy efficiency of pure SSB, it does still retain SSB's spectrum conserving property. In addition, synchronous demodulation is no longer required, and an SSB-LC signal can be demodulated by an ordinary AM radio (which uses an envelope detector). For this reason SSB-LC is also called *compatible single sideband*. To see this, write the received signal as

$$R(t) = m(t)\cos(\omega_c t) \pm \overline{m}(t)\sin(\omega_c t) + A\cos(\omega_c t)$$
$$= [m(t) + A]\cos(\omega_c t) \pm \overline{m}(t)\sin(\omega_c t),$$

where A is assumed sufficiently large that $R(t)$ is never negative. Then the magnitude (or envelope) of $R(t)$ is

$$|R(t)| = \sqrt{[m(t) + A]^2 + \overline{m}^2(t)} = A\sqrt{1 + 2\frac{m(t)}{A} + \frac{m^2(t)}{A^2} + \frac{\overline{m}^2(t)}{A^2}}.$$

For sufficiently large A, $|R(t)|$ can be written approximately as

$$|R(t)| \approx A\sqrt{1 + 2\frac{m(t)}{A}} \approx A + m(t).$$

The output of an envelope detector is thus the transmitted baseband signal, $m(t)$, added to a dc term proportional to the carrier amplitude (this dc term can easily be removed with a capacitor in series with the envelope detector—it is called a *blocking* capacitor). To make all this work it is necessary to generally have a *quite* large carrier amplitude, which means SSB-LC is actually *less* energy efficient than even DSB-LC. Still, because of its conservation of spectrum, the video signal in television is essentially a SSB-LC signal that also allows inexpensive envelope detector demodulation at the receiver.

In the early days of development that finally gave rise to commercial AM radio as we know it now, SSB was for a time a serious candidate for adoption as the standard signal format. It lost in that competition to DSB-LC, but it later found its niche in a much less public market than is national radio. During the Second World War the ultra-super top secret encrypted trans-Atlantic radio-telephone frequency-scrambled link used by English Prime Minister Winston Churchill and President Franklin Roosevelt was a single-sideband system. This system was so secure that, to any eavesdropper attempting to intercept its transmissions, they were said to sound like the Rimsky-Korsakov violin showcase "The Flight of the Bumblebee." Since that was the theme music for the then popular radio show "The Green Hornet," that became the sobriquet by which the system became known among the intelligence cognoscenti. As Tom Clancy has one of the characters say in his 1989 novel *Clear and Present Danger*, "Single sideband super-encrypted [ultra-high frequency]. That's as secure as communications get."

The "Green Hornet" was known to its developer (The Bell Telephone Laboratories) as the "X-System," and was dubbed "SIGSALY" by its military sponsor (the U.S. Army Signal Corps). It was one of the deepest secrets of the Second World War, nearly the equal of the atomic bomb project. It was finally declassified more than thirty years after the war ended, in 1976. The "Green Hornet" was the first theoretically unbreakable electronic speech scrambler constructed, as it *digitally* encrypted the information in its signal with a one-time random binary sequence recorded on two vinyl phonograph records (one at the transmitter terminal for encoding, and one at the receiver terminal for decoding) which were destroyed immediately after a single use. This random encryption was the key to the ultra-secure nature of the "Green Hornet"— as one engineer who worked on its development said, "We were convinced that we could have dropped a terminal in Berlin and without the [random] records no one could figure it out." The need for such a secure military communication system was dramatically illustrated by the fact that as early as the fall of 1941 the Deutsche Reichspost had defeated the then standard A-3 radiotelephone scrambler. With an intercept station in the Netherlands, for example, the Germans were able to instantly descramble the A-3's transmissions (including some between Churchill and Roosevelt), and then send them on to top Nazi leaders, including Adolf Hitler himself.

Compared to the elementary scrambler circuit described in Problem 16.1 (which simply inverts the input speech spectrum) the A-3 was (at least on the surface) a sophisticated step up in complexity. It divided the input speech spectrum into five channels, inverted the spectrum of each channel separately, and then randomly permutated the channels once every 20 seconds. Since there are only 120 such permutations, however, it is not difficult to understand why the Germans were able to quickly break the A-3. The compromising of the A-3 was so well known among American military leaders that many refused to use it, e.g., Army Chief of Staff General George C. Marshall rejected the A-3 on December 7, 1941 as the means to alert Pearl Harbor, Hawaii on the possibility of an imminent Japanese attack.

Given the state of the art of electronics technology in those long-ago war years, the "Green Hornet" could not avoid being a massive system—each terminal occupied several hundred cubic feet and was once described as weighing "a couple of pounds less than a sawmill"—but even so, a dozen networked terminals were deployed around the world, including one on a ship in the Pacific Ocean. The terminals used by Roosevelt and Churchill were, respectively, in the Pentagon (with a secure link to the White House) and in the sub-basement of Selfridge's department store in London (with a secure link to Churchill's lair in the underground, bombproof Cabinet War Rooms). After Roosevelt's death, President Harry Truman continued to use the "Green Hornet" to talk with Churchill; on April 25, 1945, for example, the two men discussed (and eventually rejected) the offer by Nazi SS and Gestapo head Heinrich Himmler to surrender German forces fighting the British and the Americans, while not surrendering other German forces fighting the Russians on a different front.

Unlike the A-3, the "Green Hornet" was never compromised, and you can read more about that amazing gadget in the paper by William R. Bennett, "Secret Telephony as a Historical Example of Spread-Spectrum Communication," *IEEE Transactions on Communications,* January 1983, pp. 98–104. There is a hilarious description of one of Churchill's scrambled (in every sense of the word) conversations in the war memoir by Fitzroy Maclean, *Escape to Adventure,* Little, Brown 1951, pp. 343–345.

Notes

1. *Analytic signal* is a term from the mathematical theory of complex functions,e.g., see my book *An Imaginary Tale: the story of $\sqrt{-1}$*, Princeton 1998, p. 193.

2. See, for example, Donald K. Weaver, Jr., "Design of RC Wide-Band 90-Degree Phase-Difference Network," *Proceedings of the Institute of Radio Engineers* 42, April 1954, pp. 671–676. Weaver's paper presents a design example in which he shows how to build such a box that, over the interval $f_1 = 300$ Hz to $f_2 = 3$ kHz, provides two outputs with a 90° phase difference held to within $\varepsilon = 1.1°$, using just six resistors and six capacitors (all with reasonable values). An Electronics Workbench simulation of this circuit is shown in Figure 20.5 which indicates a phase shift of 89.67° at 1.318 kHz. At the "edges" of the frequency interval the phase difference is still very near 90° (88.99° at 302 Hz and 88.84° at 3.02 kHz). Weaver's general design equations allow f_1, f_2, and ε to be freely specified, and they are quite easy to use. Weaver mentions that he built such boxes with ε less than 0.2°. Also good reading is the earlier paper by IEEE Medal of Honor winner Sidney Darlington (1906–1997) (my own colleague for more than twenty years at the University of New Hampshire), "Realization of a Constant Phase Difference," *The Bell System Technical Journal* 29, January 1950, pp. 94–104.

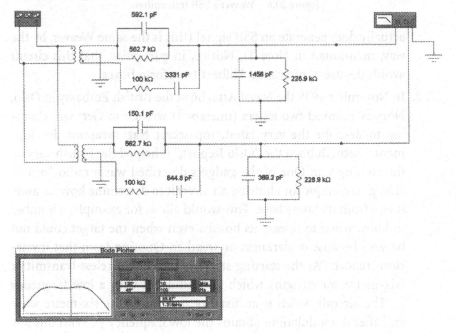

Figure 20.5   Weaver's wideband 90° phase-difference network for use in SSB.

Problems

20.1. Just to show that one can never be sure that a well-studied topic has really been exhausted (even after *decades*), consider the circuit shown in Figure 20.6. It shows Weaver's SSB transmitter, published in 1956 more than 40 years after Carson's pioneering patent application. Verify, by simply sketching spectrums, that Weaver's circuit

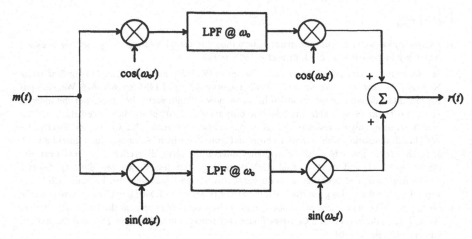

Figure 20.6 Weaver's SSB transmitter.

actually does generate an SSB signal (this is the same Weaver, by the way, mentioned in Note 2). Notice, in particular, that this circuit avoids the use of noncausal Hilbert transform boxes.

20.2. In November 1939 the Naval Attaché at the British Embassy in Oslo, Norway received two letters (unsigned) written in German, claiming to describe the very latest, top-secret Nazi weapons developments. Soon dubbed the "Oslo Report," it helped the British survive the coming war. One of the gadgets described was a radio "beam-riding" technique for allowing an aircraft to determine how far away it was from its home base. This would allow, for example, a bomber to know when to release its bombs even when the target could not be seen because of darkness or weather. Quoting from that mysterious report: "At the starting station there is a wireless transmitter (six-metre wavelength) which is modulated with a low frequency f. The aircraft which is at distance a, receives the six-metre wave and after demodulation, obtains the low frequency f. With this low frequency it modulates its own transmitter, which has a somewhat different wavelength. The thus modulated wave of the aircraft is received at the starting station and demodulated. The low frequency f thus obtained is compared with the local frequency f. Both differ by the phase angle... where a is the distance of the aircraft and c is the speed of light. By measuring the phase it is thus possible to compute the distance of the aircraft and to give the aircraft its position. In order that the measurement is free from ambiguity the phase must be [no more than] 2π."

(a) Why does the aircraft transmit a "somewhat different wavelength" than does the starting station?

(b) What is the formula for the phase angle?

(c) What frequency f gives a maximum unambiguous distance of 1000 km (about 600 mi)?

(d) Compare your answer of part c with the frequency of a 6-m wave.

Note: you can read more about the Oslo Report, including the astounding story of how its author (who survived five different concentration camps during the war) was discovered in 1954, in R. V. Jones, *Reflections on Intelligence*, Heinemann 1989, pp. 265–337.

Denouement

After all of the previous discussion on the technical demands of synchronous demodulation receivers, it is easy to see the attractiveness of using the envelope detection Chapters 5 and 6 instead. That requires the presence of a strong (or *large*) carrier in the received signal, which neither DSB-SC (by definition) and SSB (perhaps) require, as well as a bandwidth twice that of the baseband signal (i.e., twice that required by SSB). So easy and inexpensive is the envelope detector to construct, however, that the technical virtues of those other forms of AM radio simply can't compete with DSB-LC for public, commercial radio use.

Since the beginning of commercial AM radio, then, DSB-LC has been the signal format used by all countries of the world. So, what more is there to say, technically, about AM radio? As it turns out, not much more, but what does remain represents the final, great innovation that transformed the complex radio receiver of pre-1918 into the easy-to-use radio of today. That innovation was Edwin Howard Armstrong's invention of the superheterodyne receiver, which turned Fessenden's basic heterodyne concept (recall Chapter 7) into a marvel of tuning and selectivity.

Armstrong published his description of the superheterodyne in "A New System of Short Wave Amplification," *Proceedings of the Institute of Radio Engineers* 9, February 1921, pp. 3–27. It was a duplicate, almost word-for-word, of an earlier paper he gave to the Radio Club of America at Columbia

University on December 19, 1919, entitled "A New Method for the Reception of Weak Signals at Short Wave Lengths." In both papers he acknowledges the "work" of the Frenchman Lucien Levy, and for the rest of his life Levy claimed *he* was the true inventor of the superheterodyne. The Swiss-born German Walter Schottky (1886–1976) also made similar claims for a while but, unlike Levy, came to acknowledge Armstrong's prior claim. (Schottky didn't disappear into history, however; his theoretical work in the 1930s on electric currents across metal-semiconductor junctions is the basis for the well-known "Schottky diode.") Radio engineers today generally give Armstrong credit for the superheterodyne, but there is no denying that it was an idea whose time had come by the end of the First World War, and that it would surely have been developed at about the same time by others even if Armstrong had not been on the scene. As an example of how the idea was "in the air," John Carson at AT&T (the inventor, you will recall from the previous chapter, of one form of SSB) published a paper written in 1918 in which he quite specifically discusses the demodulation of an AM signal by heterodyning it with a "locally generated wave," which is the heart-and-soul of the superheterodyne. See his "A Theoretical Analysis of the Three-Element Vacuum Tube," *Proceedings of the Institute of Radio Engineers 7*, February 1919, pp. 187–200.

For you to understand why the superheterodyne is called *super*, I need to say a few words about what had come before. Before the superheterodyne receiver there was the tuned radio frequency (TRF) receiver, and before that there was the regenerative (positive feedback) receiver. The regenerative receiver circuit (invented by Armstrong while still an undergraduate electrical engineering student at Columbia) returns some of the rf energy in the plate circuit of a triode tube back (via an adjustable, coupled inductor) to the tube's input at the grid. By *very* delicate adjustment of this feedback path (often called the "tickler circuit"), Armstrong was able to achieve enormous amplification at radio frequency, i.e., he took the tube *almost* to the point of oscillation (see Chapter 8). He successfully demonstrated the regenerative receiver to representatives of AT&T on the evening of February 6, 1914. Evening was the time selected for the demonstration because then Armstrong was able to take advantage of the long-range signal paths made possible by the cooling, settling ionosphere of nighttime which could skip-bounce low-frequency radio waves around the curvature of the earth. Armstrong was, that evening, able to tune in the Pouleson arc transmitter in Honolulu, Hawaii (operating at 50 kHz), 5000 miles away.

Armstrong later developed circuits that made the delicate "tickler" adjustment *automatically*, and such receivers were called *super-regenerative*.

They never, however, played a role in commercial AM radio. Armstrong was also the first to appreciate the use of sufficient positive feedback to *electronically* generate constant amplitude oscillations (indeed, it would have been impossible for him to have overlooked them while experimenting with his regenerative amplifier.) Positive feedback oscillations had long before been observed in telephone circuits when the mouth and ear pieces were placed too close together (thus producing a howling tone sometimes called "singing"), but it was Armstrong who first realized the significance of such oscillations in circuits using De Forest's triode vacuum tube. Later, De Forest, always ready to claim an invention of somebody else as his own, asserted *he* had priority in these matters and dragged Armstrong into court. In 1934 the Supreme Court ruled in De Forest's favor, to the utter astonishment of radio engineers. So despondent was Armstrong over this incredible injustice that he attempted to return the Medal of Honor which the Institute of Radio Engineers had awarded him in 1918 in recognition of his work on both regenerative and oscillating circuits. Engineers (if not judges) knew who the real inventor was, however, and the Institute refused to take back the medal. Indeed, the IRE took the extraordinary action of publicly reaffirming its recognition of Armstrong's achievements. Armstrong later received the 1941 Franklin Medal and the 1942 Edison Medal (from the American Institute of Electrical Engineers), both awarded specifically for his invention of the oscillating and regenerative electronic circuits.[1] In 1946 De Forest also received the Edison Medal, but the citation mentions only the triode *tube*; the absence of any reference to alleged specific uses of the tube by De Forest (e.g., the development of oscillation circuitry) is impossible to overlook.

Regenerative receivers were sold commercially throughout the 1920s, but their operation was plagued by several serious faults. Most irritating was a tendency to cross the line from very-high-gain amplifier to oscillator. Then the receiver would emit loud howls and shrieks as it turned itself into a miniature transmitter (which then interfered with nearby receivers). Somewhat less irritating was the wandering nature of the tuning. A favorite station would not consistently appear at the same spot on the tuning dial because the regenerative feedback loop was quite sensitive to multiple variables (e.g., the power supply voltage, the exact coupling of the tickler back into the grid, etc.). In addition, the art of producing a good, so-called hard vacuum in a tube was still young, and so tubes often contained enough residual gas to significantly influence their characteristics. So, each tube was different (De Forest incorrectly thought the presence of gas *necessary* for triode operation!) and, in particular, if the tube in a regenerative receiver burned out, then its replacement *ensured* that all the station dial settings would change.

An alternative to using positive feedback to achieve high gain in a single stage of amplification at rf is to simply cascade several stages, each of relatively low gain. Each amplifier stage was independently tuned to the frequency band of interest. This eliminated all obvious tickler feedback loops and thus largely (but not completely) avoided the oscillation problem; but it also introduced a major new problem. These so-called TRF receivers offered their users an unhappy proliferation of knobs, one for each adjustable capacitor and/or inductor in each stage of amplification. With several knobs to individually and independently adjust, the act of tuning in any particular station became a frustrating one, indeed.[2] And even *more* knobs would commonly appear to control the filament currents to each of the tubes, a rather direct way to control the volume of the final output signal delivered to the loudspeaker. The control panels of early radio receivers were quite literally *infested* with knobs.

And there was *still* the oscillation problem, which had not been entirely eliminated with the TRF receiver. This problem, for the TRF design, is due to a far less obvious reason than the gross positive feedback path in a regenerative receiver. While it is not too hard to build a stable (nonoscillatory) amplifier that gives good gain at any *particular* frequency, it is much more difficult to do so for a *tuneable* amplifier. There are almost certain to be some frequencies at which the amplifier will oscillate. An explanation for this perplexing phenomenon of conditional stability in amplifiers didn't come until 1932, with the work of Harry Nyquist (1890–1976) at the Bell Telephone Laboratories. Nyquist was motivated to study this problem when his colleague Harold S. Black (the inventor of the negative feedback amplifier) found he could build electronic circuits that didn't obey the Barkhausen criterion (see Chapter 8). Today all electrical engineering students learn how to generate a "Nyquist plot" for an electronic circuit to determine its stability.

For a while, in the mid-1920s before the superiority of the superheterodyne design finally eliminated all competition (and while RCA refused to license other manufacturers to use the superheterodyne patents it had purchased from Armstrong), Armstrong's friend Louis Hazeltine made a fortune with his "neutrodyne" receiver. This design used *negative* or degenerative feedback to neutralize the inherent instability of tuneable rf amplifiers. Hazeltine's receiver lost its special status in 1927, however, with the introduction of tetrode vacuum tubes that greatly reduced the inherent positive feedback (at high frequencies) in triodes. So-called second generation TRF receivers using tetrodes were sold into the early 1930s as inexpensive alternatives to the superheterodyne (Philco's famous TRF Model 20 "Baby Grand" sold for $49.50, "less tubes"), but after 1932 the superheterodyne reigned supreme.

The fundamental problem facing Armstrong in building rf amplifiers was in the tube, itself. The interelectrode capacitances in the early tubes were large (where a few *pico*farads—a picofarad is a million times smaller than a microfarad—is the measure of *small*), and so at high frequencies the grid-to-cathode and the plate-to-cathode capacitances tended to short-circuit the input and the output signals, respectively. Both effects obviously reduce the gain of the tube at high frequency. And even worse, the plate-to-grid capacitance served as an unintentional regenerative feedback path from output to input, and so at high frequency the early tubes would often suddenly break into oscillation. That, of course, utterly ruined the tube's usefulness as an amplifier. In the mid-1920s, however, scientists at General Electric added another electrode to vacuum tubes. This new electrode was placed between the plate and the grid and served as an electrostatic shield that greatly reduced the capacitive coupling of the plate to the grid. Plate-to-grid capacitance in such tubes was in the millipico (i.e., femto) farad range. This new electrode was called, appropriately, the *screen grid*, and the resulting four-electrode tube was called a *tetrode*.

As with so many other developments in engineering and technology, the superheterodyne was born in war. During the First World War, Armstrong served in France with the American Expeditionary Force as an officer in the U.S. Army Signal Corps. He was charged with the general problem of intercepting German radio transmissions, particularly those at high frequencies. Related to that was the intriguing idea of building an early warning receiver to detect the approach of enemy aircraft via the electromagnetic radiation emitted by ignition systems (i.e., engine spark-plug firings). The great difficulty facing such ambitious goals for those days was that it was not possible to build amplifiers that could directly amplify very much at the frequencies involved (up to perhaps 3 MHz). It was this problem that caused Armstrong to recall Fessenden's heterodyne circuits from nearly two decades before. That is, Armstrong decided to simply move the received high-frequency signals he was trying to detect down to the lower frequencies where good amplification *could* be achieved. By the end of the war this idea had been expanded by Armstrong into the superheterodyne receiver, which revolutionized commercial radio.

The brilliant concept of the superhet was the realization that if having a *tuneable* amplifier is the root of the instability problem, then the solution is to *not use* tuneable amplifiers. (If it hurts to bang your head on the wall, then stop banging your head on the wall!) In Figure 21.1, then, we have the block diagram of what Armstrong was fond of calling the "Rolls Royce" of radio receivers. It is the circuit found in all AM radios today, and what follows is how it works.

To begin, let me first quickly describe the spectrum environment in which the everyday AM superheterodyne operates. In commercial AM broadcast radio each station is assigned a "chunk" of spectrum 10 kHz wide. Centered on each such chunk is the station's assigned carrier frequency, which is selected from the interval 540 to 1600 kHz. Thus, a station with a carrier frequency of 640 kHz (like KFI, Los Angeles) can transmit its signals over the interval 635 to 645 kHz, and a station with a carrier frequency of 1270 kHz (like WTSN in Dover, New Hampshire) can transmit its signals over the interval 1265 to 1275 kHz. This allows each AM station to transmit a DSB-LC signal carrying a baseband signal with a bandwidth of up to 5 kHz. This is wide enough for good voice and music reproduction (but not for high-fidelity broadcasting, a market niche filled by Armstrong's development of wideband FM, which allows each

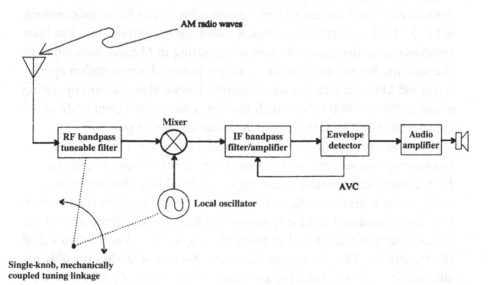

Figure 21.1   The AM superheterodyne receiver.

FM station to use ±75 kHz around its assigned carrier frequency—but that's another story).

By law, the holder of a commercial AM broadcast license in the United States must stay on his assigned carrier frequency with a maximum allowed deviation of ±20 Hz. Carrier frequencies are spaced 10 kHz apart, but two stations operating in the same geographical area would not be assigned adjacent carrier frequencies. That could result in what is called *adjacent channel* interference, which is discussed later in this section. Very-high-power stations (like KFI, which transmits with the legal maximum of 50,000 W) could conceivably be received anywhere in the entire country, and such stations originally received carrier frequency assignments duplicated *nowhere*. Such stations were called *clear channel* stations, not because they necessarily were always received clearly, but because their carrier frequency had been cleared for their exclusive use. Today, with thousands of AM broadcast stations, there are simply not enough frequencies available in the AM band to have a totally cleared channel for all of the so-called clear channel stations. There are, in fact, now over 90 such "clear channel" stations (all transmitting with the legal maximum of 50,000 W).

Those stations that do share a common carrier frequency are, however, at least widely separated by geography. For example, there are two clear channel stations at 1030 kHz: one is WBZ in Boston, and the other is KTOW in the rather distant town of Casper, Wyoming. In the United States, station carrier frequencies end with a "0," but outside the United States it is not uncommon to find stations with carrier frequencies ending with "5." In the United States, this is called a *split channel*. This has been troublesome in the past with stations operating in Mexico, with adjacent channel interference resulting (e.g., a high-power Mexican station operating at 995 kHz will interfere with a nearby United States station operating at either 990 or 1000 kHz)—such foreign stations have been aptly called "border blasters"! How this occurs is discussed later in this chapter.

When you tune your AM superheterodyne receiver, the knob you turn is adjusting two separate and distinct parts of the circuit simultaneously. First, a front-end *tuneable* radio frequency bandpass filter connected directly to the antenna is adjusted. You are moving the center frequency of this filter's passband until it is aligned with the carrier frequency of the station that you want to select from all others (this filter is often called the *preselector*). There is no problem with this tuneable filter possibly oscillating (as with the tuneable amplifier stages in the old TRF receivers) because it is either a relatively low-gain device, or even simply passive with *no* gain at all. In the latter case the rf filter *attenuates* the signal centered in its passband *less* than it attenuates signals that are displaced from the center frequency.

A less obvious role for the preselector rf filter is that it isolates the local oscillator (which is the second piece of circuitry adjusted by the tuning knob) from access to the antenna, i.e., the presence of the rf filter "unilateralizes" the receiver. (This is important because the local oscillator can, itself, radiate its rf signal into space and so interfere with other, nearby receivers. More will be said about the local oscillator, soon.) The passband of the rf filter does not, in fact, have to have a very sharp cutoff, i.e., its skirts do not have to even come close to being vertical. The filter's frequency response curve can actually be pretty "sloppy" in its rate of roll-off, and just how sloppy we'll see in just a bit. (This is good, because if is *very* difficult to build a *tuneable* filter with a sharp roll-off at its passband edges!) Notice that this means if two stations have carrier frequencies that are not very far apart then *both* signals will get through the rf filter pretty much the "same for wear." This might appear to be the beginning of a bad case of adjacent channel interference, but you'll soon see how the superheterodyne neatly sidesteps this.

Now, continuing to move through Figure 21.1, the local oscillator shifts (with the aid of the multiplier, or "mixer") the selected chunk of spectrum that has gotten through the rf preselector filter into the input of a bandpass filter with a center frequency *fixed* at 455 kHz. This is an *active* filter, in fact, and as it is also an amplifier it is called the *intermediate frequency amplifier* (or "IF amp"). The intermediate frequency (the 455-kHz center frequency) gets its name from the fact that $f_{IF} = 455$ kHz is a frequency *below* the high frequency of any AM broadcast band carrier and *above* the low frequencies of the baseband information signal that modulates the carrier. The IF amplifier is a high-gain bandpass filter with a bandwidth of 10 kHz (the bandwidth of the standard AM DSB-LC signal, which of course is twice the bandwidth of the bandpass signal) and with *very* steep skirts. Since the IF amp is *not* tuneable it is relatively easy to build such a sharp cutoff bandpass filter/amplifier that is stable (that won't oscillate). The sharp cutoff of the IF amplifier eliminates any adjacent channel signal interference (the previously mentioned case of two stations with carrier frequencies close together) that has gotten through the sloppy rf filter and to the IF amp input (along with the desired channel). Only the desired channel can survive the passage *through* the IF amp.

From the spectrum shifting (or heterodyne) theorem we know that, when a signal is "mixed" (multiplied) with the local oscillator signal, the spectrum of the signal will be translated in frequency by the value of the oscillator frequency. Thus, there are *two* station carrier frequencies that will be shifted into the IF amplifier passband, i.e., if f_{LO} is the local oscillator frequency, and if $f_{LO} > f_{IF}$, then the spectrums around the carrier frequencies $f_{LO} \pm f_{IF}$ will *both* be shifted so as to be centered on f_{IF}. If, on the other hand, $f_{LO} < f_{IF}$ then the *two* station spectrums around the carrier frequencies $f_{IF} \pm f_{LO}$ will both be shifted to be centered on f_{IF}. In either case, we call the carriers of the two signals that could potentially end up in the IF amplifier passband *image frequencies*.

If there are two stations operating, in fact, on image frequencies, then that can be a catastrophic problem because they'll end up "on top" of each other (so to speak) in the IF amplifier! The sharp cutoff of the IF amplifiers pass band does nothing in eliminating this problem of two image frequency stations interfering with each other (remember, it is adjacent channel interference that the sharp cutoff addresses), but now, at last, you can see why the front-end preselector rf filter is present. If the rf filter is tuned to one of the image frequencies (presumable the one you want to listen to) then, by definition, the other image frequency is *not* tuned. We say that the untuned image is *rejected* and, while it may still put some energy into the IF amplifier passband, we hope it will not be much energy. Indeed, the further apart in frequency that the image frequencies are, the better the rf filter will reject the unwanted image.

So, how far apart *are* the images? In the case of $f_{LO} > f_{IF}$ the image separation is $(f_{LO} + f_{IF}) - (f_{LO} - f_{IF}) = 2f_{IF} = 910$ kHz, a constant independent of tuning (of the particular station you are listening to). In the case of $f_{LO} < f_{IF}$ the image separation is $(f_{IF} + f_{LO}) - (f_{IF} - f_{LO}) = 2f_{LO}$, a variable dependent on the setting of the local oscillator (which is, obviously, always less than $2f_{IF}$). Thus, for a given value of f_{IF}, the choice $f_{LO} > f_{IF}$ gives the better image frequency separation. And if f_{IF} is sufficiently high then even a sloppy, tuneable rf filter will give excellent image frequency rejection.

In the early days of radio f_{IF} was much smaller than today's value (since 1938) of 455 kHz because of technology limitations in building tuned IF amplifier coupling transformers that could work at high frequency. In 1924, for example, General Electric built radios for RCA using an IF frequency of just 40 kHz. The increase of the value of f_{IF} was motivated by the desire for better image rejection by the preselector rf filter. Still, at carrier frequencies significantly above the commercial AM broadcast band, even $2f_{IF} = 910$ kHz is insufficient separation between image frequencies to allow the rf preselector filter to provide much rejection of the unwanted image. At a carrier frequency of 25 MHz, for example, such a frequency separation is less than 4%, and a sloppy rf preselector filter's responses to both image signals would be essentially identical. One could increase f_{IF}, of course, but this requires the envelope detector to operate at the elevated frequency, as well as complicating the design of the IF amplifier (it still must have a bandwidth nominally no more than 10 kHz, which becomes ever more difficult to achieve as the center frequency of the passband, f_{IF}, increases). Or, one could use an ingenious approach called *dual conversion*. This enhancement to the basic superheterodyne receiver is explored in Problem 21.3.

The final technical question facing us is which one of the two image carrier frequencies $f_{LO} \pm f_{IF}$ should our rf filter tune to? In practice, AM superheterodynes tune to the carrier at $f_{LO} - f_{IF}$ (you'll see why, soon), but in fact tuning to $f_{LO} + f_{IF}$ would work just as well. If f_c denotes the carrier frequency, then we have $f_{LO} = f_c + f_{IF}$ and so the local oscillator is always 455 kHz *above* the desired carrier frequency. This simply means that when the common mechanical shaft turned by the tuning knob adjusts the rf preselector filter to be centered on f_c, it is simultaneously adjusting the local oscillator to $f_c + 455$ kHz. You can easily verify this for yourself by taking two superheterodynes and tuning one to a frequency at the high end of the AM band, say 1270 kHz. Then, holding the other radio close by, tune it from the low end to the high end of the AM band. When you reach 810 kHz on the dial you will suddenly hear a 5 kHz tone from the first radio. That's because the second radio's local oscillator is operating at $810 + 455$ kHz $= 1265$ kHz, and that signal, while weak, is still strong enough to get from the second radio to the antenna of the first one and from there into the high end of that radio's IF amplifier's passband. Then, *slowly* continue to increase the dial setting of the second radio which will *slowly* sweep its local oscillator signal across the first radio's IF passband. You will hear the tone frequency *decrease* as you approach the middle of the IF passband—eventually the tone frequency will reach dc and the tone will vanish (when the second radio is tuned to 815 kHz). Then, as you continue to increase the second radio's local oscillator frequency, the tone will again suddenly become audible and increase back up to 5 kHz. Finally, the tone will again vanish as the second radio is tuned past 820 kHz, because then its local oscillator's frequency will exceed 1275 kHz (which will put the local oscillator's heterodyned signal below the low end of the first radio's IF passband).

Since the rf filter is sloppy, you should now also see that this parallel, offset tracking of the rf filter and the local oscillator does not have to be perfect, by any means. For example, if you want to listen to KFI at 640 kHz, then you turn the tuning knob until the local oscillator is operating at 1095 kHz, which you know you've done when the station comes in clearly. But suppose the tracking of the rf filter is not quite right, and instead of being set at 640 kHz it is actually at 637 kHz? This is of no real concern because the 640 ± 5 kHz station signal will still get through the (sloppy) rf filter with no trouble, and the image station at 1550 kHz (if any) will still be suppressed. The superheterodyne idea *is* fiendishly clever! But, back to the original question—why do we tune to $f_c = f_{LO} - f_{IF}$ and not to $f_c = f_{LO} + f_{IF}$?

Consider the two possibilities. If we tune to the carrier frequency $f_c = f_{LO} - f_{IF}$ (as we actually do) then, 540 kHz $< f_c <$ 1600 kHz or 540 kHz $< f_{LO} - 455$ kHz < 1600 kHz or 995 kHz $< f_{LO} < 2055$ kHz. That is, the local oscillator must be designed to vary over a 2:1 frequency interval. On the other hand, if we tune to the carrier frequency $f_c = f_{LO} + f_{IF}$ (contrary to actual practice) then 540 kHz $< f_{LO} + 455$ kHz $< 1,600$ kHz or, 85 kHz $< f_{LO} < 1,145$ kHz. Now the local oscillator must be designed to vary over a greater than 13:1 frequency interval. This is much more difficult to do than is the first case, and that is why the historical decision was to use $f_c = f_{LO} - f_{IF}$.

The output of the IF amplifier is now envelope detected to extract the baseband signal, which is then given a final power boost by the audio amplifier before being delivered to the loudspeaker. As a final elegant touch, the instantaneous amplitude of the carrier (which is a measure of the signal strength) can be extracted at the output of the envelope detector and used as a feedback signal to control the gain of the IF amplifier (that is, if this signal *weakens* we want the IF amp gain to *increase*). This gives what is called instantaneous automatic gain control (IAGC), or what is also often called automatic volume control (AVC). AVC was invented by Harold Wheeler (1903–1996) in 1926 while working for Hazeltine, and was first used commercially in 1929. See Figure 21.2 for the circuit details of just how the AVC feedback signal was generated and used in vacuum-tube superheterodyne receivers. The details are only slightly different for today's transistor and integrated circuit designs.

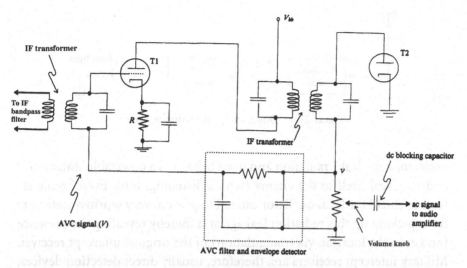

Figure 21.2 This circuit shows the generation of the automatic volume control signal in the envelope detector, from where it is fed back to control the grid-to-cathode bias (and, hence, the gain) of the IF amplifier tube. The output of the IF bandpass filter (not shown) is applied (via a tuned circuit called an *IF transformer*) to the grid of T1, the IF amplifier tube. T1 is *partly* biased (see Chapter 8) by the dc voltage drop produced across R by the quiescent tube current. The amplified voltage at the plat of T1 is coupled (via another IF transformer) into the plate of T2, which performs the detection (rectification) operation. The voltage at the top of the variable resistance volume control (called a *potentiometer*), v, has a positive dc value proportional to the signal amplitude produced by T1. This dc value is extracted from v by the AVC filter (which doubles as part of the envelope detector) to produce the AVC signal, V. V is then fed back to the grid circuit of T1. Notice that if V *increases* then the grid-to-cathode bias voltage on T1 becomes *more negative*, which tends to reduce the IF amplifier tube current, i.e., to turn T1 off. A tendency to turn T1 off, of course, is equivalent to reducing the gain of the IF amplifier, which results in a *decrease* of V. Hence, this circuit achieves an automatic self-regulation of the signal amplitude sent on to the audio amplifier by the volume control. By 1932 essentially all radio receivers had AVC.

The receiver design described in this section is used not only in AM broadcast radio, but in FM broadcast and television receivers as well, using 10.7 MHz and 44 MHz, respectively, for f_{IF} (see Problem 21.1 for some motivation on the FM f_{IF} value). But not all receivers today are super-heterodynes. Ironically, the original problem that motivated Armstrong to develop the superhet (the detection by the military of electromagnetic radiation emitted by the enemy's equipment) is precisely what precludes the use of the superhet in certain situations. In particular, military receivers used in time of war to surreptitiously intercept enemy radio transmissions do not use the superhet design because, as I mentioned before, the local

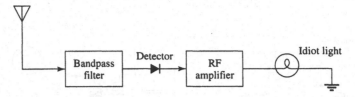

Figure 21.3   Direct energy detection receiver.

oscillator can "leak" radiation and itself behave as a detectable transmitter and so reveal itself to the enemy (who is listening, too). Even heroic efforts to shield the local oscillator can fail to prevent very sensitive detectors from picking up this radiation leakage and thereby revealing the presence (and even the location via triangulation) of the original intercept receiver. Military intercept receivers are, therefore, usually direct detection devices, with an "idiot light" display (as shown in Figure 21.3). Such a receiver is an energy detector (not an information demodulator), and the idiot light merely tells an observer if there is a signal present within the frequency interval covered by the front-end bandpass filter.

For fascinating stories on some of the occupational hazards for spies caused by "leaky" local oscillators, see Peter Wright, *Spycatcher*, Viking 1987. Wright joined M15 (the internal security part of Britain's secret service, roughly equivalent to the American FBI) in 1955 as its first scientific officer. Skilled in the art of electronic espionage, he rose to become Assistant Director by the time he retired in 1976. The term "idiot light" is from the auto industry. It refers to the 1960s replacement of expensive dashboard instrument gauges for displaying coolant temperature, oil pressure, and alternator performance, with cheap on–off lights. When the oil light illuminates, for example, that usually means you're down 3 to 5 lbs of pressure (and are probably chewing up the cylinder walls and/or the piston rings and valves even as you gaze in horror at the pretty red glow on the dash). Even an idiot can understand, *then*, that there is a problem! A personal opinion: idiot lights, on cars, are for idiots, and any electrical engineer or physicist worthy of the title should replace them on his or her personal vehicle with gauges. On rental cars, however, it's okay to tolerate them. Idiot lights and direct detection receivers are often used on military aircraft to quickly alert a pilot when he is being illuminated by a radar, possibly the weapons radar of an enemy aircraft about to launch a missile or gun attack, or the radar of a SAM (surface-to-air missile) ground-based site. Indeed, since radars that track targets use a different frequency band than do radars that are merely performing routine area surveillance, then there could be *two* idiot lights on the cockpit display: one that glows green for incident surveillance frequency energy, and one that glows red for incident tracking frequency energy. *Tracking* by enemy radar is considered by military pilots in a combat zone to be a hostile act, and maybe the same circuit that drives the red light would also honk a cockpit warning horn. These "radar-warning" receivers have saved many a pilot's life in combat. "Fuzz-buster" units used by speeding motorists to detect police radars monitoring traffic use the same general direct detection approach.

Any receiver that shifts the input signal frequencies to a new location in the spectrum is a heterodyne receiver. The prefex *super* is reserved for those receivers that also use both a tuneable front-end rf filter preselector for image rejection, and a fixed frequency sharp cutoff bandpass IF filter for adjacent channel suppression. I like to think what happened, when experimenters in the early days of radio first listened to receivers like that, is that they clapped their hands with joy, jumped up and down with eyes gleaming, and shouted "Super, by damn, now that's a *super* heterodyne receiver!" And that is how the superhet got its name.

Well, anyway, I like to think it *could* have happened that way.

Notes

1. A scholarly study of the technical details of these (and other) circuits is the paper by D. G. Tucker, "The History of Positive Feedback: the oscillating audion, the regenerative receiver, and other applications up to 1923," *The Radio and Electronic Engineer* 42, February 1973, pp. 69–80. Two lines in Professor Tucker's paper are, in particular, quite interesting: "In contrast to Armstrong's very professional and scientific approach to radio, De Forest appears almost as a fumbling amateur. In his patent specifications as in his published papers, he shows little understanding of what he is doing."

2. Two excellent articles on the problems of tuning in the early days of radio are by Arthur P. Harrison, Jr. See his "Single-Control Tuning: an analysis of an innovation," *Technology and Culture* 20, April 1979, pp. 296–321, and "The World vs. RCA: Circumventing the Superhet," *IEEE Spectrum*, February 1983, pp. 67–71. Both articles contain fascinating closeup photographs of the tuning mechanisms (some mechanically quite exotic) of early radio receivers.

3. An excellent discussion of Nyquist's work is in Chapter 16 of John G. Truxal's nifty book *Introductory System Engineering*, McGraw-Hill, 1972.

Problems

21.1. An FM superheterodyne receiver operates in the frequency band of 88 to 108 MHz. If $f_{LO} > f_{IF}$, and if we want to have all possible image frequencies fall outside the FM band, then show that f_{IF} must be at least as large as a particular minimum value (i.e., calculate it). Hint: the actual $f_{IF} = 10.7$ MHz is just slightly higher than this calculated minimum value.

21.2. Repeat the first problem for the case of the AM band (where, again $f_{IF} < f_{LO}$) and show that the actual $f_{IF} = 455$ kHz is *less* than the required minimum value to ensure all possible image frequencies fall outside the AM band. Why do you think this is so, i.e., why isn't f_{IF} at least equal to the minimum value you calculated? Hint: Consider the situation if f_{IF} were such as to be, *itself*, in (or very near) the AM band. Then the IF filter/amplifier could *directly* receive very strong AM radio signals, i.e., bypass the antenna, preselector, and mixer. This would defeat the whole concept of "tuning"! This doesn't occur in the first problem, because $f_{IF} = 10.7$ MHz is *well* outside the FM band. For AM radio, $f_{IF} = 455$ kHz is a compromise between having f_{IF} large for good image rejection, but not so large as to intrude on the AM band itself.

21.3. The front end of a so-called *dual conversion* AM superheterodyne receiver is shown in Figure 21.4. It uses the usual local oscillator/mixer/IF amplifier circuitry that ultimately shifts the desired antenna signal down to 455 kHz for envelope detection (the second IF stage in the figure). But it also uses a preliminary stage of heterodyning (the first IF stage). This design is used in superior performance, high-frequency receivers to achieve good image rejection without the need for a high value of the final intermediate frequency.

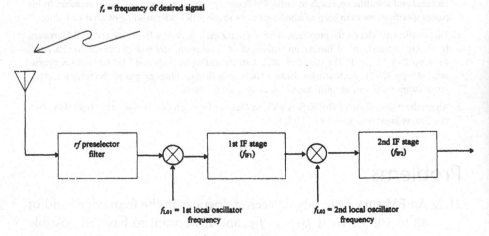

Figure 21.4   Dual-conversion superheterodyne receiver.

(You will show this, soon!) Instead of image frequencies coming in pairs, however, as they do in single conversion superheterodynes, in the dual conversion version image frequencies come in *quadruples*.

(a) Suppose f_{LO1} and $f_{LO2} > f_{IF1}$, where f_c is the carrier frequency of the desired signal. Show that there are *three* carrier frequencies (in addition to f_c) at the antenna that place energy in the middle of the passband of the second IF stage. Answers: $f_c - 2f_{IF1}$, $f_c - 2(f_{IF1} + f_{IF2})$, and $f_c + 2f_{IF2}$. Hint: Start by arguing why $f_{LO1} = f_c - f_{IF1}$ and $f_{LO2} = f_{IF1} + f_{IF2}$ (notice that f_{LO2} is a *constant* i.e., only f_{LO1} is varied during tuning). And while both IF amplifiers have steep skirts at the edges of their passbands, do *not* assume the skirts are *perfectly* vertical.

(b) Suppose that the desired $f_c = 25\,\text{MHz}$, and that $f_{IF1} = 5\,\text{MHz}$ and $f_{IF2} = 455\,\text{kHz}$. Further, as usual, suppose the nominal bandwidth of the second IF amplifier is 10 kHz. Then, to have the same relative bandwidth suppose the first IF amplifier has a bandwidth of 50 kHz. Show that the image frequencies of (1) 14.09 MHz, (2) 15 MHz, and (3) 25.91 MHz, respectively, are rejected by (1) both the preselector rf filter and the first IF amplifier, (2) the rf preselector filter only, and (3) by the first IF amplifier only.

(c) Repeat for the case $f_{LO1} < f_c$ and $f_{LO2} < f_{IF1}$.

So, at last, we have the modern radio. But where did *broadcast* radio come from? It wasn't from Marconi, whose "vision" of wireless was always limited to developing a profitable monopoly in wireless *telegraphy* to compete with the existing monopolies of the land-line and submarine cable telegraph companies in the business of sending *private* messages. It *might* have come as early as Fessenden's 1906 voice broadcast, as that event almost resulted in AT&T investing in his work. But the timing was unfortunate—less than a year later the Wall Street panic of 1907 caused AT&T to actually retrench (12,000 employees were fired). The concept of public broadcasting appears to be the original idea of Lee De Forest (at last, something he didn't "borrow" from someone else!), who transmitted opera music on a regular basis in 1907. The technology of that day was simply not up to the task, however, and matters had to wait until the early 1920s. That story is told in the epilogue.

CHAPTER

22

Epilogue

"Last week 745 [sic] human lives were saved from perishing by the wireless. But for the almost magic use of the air the *Titanic* tragedy would have been shrouded in the secrecy that not so long ago was the power of the sea... Few New Yorkers... realize that all through the roar of the big city there are constantly speeding messages between people separated by vast distances, and that, over housetops and even through the walls of buildings and in the very air one breathes, are words written by electricity."

—from the *New York Times*, April 21, 1912, in a story entitled "Wireless Crowns a Remarkable Record as Life-Saver"

The superheterodyne radio receiver arrived essentially at its final technical form in the early 1920s. It quickly started to generate big bucks; in 1924 RCA sold nearly 150,000 copies of its famous superheterodyne receiver, the *Radiola Super VIII*. Radio, itself, however, was then just beginning its nontechnical evolution. Once KDKA began its regular broadcast schedule, though, the radio business metaphorically exploded. Part of the folklore of radio history is that a prescient David Sarnoff (head of both RCA and NBC from 1930 to 1969) foretold, in 1915, the coming of the radio age of mass entertainment. That was the year (just three years after he gained worldwide attention as a twenty-one-year old radio operator who claimed to have received messages during the Titanic disaster) that Sarnoff also claimed to have written his so-called Radio Music Box memo to the management of the American Marconi Company. In that memo Sarnoff proposed to make radio a "household utility." Proposing a selling price of $75, Sarnoff suggested that the box "if manufactured in quantities of 100,000 or so could yield a handsome profit." Notice that this profit was to come from selling the *box*, itself, and not from advertising. *That* was an "advanced" concept that Sarnoff missed. Recent scholarly research[1] has both cast doubt on the date of the memo and raised questions about Sarnoff's self-promotional inclinations, but the truth of the memo's thesis (whenever it may have actually been written) cannot be denied.

> Sarnoff's claims to have heard the *Titanic's* distress calls are today thought to have been nothing but "imaginative hype." Sarnoff will, in the long run, probably be remembered mostly as the man who, time and again, derailed Armstrong's attempt to commercially develop FM radio (which Armstrong engineered in the early 1930s). Sarnoff's existing radio empire rested on AM radio, and he wanted no competition. In 1954, after years of court battles, a despondent Armstrong committed suicide. When told of Armstrong's death a shaken Sarnoff said of his one-time friend "I did not kill Armstrong," even though he hadn't been asked if he had. But if they had been asked, most radio engineers would have had a different answer about Sarnoff's role.

Where there was one licensed station in America in 1920, there were nearly 600 stations just five years later,[2] and the number of radio receivers went from thousands of crystal sets to millions of vacuum-tube circuits. In May 1922 *Radio Broadcast Magazine* declared,

> The rate of increase in the number of people who spend at least a part of their evening in listening in is almost incomprehensible. To those who have recently tried to purchase receiving equipment, some idea of this increase has undoubtedly occurred as they stood perhaps in the fourth or fifth row at the radio counter waiting their turn, only to be told, when they finally reached the counter, that they might place an order and it would be filled when possible.... It seems quite likely that before the movement has reached its height... there will be at least five million receiving sets in this country.

This would prove to be a gross underestimate.

Indeed, the radio audience reached twenty-million listeners by 1926. It's not hard to understand how that happened when you realize that at *one* New York department store, on *one* day in early 1925, 240 clerks sold 5300 five-tube receivers! And that despite a cost of $100, and a power supply involving a 90-A (!) battery for the tube filaments, enough to crank the starter of a small car engine (ac powered sets that plugged into the house current didn't go on sale until the following year). The listening audience was expanded still more when car radios appeared in 1928, under the clever marketing name of *Motorola*. Less clever was the Crosley Radio Corporation's 1930 "Roamio"—for *roaming radio*! The police, in particular, found this innovation especially useful, as did the creators of the 1930s crimebusters comic strip *Radio Patrol*.

Radio quickly penetrated into the everyday language of even the non-electronic-oriented person. For example, consider these words, spoken by the retiring President of the American Mathematical Society in September 1921: "What is the soul of mathematics, and to what wavelengths must our own souls be tuned to catch its message?" Just a few decades earlier this metaphor would have been meaningless to every being on the earth (what do "wavelengths" and "tuned to catch a message" have to do with each other?, even the great Lord Kelvin would have wondered), but by 1921 "even mathematicians" knew! See David Eugene Smith, "Religio Mathematici," *American Mathematical Monthly*, October 1921.

So excited were the early listeners (up to 1922) to be hearing anything at all that they seemingly cared little for the details of *what* they heard over their crystal sets. Indeed, unless your interests ran heavily to the National Bureau of Standards' time signals over WWV-Arlington, Virginia, or to the endless playing of phonograph records, it was the radio gadget itself that captivated, not the inherent entertainment content (or the lack of it) in the sound it generated. Station owners soon realized that such a narrow appeal could only be a fad, however, and not the basis for longterm profitable success, and local experts on various topics began to be invited to present "radio lectures" to the listening audience. One such authority amusingly described his attempt to introduce himself to the page, who greeted him at the station's entrance, as being curtly brushed aside—the young man already knew who he was: "You're one of those broadcasting guys, a—a regular scientific gent that comes up here to give the radio fans highbrow stuff."[3]

But not everybody thought radio "highbrow." As another writer wrote of early radio fans, "They are aware that the highbrow sneers at their tribe because so little matter of high intellectual content is broadcast, a criticism which, of course, does not apply to music, since nightly the finest compositions of great composers are put on the air, as well as the worst vulgarities of the jazz barbarians." See Bruce Bilven, "The Legion Family and the Radio: What We Hear When We Tune in," *The Century Magazine*, October 1924. A few months before the same author had written ("How Radio is Remaking Our World," *The Century Magazine*, June 1924) "Here is the most wonderful medium for communicating ideas the world has ever been able to dream of, yet at present the magic toy is used in the main to convey outrageous rubbish, verbal and musical, to people who seem quite content to hear it."

As another example of this skeptical attitude toward commercial radio, consider Lee De Forest's famous acidic criticism to an audience of radio executives:

Why should anyone want to buy a radio or new tubes for an old set when nine-tenths of what one can hear is the continual drivel of second-rate jazz, sickening crooning by degenerate sax players, interrupted by blatant sales talk, meaningless but maddening station announcements, impudent commands to buy or try, actually imposed over a background of what might alone have been good music?[4]

Taking a rather self-righteous position for a man who spent a significant part of his career in court trying to break the patents of others, in endless efforts to make money, De Forest also declared to the radio executives that

The radio was conceived as a potent instrumentality for culture, fine music, the uplifting of America's mass intelligence. You debased this child [De Forest was fond of calling radio "his child," and himself the "father of radio" and the "grandfather of television," claims, if you remember Maxwell and Hertz, with no merit], you have sent him out in the streets in rags of ragtime, tatters of jive and boogie-woogie, to collect money from all and sundry.

De Forest was tireless in proclaiming himself to be the "Father of Radio"; in his 1930 Inaugural Address as President of the Institution of Radio Engineers he described himself with that very phrase. He used it again as the title for his 1950 autobiography, which in 1953 sold

> 54 copies. A planned book to be written by his wife, to be called *I Married a Genius*, was never completed. "Father of Radio" is a title probably just a bit too grandiose to apply to any single individual, and in any case De Forest was actually less entitled to it than others that history has mostly forgotten. One such person was the Dutchman Hanso Idzerda (1885–1944) who was the first to develop and build transmitters specifically intended for broadcasting music and speech to an *international* audience. By late 1919 his transmissions from The Hague, at a wavelength of 1030 meters with 75 watts of power, were reaching England. He was shot by the Nazis after being discovered in a restricted rocket launch area, and has faded from radio history. You can read more about Idzerda in Pat Hawken's "Birth of Broadcasting," *Electronics World*, May 1997.

De Forest is now generally regarded by historians as a shameless self-promoter with dubious ethics. In February 1902, for example, a month after the De Forest Wireless Telegraph Company was formed with a one-million-dollar stock offering, he wrote in his personal journal "Soon, we believe, the suckers will begin to bite!" (and by the narrowest of margins he avoided being convicted and sent to prison in 1914 for yet another radio stock fraud scheme). De Forest's technical reputation has also been in sharp decline for some years as more is learned about how little he actually understood of the science of radio (and, indeed, how little he understood the physics of even his own lucky invention, the vacuum-tube triode). Still, there can be no doubt he had a valid point about the crass, commercial side of early American radio.

Shortly after De Forest's pompous lecture to the radio executives, a writer chronicled his own dismal experience with radio.[5] After first declaring radio "one of the most significant and marvelous inventions of this mechanical age," and referring to it as the "Tenth Muse," he observed that much of American radio advertising *was* pretty hard to take. He offered this as a typical example:

Announcer: Miss Edna W. Hopper is going to tell you how she managed to have all her teeth, at sixty... Miss Hopper...

Author: A voice, brutal, merciless, aggressive, like a whip, like a machine gun, like a revivalist, crashes upon the air. No one who tuned in on it would fail to stop and listen. It roams up and down a staircase of inflection, charges around corners, and chins itself on a significant word.

Lady: I spoke over the air about my own TEETH....I have all my
own TEETH....I'll tell you how you can have a dazzling
SMILE....I not only have white sparking teeth but I have
kept *all* my own TEETH....the way to have white *teeth* is
QUINDENT...TEETH...QUINDENT...TEETH...

It was exactly this sort of electronic babble-talk, of course, that helped
launch the modern consumer society. In a speech given February 1922
at the first National Radio Conference, Secretary of Commerce Herbert
Hoover declared "It is inconceivable that we should allow so great a possibility for service, for news, for entertainment, for education, and for vital
commercial purposes [as is radio] to be drowned in advertising clutter."
Well, of course, as we all know today, history hasn't quite evolved like Mr.
Hoover thought it would.

> To better understand De Forest's complaint, read about the finan-
> cial pressures on early radio stations in H. Le B. Bercovici's humor-
> ous "Station B-U-N-K," *The American Mercury*, February 1929. This
> article also tells us that B-U-N-K's rivals included stations J-U-N-K and
> B-O-S-H. More scholarly is the book by Susan Smulyan, *Selling Radio:
> The Commercialization of American Broadcasting 1920–1934*, Smith-
> sonian Institution Press, 1994.

The level of programming at the BBC, however, was certainly on a consistently higher plane than that of the Americans, even in the beginning—
when Mozart's *Magic Flute* was broadcast live from the Covent Garden
Opera House in January 1924, for example, it was talked about long after and such acclaim was not unusual. The British also took the next big
step in radio programming with the "invention" of the radio drama. This
was actually quite revolutionary, as it meant the story had to rely totally on
speech and sound effects (and the listener's imagination). On January 15,
1924, for example, *A Comedy of Danger*, the story of a rescue from a coal
mine accident (in which literally everything takes place in the dark) was
broadcast over 2LO, to much acclaim. Curiously, when the play's author
was in America a few months later, he found that radio executives there

...rejected the whole idea [of radio stories]. That sort of thing might
be possible in England, they explained, where broadcasting was a
monopoly and a few crackpot highbrows in the racket could impose
what they liked on a suffering public. But the American setup was

different: it was competitive, so it had to be popular, and it stood to reason that plays you couldn't see could never be popular.[6]

Never have the so-called experts been so wrong! Still, it took over four more years before adult storytelling began to appear on nationwide American radio, starting with "Real Folks" on NBC, August 6, 1928, and then "Amos n' Andy" a year later (but it had been on Chicago's local radio station, WGN, as Sam n' Henry," as early as January 1926). And then the floodgates opened and during the following years radio gave the American public "Just Plain Bill," "The Romance of Helen Trent," "Ma Perkins," "John's Other Wife," "Pepper Young's Family," "Our Gal Sunday" ("Can a coal miner's daughter find happiness married to England's richest lord?") "Stella Dallas," "Young Dr. Malone," "When a Girl Marries," "Backstage Wife," "Young Widder Brown," "The Second Mrs. Burton,"—on and on goes the list.[7]

These shows were *serials*, continuously evolving, fifteen-minute daily presentations of "success stories of the unsuccessful," as Robert West described them in his 1941 book *The Rape of Radio*. Sponsored mostly by the manufacturers of soaps and cleansing agents (in 1936 the top radio advertiser *by far* was Procter and Gamble), these programs became known as "soap operas."[8] So popular were they that Philco and Paramount Pictures collaborated on the 1932 film *The Big Broadcast* to bring many of the radio celebrities to the screen, to be seen and not just heard (as well as to include Philco radios as props in the movie). Critics, however, found such programs easy targets. One, the American humorist James Thurber, defined the soaps as

> A kind of sandwich, whose recipe is simple enough, although it took years to compound. Between thick slices of advertising spread twelve minutes of dialogue, add predicament, villainy, and female suffering in equal measure, throw in a dash of nobility, sprinkle with tears, season with organ music, cover with a rich announcer sauce, and serve five times a week.

Directed specifically at the millions of stay-at-home women of the decades from the 1920s through the 1950s, the soaps were both enormously popular *and* very profitable. But the coming of television, and changing economic forces that encouraged the mass departure of adult women from the house and into the labor force, combined to spell the demise of these shows after three decades of success. When "Ma Perkins" said her final goodbyes on Friday, November 25, 1960, after 7065 broadcasts, it was the end of the radio serial (the soap opera itself, of course, simply moved to television where it has thrived and multiplied like bacteria on a dirty sock).

> The large number of soaps on radio obviously required a huge and steady production of words, and so a new job category was created— "radio soap-opera script writer." When she died, The New York Times headlined Irna Phillips' (1901–1973) obituary with the title "Queen of Soapland":
>
> > For some 40 years Miss Phillips fashioned for radio and television thousands of tales of divorce, desecration, disease, and romance (often illicit), which drew an audience of millions of housewives and crowned her undisputed queen of soapland.... In the early nineteen-forties Miss Phillips had five dramatic serials running on radio. During those years she was dictating her dramas to secretaries for six or eight hours a day at the rate of over 2-million words annually. In a good year she earned $250,000.
>
> Phillips survived the arrival of television; in 1949 she wrote the first video soap, NBC's "These Are My Children." And in 1964 ABC hired her as a consultant to "Peyton Place," about what really goes on behind closed doors in a small New England town.
>
> Many other women, less well known, were also successful at soap opera writing, and their imaginary tales of heroines surviving adversity often paralleled their own real lives. For example, Jane Crusinberry (1892–1984) was a divorced mother selling cosmetics door-to-door in 1934 Chicago when she created and wrote "The Story of Mary Marlin" for WMAQ. It was soon picked up for national broadcast in 1935 by NBC, and it became one of the big hits of the 1940s war years. The storyline was a bit preposterous (Mary had assumed her husband's seat in the U.S. Senate while he wandered through Asia in an amnesic fog after a plane crash!), but perhaps it was no more unlikely than Crusinberry's own true success story.

Politicians discovered radio quickly; on June 21, 1923, Warren G. Harding became the first President of the United States to be heard over the radio (on WEAF-New York), less than 2 months before his death. A few months later, in November, ex-President Woodrow Wilson reached a wider audience through a "network" of three stations (WEAF, WCAP-Washington, D.C., and WJAR-Providence, RI); as did Harding, Wilson died just a few months after his first appearance on radio. Not at all discouraged by this unpromising track record of "broadcast-and-die," Harding's successor took to radio eagerly; Calvin Coolidge's 1925 inauguration was broadcast over a coast-to-coast network of 27 stations. He broke that record when his February 22, 1927 address to a joint session of Congress was transmitted over a network of 42 stations, from WCSH-Portland, Maine to KPO-San Francisco, California, reaching an audience of twenty million. In addition, it was the first international political broadcast, as both WGY-Schenectady, NY and KDKA transmitted the President's speech via short-wave to London (the BBC then rebroadcast it, over 2LO, to all of England), Paris, and South Africa. This political use of the new medium was brought to its highest form (ever) with Franklin Delano Roosevelt's "fireside radio chats" before and during the Second World War.

These informal networks of stations, temporarily formed to broadcast particular, special events, quickly evolved into the formal network corporate structures of today. With their ability to guarantee multiple program outlets across broad regions of the country, the networks could also guarantee an ability to devote a high level of financial support for new productions, as well as allowing station affiliates to share each others programs. And so the networks helped radio grow. In 1926 Sarnoff at RCA founded the National Broadcasting Company (NBC), leasing AT&T's long-distance telephone lines for $1 million a year to transmit programs from one station to another for local broadcasting. In fact, there were actually *two* NBC networks: the Red Network, so-called because AT&T marked its circuit-route diagrams in red to distinguish the radio circuits from regular telephone circuits, and the Blue Network (marked in blue) when new circuits were added when more capability than Red could provide was required. By the start of 1927 both NBC/Red and NBC/Blue were in operation. The Columbia Broadcasting System (CBS) began its operations soon after, in September 1927. Later, in 1943, when concern about broadcasting monopoly grew, the FCC forced RCA to divest itself of one of its networks—it sold the Blue Network, which then became the American Broadcasting Company (ABC).

Running neck-and-neck with the politicians in the use of radio were the radio priests, the so-called Bible-thumpers who were the ancestors of today's television evangelists. Indeed, two months to the day after KDKA carried the Harding–Cox election results, the same station broadcast the January 2, 1921 sermon of the pastor of Pittsburgh's Calvary Episcopal Church (a broadcast arranged by one of the station's engineers, who sang in the choir). The joining of radio and religion quickly grew from this simple beginning to what became known by such terms as the "Electric Gospel," the "Electric Pulpit," and the "Invisible Church." These phrases describe what one writer called "the promise of GE & Jesus walking hand-in-hand to make radio [a] rousing commercial success."[9]

The Paulist Fathers of the Catholic Church, in particular, very early on recognized the power of radio. As their monthly magazine stated, little more than a year after KDKA's pioneering broadcast,[10]

Behold, now is the acceptable time for the Catholic Church to rise to this great and unique occasion, before the privilege is entirely pre-empted by those outside the Faith, and not allow the wireless telephone...to be used as the medium of heresy. *The Catholic Church should erect a powerful central wireless transmitting station...*

And then the author really worked himself up into an excited state, declaring that such a transmitting station would allow the Church to

reach untold millions at the very poles of the world. It would be the Super International Catholic Truth Society.... The burning sands of the Sahara, the frozen steppes of Alaska, the jungle fastness of India, the inescapable gorges of the Himalayas, the serene calm of the mountain shepherd hut, as well as the far-flung congregations aboard ocean liners, lashed by the angry seas, could all be put in touch with Christ's truth instantaneously and simultaneously since the wireless telephone leaps over all barriers of time and space.

Three years later the Paulist Fathers announced they were about to follow this spirited advice and establish a broadcasting station.[11] In an anticipatory echo of De Forest, the Fathers stated their hope to be able to "present a program that will be a relief...from too much 'Hot Mammy' on the saxophone"!

On a more serious note, in Rome the church began operation of Vatican Radio HVJ in February 1931. As Marconi said at the inaugural ceremonies, this powerful (25 kiloW) shortwave station made it possible for "the first time in history that the living voice of the Pope will have been heard simultaneously in all parts of the globe." Two years later, almost to the day, Pius XI had additionally at his disposal a fifteen mile microwave radiotelephone link (at 500 MHz), connecting Vatican City and the Pope's summer residence. These radio facilities freed the Catholic Church in Rome from total dependency on the Italian telegraph system, which was controlled by Mussolini's Fascist government. This was important, as it was just at this time that Pius XI was writing his famous anti-Fascist encyclical.

Radio came just a bit too late for such masters of religious fervor in revival tents as Billy Sunday, but others like Father Charles Coughlin, Charles Fuller, and Aimee Semple McPherson[12] became celebrities, even cult figures, to millions of listeners in the Great Depression of the 1930s. During his entire career, for example, it has been estimated that Billy Sunday was perhaps heard by several million people. An impressive total, to be sure, but Father Coughlin was heard, in his peak years, by millions of radio listeners with each Sunday afternoon broadcast. When WCAU-Philadelphia polled its audience for its preference—the New York Philharmonic or Father Coughlin—the priest beat the musicians by more than fifteen to one![13] The early audiences simply loved religious radio, even more than baseball, e.g., WEAF-New York voluntarily gave up an opportunity to carry a Sunday game of the 1924 World Series (with the Giants in it!) to instead broadcast a church service.

In addition to the "adult" soaps, early radio introduced yet another type of program; the child's crime-and-horror show. As one writer of the time (Worthington Gibson, "Radio Horror: For Children Only," *The American Mercury*, July 1938, pp. 294–296) said of those productions,

> Come five o'clock every weekday afternoon, millions of American children drop whatever they are doing and rush to the nearest radio set. Here, with feverish eyes and cocked ears, they listen for that first earsplitting sound which indicates that the Children's Hour is at hand. This introductory signal may be the wail of a police siren, the rattle of a machine gun, the explosion of a hand grenade, the shriek of a dying woman, the bark of a gangster's pistol, or the groan of a soul in purgatory. Whatever it is, the implication is the same: Radio has resumed its daily task of cultivating our children's morals—with blood-and-thunder effects.

The writer concluded by suggesting that one way "to end this curse upon your children's minds" was simply "to toss your radio set out the window." When one considers the modern-day debates on the violence on television and in films, and on possibly censoring the Internet, it is clear that there is, indeed, very little that is new under the sun.

The 1930s and 1940s children's radio crime-horror-adventure programs did have one positive feature; a tendency of their producers to use classical music for their themes. A cynic might say that was because such music was in the public domain, and thus free, but so what?—the music *was* good. Who, even today, can listen to the overture from Rossini's "William Tell" and not instantly think of the *Lone Ranger*? *The Green Hornet* flew into the homes on the frantic violin notes of Rimsky-Korsakov's "The Flight of the Bumblebee," and *The Shadow* and *I Love A Mystery* crept into darkened bedrooms on the eerie sounds of Saint-Saen's "Omphale's Spinning Wheel" and Sibelius' "Valse Triste," respectively. The wonderfully horrific *Escape* used Mussorgsky's "Night on Bald Mountain," and the adventure program *Arabesque* introduced its young fans to Rimsky-Korsakov's "Scheherazade." And the no-nonsense *The FBI in War and Peace* did its thing to the notes of the march from Prokofiev's "The Love for Three Oranges." This intellectual aspect of the Children's Hour actually carried over to the early television adventure shows, as well, e.g., on radio *The Cisco Kid* relied (and none too well, at that) on hokey organ music, but on TV the Kid and Pancho rode hard to the blood-pumping opening of Khachaturian's ballet *Gayane*, "The Sabre Dance."

The children who listened to this music in the 1930s and the 1940s did not (for the most part) become the serial killers that Worthington Gibson feared. Rather, they grew-up to become the middle-class consumers who fueled the 1950s craze for hi-fi sets and long-playing recordings of the classical music they had first heard on radio, during the Children's Hour.

But even when such programs were in their heyday, much more was going on with radio in the 1930s than simply soap operas, church services, crime shows, and baseball games. In 1926, for example, the U.S. Department of Agriculture started its Radio Service, to broadcast educational programming to farmers. That same year saw the start of the U.S. Bureau of Home Economics' "Aunt Sammy" broadcasts (so-called as the star of the show was supposed to be the wife of Uncle Sam!) These programs, which continued until 1944, carried valuable information to America's homemakers on such topics as nutrition, sanitation, child care, and emergency plumbing repairs. It was an enormously popular production, with a faithful audience of five million. When the meal plans that had been broadcast were compiled into a book (*Aunt Sammy's Radio Recipes*), the Government Printing Office received over a million requests. Not until Julia Child's famous *The Joy of Cooking* appeared decades later would there be a more widely read cookbook.

Early radio could be technically inventive, too. In the late 1930s, for example, an amazing technology briefly made an appearance in broadcast radio—*facsimile* radio. This was a radio that could receive and print pictures. Facsimile over telegraph lines had been demonstrated as early as

1843 by the Scottish inventor Alexander Bain (1810–1877), and then later over telephone lines by the German-born electrical engineer Arthur Korn (1870–1945). Facsimile by radio, however, was pioneered by the English-born inventor William Finch (1897–1990). It was Finch's system that was used by WLW-Cincinnati (owned by the Crosley family) to broadcast news pictures and stories and weather maps with its regular AM transmitter, starting in 1938. Anyone with a proper receiver (the Crosely "Reado"—for "read and radio") could, for the purchase price of $79.50, print these images at home. Facsimile radio never caught on, however, and almost totally faded away. But not completely. Since 1990 an abbreviated version of the *New York Times* newspaper has been available daily as a worldwide facsimile broadcast. Called the *Times* Fax, it is popular reading at such remote locations as resorts and cruise ships at sea.

Finch's reduction of a two-dimensional black-and-white image (which could be text as well as a photo) to a one-dimensional time-varying signal that could then be sent by AM radio (just like an audio signal) is easy to understand. The image to be sent was printed on a strip of paper four inches wide, and then slowly pulled along a horizontal direction. Above the moving paper was a pendulum, swinging across the width of the paper from one edge to the other. Mounted in the pendulum tip was a tiny light source, and also a sensitive photoelectric sensor that measured the intensity of the light reflected by the image-spot instantaneously beneath the light. The sensor output signal would vary as the light swung over white and black areas of the image. The oscillation rate of the pendulum was 100 back-and-forth swings per minute and so, with a paper speed of one inch per minute, the "horizontal scan line density" was 100 lines/inch.

At a receiver, a second pendulum swung synchronously over a strip of specially treated black paper with a white coating. A light source on the tip of this pendulum would, in accordance with the received signal, burn away more or less of the coating and so produce a black-and-white facsimile of the original image. Because of its mechanical scanning this system was slow, requiring eleven minutes to send a single page one-half the width of a sheet of standard paper.

Governments had discovered, by the 1930s a two-pronged truth about radio—it is a powerful instrument for controlling the local population, and it is an equally powerful weapon of reverse propaganda in the hands of foreign governments. International radio became a game of thrust and counterthrust. When one government would build a transmitter on its border, its neighbor would build one nearby. The opening broadcast of the first station was the cue of the second station to transmit, too—on the same frequency. (Was it simply a coincidence, for example, that the ring of maritime radio navigation beacons along England's coast just happened to operate on the same frequency as did Radio Moscow?) A transmitter-power race started and, as one writer drew the military analogy, as powerful radio stations in the 150 to 200-kW range appeared it was "as startling as if gun calibers were doubled in the naval world."[14]

The military uses of radio (and of electronics, in general) hardly needs to be explained today, with the spectacular images of high-tech weaponry that all the world saw daily during the 1991 Gulf War and the 1999 NATO air war in Yugoslavia. Indeed, those images, themselves, were powerful testimony to the way radio and its descendent, television, have made the worlds at the opposite ends of the twentieth century as different as are the surfaces of the Earth and the Sun. The impact of radio on war is, however, as is nearly everything about radio, not new. Radio-telegraphy made itself felt almost immediately in the First Word War (the U.S. Navy, as mentioned at the start of Chapter 7, took over radio during that war and in 1917 banned all amateurs from the air until an Act of Congress reversed matters two years later).

Perhaps even more important than its direct use by the military was the impact of radio, in the early days of the Second World War, with its nearly instantaneous spreading of information around the world. This informational role was illustrated, better than by any words I could write, by a 1941 drawing in *Punch*. War had come again to Europe, but now ordinary AM radio was there, too, and it played a major role in helping to combat one of the greatest evils the world has ever known. In just over twenty years radio had changed from being just a dream to being the "secret hope," an everyday gadget of information that helped defeat those who tried to enslave the world. And when D-Day (June 6, 1944) came, and when the countdown to the defeat of Nazi Germany began with the Allied invasion of Europe, radio was there to report it "live."

See John McDonough, "The Longest Night: Broadcasting's First Invasion," *American Scholar*, March 1994, for a detailed description of the radio converge of the Normandy invasion. I call that coverage "live" because it was actually broadcast in the form of delayed recordings (remember, there was no CNN in 1944!) That actually represented a big change for radio because, before D-Day, the radio networks had barred recordings of any kind, claiming they would be a deception on the public. The *real* reason for the ban, however, was that the

THE SECRET HOPE

Figure 22.1   Resistance fighters gather to hear the latest radio news about the war against the Nazis.

use of recorded programs would free independent, local stations from minute-by-minute dependency on the networks for access to national news and big-name entertainment. Once the radio audience heard actual "live action" battle reporting, however (including multiple strafing attacks by a German war plane on an invasion ship, whose gun crews eventually shot it from the air as listeners sat glued to their seats), the networks were forced to drop the ban on recordings. The public *demanded* it. Other forces, too, were at work to suppress news coverage by radio, recorded or live. During the first decade or so of radio, newspapers were generally quite hostile. See George E. Lott, Jr., "The Press-Radio War of the 1930s," *Journal of Broadcasting* 14, Summer 1970, pp. 275–286.

American radio listeners had, actually, already gotten a taste of the thrill of real-time radio coverage *before* the start of the Second World War. Filled with a vast amount of hydrogen, the joint National Geographic Society/U.S. Army Air Corp balloon *Explorer I* began a near fatal ascent on the morning of July 28, 1934, from a site just outside of Rapid City, South Dakota. At about 60,000 feet the balloon began to tear, and the three man crew decided to terminate the flight. All the way down, with a total and catastrophic failure at any instant possible, the radio conversations between the crew and the ground were broadcast live by NBC—and millions tuned in. (Compare this to the total blackout by NASA of the Apollo 13 moon mission explosion in 1970.) When just a few thousand feet above ground, the balloon finally did explode and the crew dramatically jumped to safety through a hatch opening (all had parachutes!) You can find stunning photographs of this now nearly forgotten, yet incredible adventure in *The National Geographic Magazine*, October 1934, pp. 397–434. (In 1935 *Explorer II* reached over 72,000 feet and was a complete success.) The famous radio broadcast of the 1937 explosion of the hydrogen-filled German dirigible *Hindenburg* was not a live broadcast, but rather a recording played the next day.

Nothing stays the same in our technical world, of course. Each new decade since the end of the second World War has brought forth new gadgets that have *seemed* to be on the verge of killing off radio. In the 1950s and 1960s it was television in the home, and in the 1970s it was audio-tape cassettes in the automobile. The 1980s brought laser-read CDs, and with today's talk of "information superhighways" pouring data into homes via optical links and computer networks the demise of radio is again a popular prediction. I, however, simply don't believe it. Even today, households are more likely to contain a radio than they are to have a television, a telephone, or a computer. As it has from its earliest days, radio speaks to the masses as no other technology can.

As for the precise technical future of radio, who can predict what will come next? The advent of digital radio, for use as both a visual display of news and automobile navigational information from signals transmitted by orbiting satellites, is just one of the developments that even the genius of Armstrong would have found startling.[15] Radio has a bright future! In the very long run, however, I like the imagery of Maxwell himself, who wrote an 1878 poem (less than a year before his death) that reads astonishingly like a description of all the old radio shows that even now are on their way to the next galaxy and beyond. It speaks to the staying power of radio (even though there *was* no radio when Maxwell wrote). Titled "A Paradoxical Ode," at one point Maxwell's poem says

> Till, in the twilight of the Gods,
> When earth and sun are frozen clods,
> When, all its energy degraded,
> Matter to ether shall have faded;
> We, that is, all the work we've done,
> As waves in ether, shall for ever run
> In ever widening spheres through heavens beyond the sun.[16]

Maxwell's poem may well have been the inspiration for Sir Arthur Eddington, who decades later wrote of his vision of the end of the world:

> it would seem that the universe will finally become a ball of radiation... The longest waves are Hertzian waves of the kind used in broadcasting. About every 1500 million years this ball of radio waves will double its diameter; and it will go on expanding in geometrical progression for ever. Perhaps then I may describe the end of the physical world as—one stupendous broadcast.[17]

Alas, there will be no one left but the angels to listen to Eddington's final ball of radio waves; but on the bright side, there will also be no new commercials! Do I hear De Forest's ghost applauding?

Notes

1. Louise M. Benjamin, "In Search of the Sarnoff 'Radio Music Box' Memo," *Journal of Broadcasting & Electronic Media* 37, Summer 1993, pp. 325–335.

2. This number, and the rest of the numbers quoted in this paragraph, are taken from Leslie J. Page, Jr., "The Nature of the Broadcast Receiver and Its Market in the United States from 1922 to 1927,"

Journal of Broadcasting 4, Spring 1960, pp. 174–182. Also very helpful was the unpublished doctoral dissertation by Lawrence W. Lichty, "The Nation's Station," *A History of Radio Station WLW*, The Ohio State University 1964.

3. Ford A. Carpenter, "First Experiences of a Radio Broadcaster," *The Atlantic Monthly*, September 1923.

4. From James Rorty, "The Impending Radio War," *Harper's*, November 1931.

5. Robert Littell, "A Day with the Radio," *The American Mercury*, February 1932.

6. This and other amusing radio anecdotes are told by Richard Hughes in "The Birth of Radio Drama," *The Atlantic Monthly*, December 1957. Several American stations and the BBC, too, in fact, had broadcast plays well before Hughes' drama was presented.

7. George A. Willey, "End of an Era: The Daytime Radio Serial," *Journal of Broadcasting* 5, Spring 1961, pp. 97–115. As Willey observes, there were earlier attempts at "radio romance" shows in 1927 and 1928 (e.g., "True Romances" and "Romance Isle") but they were very limited efforts. At WLW-Cincinnati, Ohio, radio stories were given the curious name of "radarios" (a contraction of "radio scenario"). A marvelous book that describes literally *every* network show broadcast during America's "golden age of radio," including cast lists and photographs, is by Frank Buxton and Bill Owen, *The Big Broadcast, 1920–1950*, Viking, 1972.

8. For how the radio soap operas were perceived *at the time*, see Whitefield Cook, "Be Sure to Listen In!," *The American Mercury*, March 1940; Merrill Denison, "Soap Opera," *Harper's*, April 1940; and Max Wylie, "Washboard Weepers," *Harper's*, November 1942. A few years later James Thurber wrote a five-part series of long, informative essays full of low-key humor for *The New Yorker* magazine, collectively titled "Soapland" (see the issues of May 15, May 29, June 12, July 3, and July 24, 1948).

9. Dave Berkman, "Long Before Falwell: Early Radio and Religion—As Reported by the Nation's Periodical Press," *Journal of Popular Culture* 21, Spring 1988, pp. 1–11. As Berkman observes, radio and religion were a natural combination; "both were built on words and music."

10. Thomas F. Coakley, "Preaching the Gospel by Wireless," *Catholic World*, January 1922.

11. Editorial comment, *Catholic World*, April 1925.

12. In 1924 McPherson started her own radio station, KFSG, just the third station in Los Angeles. Its call letters stood for "Kall Foursquare Gospel."

13. Marshall W. Fishwick, "Father Coughlin Time: The Radio and Redemption," *Journal of Popular Culture* 22, Fall 1988, pp. 33–47.

14. Herber Blankenhorn, "The Battle of Radio Armaments," *Harper's*, December 1931. The international radio wars had found their way into pulp magazine fiction by the late 1930s, as in Eric Frank Russell's "The Great Radio Peril," *Astounding Stories*, April 1937.

15. Two other new developments in radio are Internet radio (or "webcasting"), and the recent decision by the FCC (January 2000) to license hundreds of "low-power radio" stations operating at 100 watts.

16. Lewis Campbell and William Garnett, *The Life of James Clerk Maxwell*, Johnson Reprint Corporation 1969 (first published 1882, 1884), p. 650.

17. *New Pathways in Science*, Macmillan 1935, p. 71.

Technical
Appendices

The mathematically quite-technical appendices that follow are included for two reasons. First, they may be useful for review and/or for occasional consultation as you work your way through the book. Their formal content is at the advanced freshman level, but the presentation is somewhat (but not much) more advanced. Second, you may find familiar topics discussed in an unfamiliar way. It's always good to know how to do the same thing in more than one way. The approach is mostly the classical, precomputer age one of mathematical analysis. The modern tool of writing and running a computer code that can simulate extraordinarily complicated circuits is, however, included in the main text. I strongly urge any reader who is serious about studying electrical engineering to learn how to use one of the several commercially available codes as soon as possible. At the University of New Hampshire, first-semester sophomores start right in with Electronics Workbench, the circuit simulation code used throughout this book.

Complex Exponentials

Recall, from freshman calculus, the following power-series expansion for e^x around $x = 0$ (the so-called Maclaurin series, a special case of the more general Taylor series).

$$e^x = 1 + x + \frac{x^2}{2!} + \frac{x^3}{3!} + \cdots + \frac{x^n}{n!} + \cdots = \sum_{n=0}^{\infty} \frac{x^n}{n!}.$$

This series expansion is usually derived, formally, for the case of x a real quantity, and the series converges for all real x, i.e., for $-\infty < x < \infty$. When you use an electronic calculator to evaluate exponentials, this is the algorithm that is invoked when you push the EXP button (the summation is preprogrammed right into the calculator's processing chip). Now, without concern about questions of convergence if x is allowed to more generally be a complex-valued quantity, let's just assume the series continues to work and see what happens. Thus, to be specific, replace x with jx (where $j = \sqrt{-1}$) and write

$$e^{jx} = 1 + (jx) + \frac{(jx)^2}{2!} + \frac{(jx)^3}{3!} + \frac{(jx)^4}{4!} + \cdots$$

$$= 1 + jx - \frac{x^2}{2!} - j\frac{x^3}{3!} + \frac{x^4}{4!} + \cdots$$

or, collecting real and imaginary terms together,

$$e^{jx} = \left(1 - \frac{x^2}{2!} + \frac{x^4}{4!} \cdots\right) + j\left(x - \frac{x^3}{3!} + \cdots\right).$$

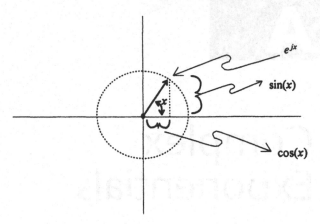

Figure A.1   The complex-valued vector in polar form (e^{jx}) and its rectangular form
$\cos(x) + j\sin(x)$.

But, again from freshman calculus, you should recognize the two power-
series expansions in the parentheses as those for $\cos(x)$ and $\sin(x)$, respectively, i.e., we have, for all real-valued x,

$$e^{jx} = \cos(x) + j\sin(x).$$

This result (displayed in Figure A.1) is called Euler's identity, after the
great Swiss mathematician Leonhard Euler (1707–1783), but it was actually published first (in disguised form) by the Englishman Roger Cotes
(1682–1716) in the pages of the *Philosophical Transactions of London* in
1714 (when Euler was just 7 years old).[1] With Euler's identity we can do
many astonishing calculations and, as you'll see by the time you finish this
book, without it radio is a lot harder to understand. But first, and most
immediately, it provides us with the geometric interpretation of e^{jx} shown
in Figure A.1, as a vector in the complex plane with real part $\cos(x)$ and
imaginary part $\sin(x)$. The angle this vector makes with the real axis is x
(in radians), and from the Pythagorean theorem and the trigonometric
identity $\cos^2(x) + \sin^2(x) = 1$ we see that the vector has unit length. As
x varies, e^{jx} rotates in the complex plane, with its tip always on the circle
with unit radius centered on the origin.

One of the more utilitarian applications of Euler's identity is the quick
and easy derivation of trigonometric identities. For example, since

$$e^{j(x+y)} = e^{jx}e^{jy},$$

then

$$\cos(x + y) + j \sin(x + y) = \{\cos(x) + j \sin(x)\}\{\cos(y) + j \sin(y)\}$$
$$= \cos(x) \cos(y) + j \cos(x) \sin(y)$$
$$+ j \sin(x) \cos(y) - \sin(x) \sin(y).$$

Equating real and imaginary parts, respectively, on each side of the equality sign gives us the trigonometric addition formulas

$$\cos(x + y) = \cos(x) \cos(y) - \sin(x) \sin(y),$$
$$\sin(x + y) = \cos(x) \sin(y) + \sin(x) \cos(y).$$

These formulas will prove to be most useful in the discussion of Chapter 20 on single-sideband AM radio.

Now, what about my earlier claim of "astonishing" things we can calculate with complex exponentials? Consider, for example, the question of the value of the strange-looking expression

$$(j)^j = ?$$

To answer this, write Euler's identify for $x = \pi/2$ to arrive at

$$e^{j\pi/2} = \cos\left(\frac{\pi}{2}\right) + j \sin\left(\frac{\pi}{2}\right) = j.$$

Then,

$$(j)^j = (e^{j\pi/2})^j = e^{j^2(\pi/2)} = e^{-\pi/2} = 0.20788\ldots.$$

Who would have even dreamed such a statement before Euler's identity? The imaginary power of an imaginary number not only has meaning, but in fact can be *real*! (This wonderful result was first stated by Euler in a 1746 letter.) This isn't quite all there is to e^{jx}, however. Using the geometrical interpretation of e^{jx}, we realize that it (a vector) returns to any given position in the complex plane after a further rotation of 2π radians (more precisely, after a rotation in *either* direction of any *integer* number of 2π radians). Thus, we should really write

$$j = e^{j(\pi/2 \pm 2\pi k)}, \qquad k = 0, 1, 2, 3, \ldots$$

and so

$$(j)^j = (e^{j(\pi/2 \pm 2\pi k)})^j = e^{-(\pi/2 \pm 2\pi k)} = e^{-\pi(1/2 \pm 2k)}$$

where k is any integer, not just zero. $(j)^j$ has an *infinity* of real values. That is *really* surprising!

As another special case, write $x = \pi$ and so

$$e^{j\pi} = \cos(\pi) + j \sin(\pi) = -1$$

or, rearranging,

$$e^{j\pi} + 1 = 0.$$

This incredible (I only just barely hesitate to use the word *mysterious*) expression relates five of the most important numbers in mathematics. It has been called, perhaps with only slight exaggeration, one of the most remarkable statements in all of mathematics.

In your study of radio in this book, we will specifically deal with sinusoidally varying time functions. If the angular frequency is denoted by ω (in units of radians per second), then

$$e^{j\omega t} = \cos(\omega t) + j \sin(\omega t)$$

represents a rotating vector of unit length making angle $\theta = \omega t$ with the real axis. The vector is rotating because the angle θ is changing with time, i.e.,

$$\frac{d\theta}{dt} = \omega.$$

If $\omega > 0$, then the vector rotates counterclockwise (CCW), and if $\omega < 0$ (a *negative* frequency!) then the vector rotates clockwise (CW). See Figure A.2. Here then is another wonderful insight complex exponentials give us; a negative frequency is not science fiction, but just the rotation rate of a vector spinning in the opposite sense from that of a positive frequency vec-

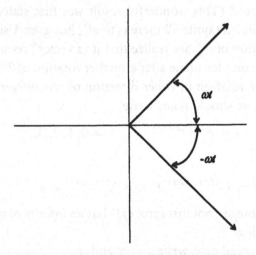

Figure A.2 Counter-rotating complex vectors with mutual cancellation of their imaginary components.

tor. Since one complete rotation is 2π radians, then the rotation frequency in rotations per second (or cycles per second, or the modern *hertz*, abbreviated Hz), denoted by the symbol f, is given by the simple expression all physicists and electrical engineers should know as well as their own name,

$$f = \frac{\omega}{2\pi}.$$

Now, we also have

$$e^{-j\omega t} = e^{j(-\omega t)} = \cos(-\omega t) + j\sin(-\omega t) = \cos(\omega t) - j\sin(\omega t),$$

because the cosine and the sine are even and odd functions, respectively. [Recall that a function $g(t)$ is said to be even if $g(-t) = g(t)$ for all t, and that it is said to be odd if $g(-t) = -g(t)$ for all t. The properties of evenness and of oddness are quite restrictive; practically all functions have neither property. And yet any function can be written as the sum of an even function and an odd function (see Chapter 15).] Continuing we have the highly useful results, from adding and subtracting the expressions for $e^{j\omega t}$ and $e^{-j\omega t}$,

$$\cos(\omega t) = \frac{1}{2}[e^{j\omega t} + e^{-j\omega t}],$$

$$\sin(\omega t) = \frac{1}{2j}[e^{j\omega t} - e^{-j\omega t}].$$

There are simple geometrical interpretations of these two expressions. For $\cos(\omega t)$, for example, we have two vectors rotating *opposite* to each other in such a way that at each instant of time the imaginary components cancel and the real parts add. That is, this sum of two complex exponentials is a sinusoidal oscillation completely confined to the real axis. A similar conclusion follows for $\sin(\omega t)$, with the oscillation confined to the imaginary axis.

As an amusing example of the power of complex exponentials to reduce difficult problems to routine problems, consider this question. A man stands at the origin of the complex plane, and walks forward along the positive real axis for unit distance. He then pivots on his heels in a CCW way through angle θ and walks forward for a one-half unit distance. He then pivots again through a CCW rotation of θ and moves forward for a one-quarter unit distance. He continues doing this for an infinity of equal rotations and ever-decreasing distances (each one-half of the previous distance). Where does he end up in the complex plane, and for what angle θ is he the farthest away from the real axis? A sketch of this process looks like

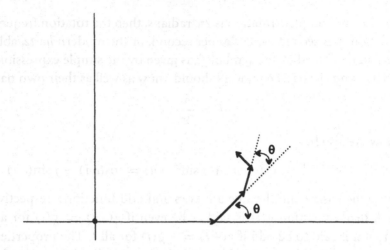

Figure A.3   A walk in the complex plane.

Figure A.3 (where I've assumed a value of $\theta < 90°$ simply for the purpose of drawing a clear diagram).

This "walk in the complex plane" is mathematically described by a sum of vectors, i.e., after the $(n+1)$st step the vector sum $S(n+1)$ points to the man's location in the plane:

$$S(n+1) = 1 + \frac{1}{2}e^{j\theta} + \frac{1}{4}e^{j2\theta} + \cdots + \frac{1}{2^n}e^{jn\theta}.$$

The man's distance from the real axis is the imaginary part of the $S(n+1)$. Thus, what we want to do is find the θ that maximizes the imaginary part of $S(\infty)$. Writing out $S(\infty)$ we have

$$S(\infty) = 1 + \frac{1}{2}e^{j\theta} + \frac{1}{4}e^{j2\theta} + \frac{1}{8}e^{j3\theta} + \cdots.$$

Recognizing this as a geometric series, with the common factor between any two adjacent terms as $(\frac{1}{2})e^{j\theta}$, we use the standard trick of multiplying through by the common factor to get

$$\frac{1}{2}e^{j\theta}S(\infty) = \frac{1}{2}e^{j\theta} + \frac{1}{4}e^{j2\theta} + \frac{1}{8}e^{j3\theta} + \cdots.$$

Subtracting, we have

$$S(\infty) - \frac{1}{2}e^{j\theta}S(\infty) = 1 = S(\infty)\left[1 - \frac{1}{2}e^{j\theta}\right],$$

and so

$$S(\infty) = \frac{1}{1 - (\frac{1}{2})e^{j\theta}} = S_r + jS_i$$

where S_r and S_i are the real and imaginary parts of $S(\infty)$, respectively. To get our hands on explicit expressions for S_r and S_i, we use another standard trick; multiplying through top and bottom of a ratio of complex quantities by the conjugate of the bottom. (Recall that the conjugate of a complex quantity is found by replacing every occurrence of j with $-j$.) Thus,

$$S(\infty) = \frac{1}{1 - (\frac{1}{2})e^{j\theta}} \frac{1 - (\frac{1}{2})e^{-j\theta}}{1 - (\frac{1}{2})e^{-j\theta}}$$

$$= \frac{1 - (\frac{1}{2})e^{-j\theta}}{1 - (\frac{1}{2})e^{j\theta} - (\frac{1}{2})e^{-j\theta} + \frac{1}{4}}$$

$$= \frac{1 - (\frac{1}{2})e^{-j\theta}}{\frac{5}{4} - \cos\theta}.$$

Expanding the numerator with the aid of Euler's identity we immediately have

$$S(\infty) = S_r + jS_i = \frac{1 - (\frac{1}{2})\cos\theta}{\frac{5}{4} - \cos\theta} + j\frac{(\frac{1}{2})\sin\theta}{\frac{5}{4} - \cos\theta}.$$

Thus, after an infinite number of steps the man's distance from the real axis is

$$S_i = \frac{(\frac{1}{2})\sin\theta}{\frac{5}{4} - \cos\theta}.$$

We can answer the second question about θ in the usual way, by setting $dS_i/d\theta = 0$. If you do that you'll find $\cos(\theta) = 0.8$, or $\theta = 36.87°$ (and so my guess of $\theta < 90°$ for the previous sketch was correct), and the maximum value for S_i is $\frac{2}{3}$. Without complex exponentials I think this would be a very awkward problem to analyze.

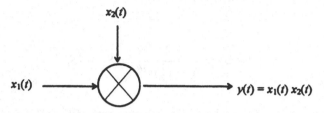

Figure A.4   An analog multiplier.

As a final example of the power of the complex exponential representations of sinusoids, consider this little problem. Suppose we have a box that has two inputs and one output. The signal at the output is the product of the two input signals (never mind how such a box could be made, but be assured that it is possible and that you'll find out how it is done in AM radio in Section 3). I'll represent this box with the symbol shown in Figure A.4.

Now, suppose we have four such boxes wired in a sequential chain, as shown in Figure A.5, with $x(t) = \cos(\omega_0 t)$ as the input to the chain and $y(t)$ as the output. It should be clear that $y(t) = \cos^{16}(\omega_0 t)$, and that this represents a periodic signal of narrow positive pulses as shown in the MATLAB plot of Figure A.6. Indeed, by using even more multipliers in the chain we can make the output pulses as narrow as we'd like. This results from the fact that $|\cos(\omega_0)t)| \leq 1$, and the fact that the square of any number with absolute value less than one is even smaller in magnitude. Now, what are the various frequency components present in $y(t)$, and what are their amplitudes? This is a question that is generally answered with a Fourier series analysis (as I'll discuss in Chapter 9), but this particular problem is easily handled with just Euler's identity and the binomial theorem.

The binomial theorem is the infinite series

$$(1 + x)^n = 1 + nx + \frac{n(n-1)}{2!}x^2 + \frac{n(n-1)(n-2)}{3!}x^3 + \cdots$$

which converges for $|x| < 1$. If n is a nonnegative integer the series actually has only a finite number of nonzero terms, because after the $n+1$st term there is a zero factor in each coefficient. If a and b are any two quantities and n is any nonnegative integer, then

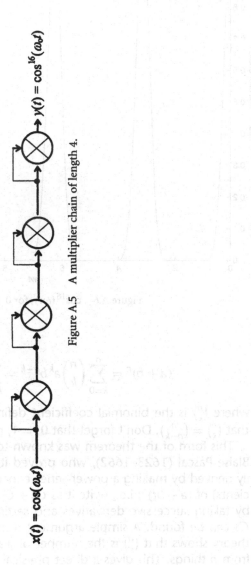

Figure A.5 A multiplier chain of length 4.

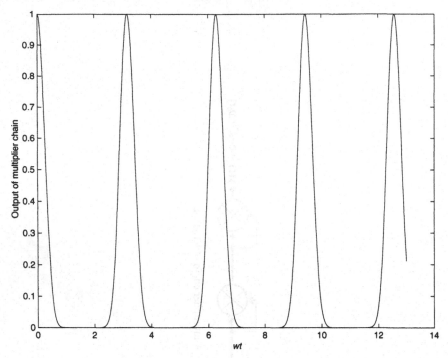

Figure A.6 $\cos^{16}(\omega t)$, for $0 < \omega t < 4\pi$.

$$(a+b)^n = \sum_{k=0}^{n} \binom{n}{k} a^k b^{n-k} = \sum_{k=0}^{n} \binom{n}{k} a^{n-k} b^k$$

where $\binom{n}{k}$ is the binomial coefficient defined as $n!/(n-k)!k!$. Notice that $\binom{n}{k} = \binom{n}{n-k}$. Don't forget that $0! = 1$, not zero! (Pun intended.)

This form of the theorem was known to the French mathematician Blaise Pascal (1623–1662), who proved it in 1654. This result is easily derived by making a power-series expansion (with unknown coefficients) of $(a + bt)^n$; i.e., write it as $C_0 + C_1 t + C_2 t^2 + \cdots + C_n t^n$. Then, by taking successive derivatives and setting $t = 0$, the values of the Cs can be found. A simple argument from elementary combinatorial theory shows that $\binom{n}{k}$ is the number of ways k things can be selected from n things. This gives a direct physical reason for why mathematicians define $\binom{n}{k} = 0$ when $k > n$, e.g., there are zero ways to take six resistors from a pile of five resistors! Sometime in the years 1665–1666, Isaac Newton (1642–1727) extended the theorem to rational values of n (even negative ones), although his "proof" was mostly intuitive. It wasn't until 1826 that the binomial theorem was *really* proven, for arbitrary n, by the Norwegian mathematician Niels Henrik Abel (1802–

1829). By assigning specific values to a and b, some interesting and useful identities involving the binomial coefficients can be easily derived. For example, if $a = b = 1$, then

$$\sum_{k=0}^{n} \binom{n}{k} = 2^n.$$

Now, what if instead of being real we have a and b complex? For example, suppose $a = e^{i\pi/4}$ and $b = e^{-i\pi/4}$. (Notice that now I am using $i = \sqrt{-1}$, rather than j; it's done both ways!) Then, $a + b = 2\cos\left(\frac{\pi}{4}\right) = \sqrt{2}$ and so

$$2^{n/2} = \sum_{k=0}^{n} \binom{n}{k} e^{i(n-k)\pi/4} e^{-ik\pi/4} = \sum_{k=0}^{n} \binom{n}{k} e^{in\pi/4} e^{-ik\pi/2}.$$

Suppose next that n is even in this last expression, i.e., for m any non-negative integer suppose that $n = 2m$. Then

$$2^m = \sum_{k=0}^{2m} \binom{2m}{k} e^{im\pi/2} e^{-ik\pi/2}$$

or,

$$2^m e^{-im\pi/2} = 2^m \left\{ \cos\left(\frac{m\pi}{2}\right) - i\sin\left(\frac{m\pi}{2}\right) \right\} = \sum_{k=0}^{2m} \binom{2m}{k} e^{-ik\pi/2}.$$

Writing out the summation term by term we obtain

$$2^m \left\{ \cos\left(\frac{m\pi}{2}\right) - i\sin\left(\frac{m\pi}{2}\right) \right\} = \binom{2m}{0} - i\binom{2m}{1} - \binom{2m}{2}$$
$$+ i\binom{2m}{3} + \binom{2m}{4} - \cdots.$$

Equating real and imaginary parts on both sides gives us the two remarkable identities

$$\binom{2m}{0} - \binom{2m}{2} + \binom{2m}{4} - \cdots = 2^m \cos\left(\frac{m\pi}{2}\right)$$
$$\binom{2m}{1} - \binom{2m}{3} + \binom{2m}{5} - \cdots = 2^m \sin\left(\frac{m\pi}{2}\right).$$

Using different complex values for a and b will give yet other identities.

Now, writing

$$\cos(\omega_0 t) = \frac{e^{j\omega_0 t} + e^{-j\omega_0 t}}{2}$$

and making the associations of (see the above shaded box)

$$a = \frac{1}{2}e^{j\omega_0 t}, \quad b = \frac{1}{2}e^{-j\omega_0 t},$$

we have

$$y(t) = \sum_{k=0}^{16} \binom{16}{k} \left(\frac{e^{j\omega_0 t}}{2}\right)^k \left(\frac{e^{-j\omega_0 t}}{2}\right)^{16-k} = \frac{1}{2^{16}} \sum_{k=0}^{16} \binom{16}{k} e^{j(2k-16)\omega_0 t}.$$

You can now literally write down the answers by inspection. For example, for $k = 8$ we get the only term which is independent of t:

$$\frac{1}{2^{16}} \binom{16}{8} = \frac{16!}{(8!)^2 2^{16}} = 0.1964.$$

That is, this is the dc (or average) value of $\cos^{16}(\omega_0 t)$. A pretty result for so little work! In the same way, we can calculate (for example) the peak amplitude of the term that represents the output component at frequency $6\omega_0$. We do this by observing that the terms for $k = 5$ and $k = 11$ sum to

$$\frac{1}{2^{16}} \binom{16}{5} e^{-j6\omega_0 t} + \frac{1}{2^{16}} \binom{16}{11} e^{j6\omega_0 t} = \frac{1}{2^{16}} \binom{16}{5} 2\cos(6\omega_0 t)$$

because $\binom{16}{5} = \binom{16}{11}$. That is, the peak amplitude of the output component at frequency $6\omega_0$ is

$$\frac{1}{2^{15}} \binom{16}{5} = \frac{16!}{(5!)(11!)2^{15}} = 0.1333.$$

Note

1. You can read much more about Euler and Cotes, and their work with complex numbers, in my book *An Imaginary Tale: the story of $\sqrt{-1}$*, Princeton 1998.

Problems

A.1. What value of θ maximizes $S_r + S_i$ in the "walk in the complex plane" problem? What value of θ maximizes $\sqrt{S_r^2 + S_i^2}$?

A.2. What is the dc value of $\{\cos(\omega t) + \sin(\omega t)\}^n$, as a function of the nonnegative integer n? Partial answer: for $n = 10$ the dc value is 7.875, while for $n = 50$ the dc value is pretty close to 3,767,330.

A.3. Write $j^{(j^j)}$ as a complex number in polar form. Don't forget that $j = e^{j(\frac{\pi}{2}+2\pi k)}$ where k is any integer, so compute *three* answers using $k = 0, -1,$ and $+1$.

A.4. Derive the following identity:

$$(e^{jx})^{e^{jx}} = e^{-x\sin(x)}\left[\cos\left\{x\cos(x)\right\} + j\,\sin\{x\cos(x)\}\right].$$

A.5. Derive the identity

$$\frac{1}{2} + \cos(t) + \cos(2t) + \cdots + \cos(nt) = \frac{\sin\left\{\left[n+\frac{1}{2}\right]t\right\}}{2\sin\left(\frac{t}{2}\right)}.$$

Next integrate both sides from $-\pi$ to π and, since all the cosine integrals vanish and the right-hand side integrand is even, show how this leads to the integral

$$\int_0^{\pi} \frac{\sin\left(\left[n+\frac{1}{2}\right]t\right)}{\sin\left(\frac{t}{2}\right)}\,dt = \pi,$$

for any nonnegative integer n (including zero).

Hint: Begin with the sum $S = 1+e^{jt}+e^{j2t}+\cdots+e^{jnt}$. Sum this (as done in the text for the "walk in the complex plane" problem), and then set the real part of the sum equal to the real part of the original expression (which you can write using Euler's identity).

A.6. Show, for x and y real, that

$$\left|\sin(x + jy)\right| = \left|\sin(x) + \sin(jy)\right|.$$

Hint: you may find it helpful to read up on the so-called hyperbolic functions before attempting this problem. That is, consider the hyperbolic sine and cosine, which are, respectively,

$$\sinh(x) = \frac{1}{2}\left(e^x - e^{-x}\right)$$

$$\cosh(x) = \frac{1}{2}\left(e^x + e^{-x}\right).$$

B

What Is (and Is Not) a Linear Time-Invariant System (Superposition)

For a system to be linear it must possess the so-called superposition property. That is, with $x(t)$ as the system input and $y(t)$ as the system output, the input $x_1(t) + x_2(t)$ must result in the output $y_1(t) + y_2(t)$ where $x_1(t)$ alone produces $y_1(t)$ [and $x_2(t)$ alone produces $y_2(t)$]. It is not necessary for the inputs to be applied to the same part of the system.

While superposition is a necessary property for the system to be linear, it is not sufficient. One other property, that of scaling, is also required. That is, if $x(t)$ results in $y(t)$, then $Kx(t)$ must result in $Ky(t)$ where K is *any* (perhaps even complex) constant. This second property often puzzles students, as it seems to be merely a special case of superposition. That is, suppose we write $x_1(t) = x_2(t) = \cdots = x_K(t)$. Then, for a linear system obeying superposition, $y(t) = y_1(t) + y_2(t) + \cdots + y_K(t)$ will be the output if the input is $x(t) = x_1(t) + x_2(t) + \cdots + x_K(t) = Kx_1(t)$. But since $y_1(t) = y_2(t) = \cdots = y_K(t)$, then $y(t) = Ky_1(t)$ and it appears we have derived scaling from superposition. That is, we seem not to have to demand scaling as a property separate and distinct from superposition. But there is a subtle flaw here (see Problem B.2). It is, in fact, quite easy to demonstrate system functions that possess either one of the two properties of superposition and scaling, but not the other property. Such systems are *nonlinear*; to be linear a system must possess *both* of these independent properties (as shown in Figure B.1). Let me give you an example of each case.

A system that obeys superposition, but not scaling: Let $y(t) = \text{Re}\{x(t)\}$, where this means the output is the real part of the complex-valued input, $x(t)$. This system obeys superposition because

$$\text{Re}\{x_1(t) + x_2(t)\} = \text{Re}\{x_1(t)\} + \text{Re}\{x_2(t)\}.$$

But this system does not obey scaling because we can explicitly find a scaling factor, K, that fails to pass from input to output. Thus, suppose we pick $K = j = \sqrt{-1}$. Then, if we write the input as the complex time function

$$x(t) = u(t) + jv(t),$$

we have, for the system output when the input is $jx(t)$,

$$\text{Re}\{jx(t)\} = \text{Re}\{ju(t) - v(t)\} = -v(t) \neq j\text{Re}\{x(t)\}.$$

How, you may wonder, can there be such a thing as a complex-value signal? In an actual circuit, there can't, but I am being quite general here and, on paper, in theoretical analyses, complex-valued signals are very useful. In

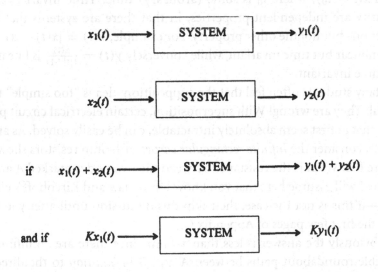

then the SYSTEM is <u>linear</u>

Figure B.1   A linear system.

Section 4, for example, there is a discussion of the so-called *analytic signal*, a complex signal extraordinarily helpful in understanding what is going on in single-sideband AM radio.

A system that obeys scaling, but not superposition: Let

$$y(t) = \frac{1}{x(t)} \left(\frac{dx}{dt} \right)^2.$$

This should be obvious to you, now, almost by inspection.

I think most would agree that $y(t) = x^2(t)$ is a nonlinear system function by inspection (but if it isn't obvious, then formally test the function for superposition and for scaling and show it fails both tests). But what about

$$y(t) = \sqrt{x^2(t)},$$

where the square-root operation is always the positive root? Can one argue that perhaps the two nonlinearities (squaring and square-rooting) "cancel"? No, because this function is equivalent to $y(t) = |x(t)|$ and you should now be able to show that the absolute value function in fact fails *both* the superposition and the scaling tests.

Finally, *time-invariant* systems are those whose outputs shift in time just as do their inputs. Thus, if $y(t)$ results from $x(t)$, then $y(t - t_0)$ results from $x(t - t_0)$, where t_0 is some (arbitrary) time. Time invariance and linearity are independent properties, in that there are systems that have either one but not the other property. For example, $y(t) = |x(t) - x(t-1)|$ is nonlinear but time invariant, while conversely $y(t) = \frac{x(t)}{(|t|+1)}$ is linear but not time invariant.

Many students often feel that the supposition idea is "too simple" to be useful. They are wrong! With superposition, certain electrical circuit problems that at first seem absolutely intractable, can be easily solved. As an example, consider the *infinite rectangular carpet* of 1-ohm resistors shown in Figure B.2. What is the resistance between the two nodes marked A and B? (Note: I will assume here that you know Ohm's law and Kirchhoff's circuit laws—if this is not the case, then skip this discussion until after you have read the first few pages of Appendix C.)

Obviously the answer is less than 1-ohm since there are an infinity of possible roundabout paths between A and B *in addition* to the direct 1-ohm path. In fact, students often claim that because of that infinity of alternate paths then the resistance "must be" zero. This ignores the fact that these roundabout paths are longer and longer (and of higher and higher

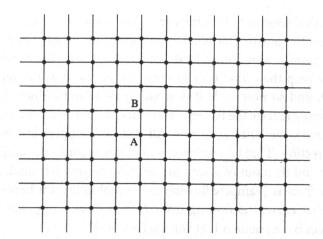

Figure B.2 An infinite carpet of 1-ohm resistors. Each segment between adjacent dots is *not* a wire, but a resistor.

resistance) the further out from A and B they loop, and so in fact the answer is actually $\frac{1}{2}$-ohm. Well, how does one show *that*? We certainly can't (or at least don't want to) write an infinite number of circuit equations! Superposition is the key.

To begin, I'll actually analyze *two* separate problems, each easy to do; then I'll argue that the sum or superposition of these two easy problems is equivalent to original hard problem. So, first imagine that we connect the positive terminal of a battery to node A and adjust the battery voltage until it forces one ampere of current into node A. Let's call the required voltage V. The negative terminal of the battery is connected to the "edge" of the carpet of resistors, at infinity. (Yes, I know, pretty hard to do in the lab, but we can *imagine* doing this.) The one ampere the battery shoves into A is thus extracted from the net at infinity and returned to the negative terminal of the battery. Now, what is the *voltage* at node B?

This question is easy to answer as soon as you notice that *by symmetry* the one ampere into A splits equally among the four resistors connected to A, i.e., there is $\frac{1}{4}$-ampere flowing from A to B. This argument works because the resistor carpet is *infinite* in extent; if it were finite a symmetry argument would depend on A being in the "center" of the carpet. For an infinite carpet, however, *every* node is at the "center" in the sense that it is as far from the infinitely distant "edge" in one direction as in any other direction. So, there is a $\frac{1}{4}$-volt drop from A to B, i.e., the voltage at B is $(V - \frac{1}{4})$ volts.

Next, let's disconnect the battery from the carpet and then reconnect it so that now the positive terminal is connected to the edge of the resistor carpet at infinity, and the negative terminal is connected to node B. As before, we keep the carpet edge grounded (i.e., use it as the zero voltage reference), and so now node B is at voltage $-V$ volts. Since all we have done is simply reverse the battery (and since node B is also in the "center" of the carpet), the battery current will still be 1 ampere but now in the reverse direction. That is, 1 ampere is now flowing into the carpet "edge" at infinity and then out of B into the negative battery terminal. Again, by symmetry there is $\frac{1}{4}$-ampere flowing in the 1-ohm resistor between A and B. Since that current is flowing from A to B then A is $\frac{1}{4}$-volt higher in voltage than B, i.e., node A is at voltage $(-V + \frac{1}{4})$.

The situation at the three critical points (A, B, and the carpet "edge") of the resistor carpet, for the two battery connections, is summarized in the following table:

	First Connection	Second Connection
node A voltage (volts)	V	$-V + \frac{1}{4}$
node B voltage (volts)	$V - \frac{1}{4}$	$-V$
voltage at carpet 'edge,' at infinity (volts)	0	0
current out of battery into A (amperes)	1	0
current out of B into battery (amperes)	0	1

Now, when we "add" these two connections (i.e., use superposition) we see that one ampere flows into A (which is at voltage $\frac{1}{4}$ volt), one ampere flows out of B (which is at voltage $-\frac{1}{4}$ volt), and there is zero current at infinity. This is precisely the original problem and, since one ampere flows from A to B when there is a $\frac{1}{2}$ volt difference between A and B, then the resistance between A and B is $\frac{1}{2}$ ohm. This was easy, but *only* because of superposition.[1]

Note

1. The curious reader may now be asking: what is the resistance between *any* two nodes in the infinite carpet? This *can* be calculated exactly, but it does require mathematics at a somewhat higher level than is in this book. The physical reason for the complexity is that we can no longer use symmetry to argue how the $\frac{1}{4}$-ampere from node A into node B splits as it leaves node B. Some students argue that there is $\frac{1}{12}$-ampere flowing out of node B towards each of the three target nodes B connects to (other than node A, of course), but that is not correct as those three target nodes are *not* in identical situations. But, not to keep you in suspense any longer, when 1 ampere flows into node A

(which is at voltage V), then the first four nodes separated *diagonally* from node A are at voltage $V - \frac{1}{\pi}$. Thus, the diagonal resistance between those nodes and node A is $\frac{2}{\pi}$-ohms. You can find it all worked out in the book by Balthazar van der Pol and H. Bremmer, *Operational Calculus Based on the Two-Sided Laplace Integral*, Cambridge University Press (2nd edition) 1955, pp. 371–372. You can find additional discussions of similar problems concerning infinite carpets of resistors in the following two papers: Francis J. Bartis, "Let's Analyze the Resistance Lattice," *American Journal of Physics* 35, April 1967, pp. 354–355, and Giulio Venezian, "On the Resistance Between Two Points on a Grid," *American Journal of Physics* 62, November 1994, pp. 1000–1004.

Problems

B.1. Show that the "linear looking" $y(t) = ax(t) + b$, where a and b are constants, is a *non*linear system function except for the special case of $b = 0$. This is a popular question on PhD oral examinations, and even Nobel laureates can go astray with it. In particular, see the footnote on p. 177 of Francis Crick's book *The Astonishing Hypothesis*, Touchstone, 1995.

B.2. Where is the flaw in the "derivation" I did of the scaling property from the superposition property? Hint: for linearity, the scaling factor K must be *any* constant; does the "derivation" make any special assumptions about K?

Two-Terminal Components, Kirchhoff's Circuit Laws, Magnetic Coupling, Complex Impedances, ac Amplitude and Phase Responses, Power, Energy, and Initial Conditions

There are three standard electrical components used in electronic circuits. These three components (resistors, capacitors, and inductors) are each characterized by being *passive*, i.e., they do not generate electrical energy, but either dissipate it as heat (resistors), or temporarily store it in an electric field (capacitors) or in a magnetic field (inductors). In addition, all three are *two-terminal* components, as shown in Figure C.1. When you take a more sophisticated course in network theory you will learn that

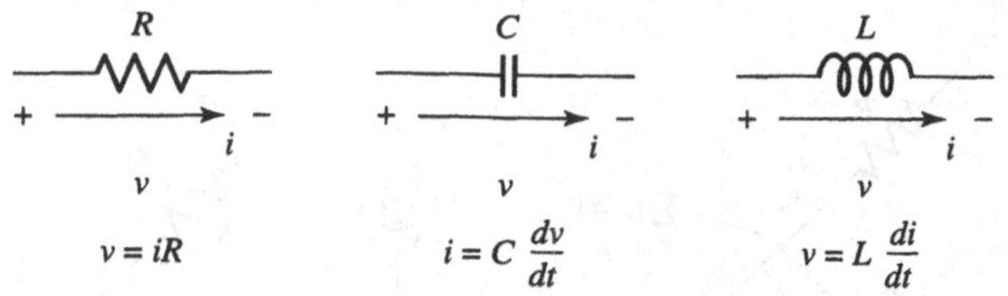

Figure C.1   The three standard electrical components used in radio circuits.

matters are actually more complex than this. When components are allowed to have more than two terminals, the zoo of components quickly expands. Three-terminal components, like triode vacuum tubes, are discussed in the main text of this book (Chapter 8). Two very useful four-terminal components are the *ideal transformer* (see the shaded boxes in this appendix) and the *gyrator* (see Problem C.1 and Appendix F).

We can formally define each of the three passive, two-terminal components by the relationship that connects the current (i) through them to the voltage drop (v) across them. If we denote the values of these components by R (ohms), C (farads), and L (henrys), and if v and i have units of volts and amperes, respectively, and if the time (t) is in units of seconds, then we have the so-called *constitutive laws*

$$v = iR,$$

$$i = C\frac{dv}{dt},$$

$$v = L\frac{di}{dt}.$$

For typical modern radio receiver circuits, the "practicality" of the various fundamental units varies. Thus, 1 ohm is a very small resistance, 1 farad is an enormous capacitance, 1 henry is a large inductance, 1 volt can be enormous, large, or typical (depending on where it is in the circuit), and 1 ampere is either an enormous or a very large current.

In any analysis of radio circuits, electrical engineers and physicists use two "laws" named after the German Gustav Robert Kirchhoff (1824–1887). These two laws (illustrated in Figure C.2) are, in fact, really the laws of the conservation of energy and the conservation of electric charge, in disguise. They are, in words

Kirchhoff's voltage law: the sum of the voltage (or *electric potential*, as physicists often call it) *drops* around any closed path in a circuit is zero. Voltage is defined to be energy per unit charge, and the voltage

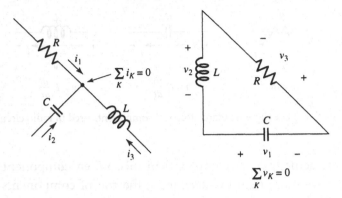

Figure C.2   Kirchhoff's two circuit laws.

drop is the energy expended in transporting a unit charge through the electric field that exists inside the component. This law physically says that the net energy change for a unit charge transported around a closed path is zero. If it were not zero, then we could repeatedly transport charge around the closed path in the direction in which the net energy change is positive and so become rich selling the energy gained to the local power company! Conservation of energy, however, says we can't do this. (Since the sum of the drops is zero, then one can also set the sum of the voltage *rises* around any closed loop to zero.)

Kirchhoff's current law: the sum of the currents into any point in a circuit is zero. This physically says that if we construct a tiny, closed surface around any point in a circuit then the charge enclosed by the surface remains constant. Whatever charge is transported into the enclosed volume by one current is transported out of the volume by other currents. Current is the motion of electric charge, Q. Mathematically, the current i at any point in a circuit is defined to be the rate at which charge is moving through that point, i.e., as $i = \frac{dQ}{dt}$. Q is measured in units of coulombs (the electron charge is 1.6×10^{-19} coulombs), and 1 ampere is equal to 1 coulomb per second.

The physical details of how an inductor "works," unlike the details for a resistor or a capacitor, will be important later in this appendix. Imagine that a coil of wire, with n turns, is carrying a current $i(t)$ as shown in Figure C.3.

The current creates a magnetic field of closed (no ends) flux lines that *encircle* or *thread through* the turns of the coil. Ampere's law,

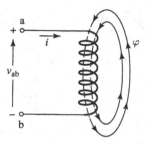

Figure C.3 A current-carrying coil, with its magnetic flux φ.

named after the French mathematical physicist André Marie Ampere (1775–1836), says that the flux produced by each turn of the coil is proportional to i, i.e., the contribution by each turn to the total flux φ is Ki, where K is some constant depending on the size of the coil and the nature of the matter inside the coil. Since the flux contributions add, then the total flux produced by the n turns is $\varphi = nKi$. Now, from Faraday's law of induction, named after the English experimentalist Michael Faraday (see Chapter 2), a *change* in the flux through a turn of the coil produces a potential difference in each turn of the coil of magnitude $\frac{d\varphi}{dt}$. Since there are n turns in series, then the total potential difference that appears across the ends of the coil has magnitude

$$v_{ab} = n\frac{d\varphi}{dt} = n\frac{d(nKi)}{dt} = Kn^2\frac{di}{dt} = L\frac{di}{dt}, L = Kn^2.$$

Notice that the so-called *self-inductance L* of an *n*-turn coil varies as the *square* of the number of turns. In the next box I'll extend the idea of magnetic flux linking the turns of a single coil to that of flux linking *two* coils. That analysis will be the prelude to developing what is called the *ideal transformer*, a tremendously useful device that is all too often dismissed by students as being both "obvious" (it is *not!*) and "boring" (it is *not!*) Beginning students tend to think of transformers as dirty-looking, heavy, ugly lumps of metal and wire. That is all wrong; transformers played a big role in early radio and have almost magical properties that continue to make them extremely important to the modern electronics circuit designer (see part c of Problem C.1).

As an illustration of the use of Kirchhoff's laws, consider the circuit shown in Figure C.4. At first the switch is open, and the current in the inductor and the voltage drop across the capacitor are both zero. Then, at time $t = 0$, the switch is closed and the signal generator [with potential difference across its terminals of $u(t)$] is connected to the rest of the circuit. Suppose we denote the resulting signal generator current by $i(t)$. We

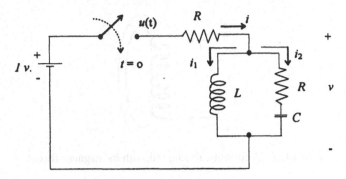

Figure C.4   A circuit with a switched input.

can derive the equation relating $u(t)$ and $i(t)$ by using Kirchhoff's two laws. Using the notation of Figure C.4, we have

$$i = i_1 + i_2,$$

$$u = iR + v$$

$$v = L\frac{di_1}{dt},$$

$$v = i_2R + \frac{1}{C}\int_0^t i_2(x)\, dx.$$

These equations completely describe the behavior of the circuit for all $t > 0$. The last term in the last equation follows by integrating the relationship $i = C\frac{dv}{dt}$ for a capacitor. If V_0 is the voltage drop across the capacitor at time $t = 0$ (the so-called initial voltage), then we have the voltage drop across the capacitor for *any* time $t > 0$ as $(\frac{1}{C})\int_0^t i(x)\, dx + V_0$, where x is, of course, simply a dummy variable of integration. In this problem it is given that $V_0 = 0$. If we differentiate the last equation (a process discussed at length in Appendix G), then we can also write $\frac{dv}{dt} = R\frac{di_2}{dt} + (\frac{1}{C})i_2$. We can manipulate and combine these equations to eliminate the variables i_1 and i_2, and thus arrive at the following second-order linear differential equation relating the applied voltage $u(t)$ to the resulting current $i(t)$

$$\frac{d^2u}{dt^2} + \frac{R}{L}\frac{du}{dt} + \frac{1}{LC}u = 2R\frac{d^2i}{dt^2} + \left(\frac{R^2}{L} + \frac{1}{C}\right)\frac{di}{dt} + \frac{R}{LC}i.$$

You should verify that this is so. It is standard practice to call the applied signal the *excitation* and the resulting signal the *response*.

To proceed in more detail, we need to be more specific now about the nature of $u(t)$. Suppose, for example, the signal generator is simply a 1-volt battery. Then, for $t > 0$ we have $u(t) = 1$ and

$$\frac{d^2 u}{dt^2} = \frac{du}{dt} = 0,$$

and the differential equation for the circuit reduces to (again, for $t > 0$)

$$2R\frac{d^2 i}{dt^2} + \left(\frac{R^2}{L} + \frac{1}{C}\right)\frac{di}{dt} + \frac{R}{LC}i = \frac{1}{LC}.$$

We solve this equation in the standard way, i.e., by noticing that $i(t)$ is the sum of the solutions for the homogeneous case (set the right-hand side equal to zero) and for the inhomogeneous case (set the right-hand side equal to the constant $\frac{1}{LC}$). The inhomogeneous solution is obvious by inspection, i.e.,

$$\frac{R}{LC}i = \frac{1}{LC}$$

or

$$i = \frac{1}{R},$$

as all the derivatives of a constant are zero.

To solve the homogeneous case we have to do a bit more work. I will *assume* the solution has the form of

$$i(t) = Ie^{st}$$

when I and s are both constants (perhaps complex). What motivates this assumption (you may be wondering)? As far as I know, the origin of this immensely clever idea is lost in the history of mathematics. The first person to think of doing this was *very* smart! Substituting this assumed solution into the homogeneous differential equation we arrive at

$$2Rs^2 Ie^{st} + \left(\frac{R^2}{L} + \frac{1}{C}\right)sIe^{st} + \frac{R}{LC}Ie^{st} = 0$$

from which we see that the common factor of Ie^{st} divides out from every term. This leaves us with simply a quadratic, *algebraic* equation for s:

$$s^2 + \frac{R^2 C + L}{2RLC}s + \frac{1}{2LC} = 0.$$

Call the two roots (perhaps complex) of this quadratic s_1 and s_2. Then, the general solution for the battery current $i(t)$ is the sum of the inhomogeneous and homogeneous solutions, i.e.,

$$i(t) = \frac{1}{R} + I_1 e^{s_1 t} + I_2 e^{s_2 t}, \quad t > 0$$

where I_1 and I_2 are constants yet to be determined. I'll show you how that final step is done at the end of this appendix, after we've discovered the way currents in inductors, and voltage drops across capacitors, can (or cannot) change instantaneously.

An enormously useful device that has more than two terminals can be created by positioning two current-carrying coils of wire (i.e., two inductors, as discussed in the previous box) near each other, so that the magnetic flux fields of each can at least partially link to the turns of the other. This results in an additional induced voltage in each coil due to the current in the *other* coil, with the coefficient of the new term called a *mutual inductance* to distinguish it from the self-inductance. That is, as shown in Figure C.5.

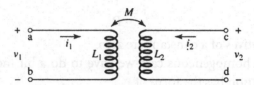

Figure C.5   Two magnetically-coupled coils.

In the notation of Figure C.5 we have

$$v_1 = L_1 \frac{di_1}{dt} + M_{12} \frac{di_2}{dt}$$

$$v_2 = M_{21} \frac{di_1}{dt} + L_2 \frac{di_2}{dt}$$

where M_{12} and M_{21} are the mutual inductances. In fact, it can be shown that $M_{12} = M_{21} = M$, but I'm not going to do it in this book; we'll simply accept it. In addition, that demonstration also shows $0 \leq M \leq \sqrt{L_1 L_2}$, with the exact value of M depending on the amount of cross-coupling flux. The so-called *coupling coefficient* is defined to be $K = M/\sqrt{L_1 L_2}$, and so $0 \leq K \leq 1$. Coils wound in air (with K typically less than $\frac{1}{2}$) are said to be *loosely* coupled, while coils wound around a common iron core (with K typically close to 1) are said to be *tightly* coupled. *If we assume* that terminals b and d can be connected together, such as to a common ground, *then* another circuit which has

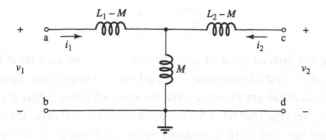

Figure C.6   Equivalent circuit of magnetically coupled coils.

the same defining Kirchhoff equations (and so it is called an *equivalent circuit*) is shown in Figure C.6.

Depending on the values of L_1, L_2 and K, it is possible for the equivalent circuit to have a negative inductance and so it could not actually be constructed. But for the purposes of mathematical analysis, on paper, this is not a problem. In Appendix E you'll see how this equivalent circuit can be used to understand some common bandpass filter circuits used in radio.

The idea of assuming a solution of the form e^{st} is a highly useful one in electronic circuit analysis. For a resistor, for example, the ratio of the voltage drop across the resistor to the current in it is *always* a constant (specifically, the ratio is R!) Because of the presence of a differentiation operation, however, this is not so, *in general*, for inductors and capacitors. Still, for a special class of time functions the voltage-current ratio *is* constant. Thus, suppose we assume that the voltage and the current for a component both vary as e^{st}. Then, for an inductor, if we write the current as $I_m e^{st} = I(s)$, the voltage drop is

$$L\frac{di}{dt} = LsI_m e^{st} = LsI(s) = V(s),$$

and so

$$\frac{V(s)}{I(s)} = sL,$$

which may be a *complex* constant since s may be complex. And similarly, for a capacitor, if we write the voltage drop as $V_m e^{st} = V(s)$, then

$$C\frac{dv}{dt} = CV_m s e^{st} = CsV(s) = I(s)$$

and so

$$\frac{V(s)}{I(s)} = \frac{1}{sC}.$$

Thus, for the special class of time functions e^{st}, we can treat inductors and capacitors just like resistors since all three components have voltage-current ratios that are constant. We do not call these ratios a resistance, however (reserving that term for the always purely real ratio for a resistor), but instead use the word *impedance* (and the symbol Z) to broadly include all three voltage-current ratios. Thus, we can write the relationship between the current $I(s)$ through an impedance $Z(s)$ (made from any connection of Rs, Ls, and Cs) and the voltage drop $V(s)$ across the impedance as $V(s) = I(s)Z(s)$.

Two resistors, R_1 and R_2, *add in series* because they are each carrying the same current (then apply Kirchhoff's voltage law). They combine in parallel as $\frac{1}{R} = \frac{1}{R_1} + \frac{1}{R_2}$ (where $R = \frac{R_1 R_2}{R_1 + R_2}$ is the effective resistance) because they each have the same voltage drop between their terminals (then apply Kirchhoff's current law). The impedances of components in parallel combine just as do the resistances of resistors in parallel, as shown in Figure C.7. Impedances connected in parallel is indicated by writing $Z = Z_1 \| Z_2$. The reciprocal of an impedance is called an *admittance*, and it is usually more convenient to work with admittances when circuit elements are in parallel.

If two impedances, Z_1 and Z_2, are in series (and so are carrying the same current), and if the total voltage drop across the two impedances is $V(s)$, then the individual voltage drops across Z_1 and Z_2 are, respectively,

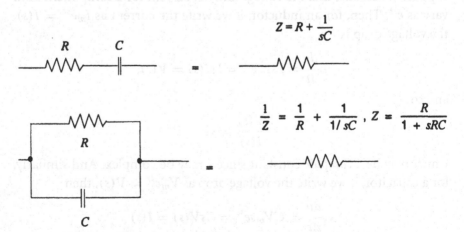

Figure C.7   The impedances of series, and of parallel, RC combinations.

$$V_1(s) = V(s)\frac{Z_1}{Z_1 + Z_2}, \ V_2(s) = V(s)\frac{Z_2}{Z_1 + Z_2}.$$

Such an arrangement is called a *voltage divider*. If the two impedances are in parallel (and so have the same voltage drop $V(s)$), and if the total current flowing in the two impedances is $I(s)$, then the individual currents in Z_1 and Z_2 are, respectively,

$$I_1(s) = I(s)\frac{Z_2}{Z_1 + Z_2}, \ I_2(s) = I(s)\frac{Z_1}{Z_1 + Z_2}.$$

Such an arrangement is called a *current divider*.

While signals of the form e^{st} are a very restricted class, for electrical engineers they include many of the signals of practical interest. Since s is, in general, complex valued, let's write it with explicit real and imaginary parts, i.e., as

$$s = \sigma + j\omega.$$

Then, we can construct the following table:

σ	ω	Nature of e^{st}
0	0	1
$\neq 0$	0	$e^{\sigma t}$
0	$\neq 0$	$e^{j\omega t} = \cos(\omega t) + j\sin(\omega t)$
$\neq 0$	$\neq 0$	$e^{\sigma t}e^{j\omega t} = e^{\sigma t}[\cos(\omega t) + j\sin(\omega t)]$

By picking various values for σ and ω, then we can model constants ($\sigma = \omega = 0$), exponentials that either increase with time ($\sigma > 0, \omega = 0$) or decrease with time ($\sigma < 0, \omega = 0$), pure constant-amplitude sinusoids ($\sigma = 0, \omega \neq 0$), or sinusoids with amplitudes that either exponentially increase with time ($\sigma > 0, \omega \neq 0$) or that decrease with time ($\sigma < 0, \omega \neq 0$). A particularly interesting special case is that of the pure sinusoid, with $\sigma = 0$. Then $s = j\omega$ and the ac impedances of resistors, capacitors, and inductors are, respectively,

$$Z_R = R, \ Z_C = \frac{1}{j\omega C}, \ Z_L = j\omega L.$$

At dc ($\omega = 0$) we see that capacitors are open circuits ($Z_C = \infty$) and inductors are short-circuits ($Z_L = 0$). $Z(j\omega)$, the ac impedance of an arbitrary connection of Rs, Ls, and Cs, is generally complex, and we write it as $Z(j\omega) = R(\omega) + jX(\omega)$. $R(\omega)$ is the *resistive* part of the impedance, and $X(\omega)$ is the *reactive* part of the impedance. Note carefully that *individual* resistors are frequency independent. But the resistive (real) part of impedances can indeed be frequency dependent. See, for example, the analysis later in this appendix for the circuit of Figure C.9.

The frequency-dependent ac impedances of capacitors and inductors (which are pure reactances) can be used to build useful circuits called *filters*. Radio circuits, in particular, would not work without filters. In Figure C.8 a simple RC filter is shown, with input signal $x(t)$ and output signal $y(t)$. In general, $x(t)$ will have energy at many different frequencies but because the filter is linear (i.e., the differential equation describing the filter circuit, and thus connecting $x(t)$ to $y(t)$, is linear) we can apply the superposition property (see Appendix B) and consider each frequency component individually. The total output, $y(t)$, will simply be the sum of the individual outputs due to each of the individual frequency components in $x(t)$.

We can extend the idea of a resistor voltage divider, as mentioned earlier, to circuits involving impedances. If in Figure C.8 we had resistor R_1 in place of R, and resistor R_2 in place of C, then we would write

$$y(t) = \frac{R_2}{R_1 + R_2}x(t).$$

For the more general case of complex ac impedances we can similarly write, for the filter of Figure C.8,

$$\frac{Z_C}{Z_R + Z_C} = \frac{\frac{1}{j\omega C}}{R + \frac{1}{j\omega C}} = \frac{1}{1 + j\omega RC}$$

as the fraction of that part of the input signal *at frequency ω* that appears in the output signal. (The fact that this fraction is *complex* has a deep significance which will be clear by the end of the next several paragraphs.) Since

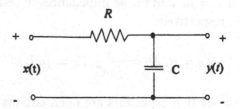

Figure C.8   An elementary low-pass filter (LPF).

the magnitude of this fraction decreases with increasing frequency, the circuit in Figure C.8 is called a *low-pass filter* (LPF), i.e., it tends to "pass" or "transfer" energy from input to output better at lower frequencies than it does energy at higher frequencies. It is easy to show that swapping the R and the C makes a high-pass filter (HPF).

The frequency-dependent ratio of output-to-input (which is generally complex) is called the *transfer function* of the filter, and is usually written as $H(j\omega)$. A particular frequency that is characteristic of the LPF of Figure C.8 is $\omega_0 = \frac{1}{RC}$, and we can write $H(j\omega)$ for that filter as

$$H(j\omega) = \frac{1}{1 + j\frac{\omega}{\omega_0}}, \quad \omega_0 = \frac{1}{RC} \text{ (radians/sec)}.$$

By convention, ω_0 is called the *cutoff* frequency of the filter.

Suppose we apply a pure sinusoid at frequency $\omega = \alpha$ to the input terminals of this filter, i.e., suppose $x(t) = \sin(\alpha t)$. Then, in fact, we are actually applying the sum of *two* signals each of the form e^{st} (where $s = \pm j\alpha$), i.e., from Euler's identity:

$$x(t) = \frac{1}{2j}[e^{j\alpha t} - e^{-j\alpha t}].$$

Thus, multiplying each complex exponential term of $x(t)$ by the filter's explicit transfer function *evaluated at the frequency of the term*, we have

$$y(t) = \frac{1}{2j}[e^{j\alpha t}H(j\alpha) - e^{-j\alpha t}H(-j\alpha)] = \frac{1}{2j}\left[\frac{e^{j\alpha t}}{1 + \frac{j\alpha}{\omega_0}} - \frac{e^{-j\alpha t}}{1 - \frac{j\alpha}{\omega_0}}\right].$$

There are a lot of js in this expression, but since a real signal applied to the input of a filter made of real hardware *must* produce a real output, then we know this complicated expression with all those js must in fact really be real. (If it isn't, that's the math saying a mistake has been made.) Indeed, you can see this is so *by inspection* if you notice that the expression inside the brackets is the difference of conjugates, and so is equal to $2j$ times the imaginary part of the first term, i.e.,

$$y(t) = \frac{1}{2j}\text{Im}\left[\frac{e^{j\alpha t}}{1 + \frac{j\alpha}{\omega_0}}\right]2j = \frac{\sin(\alpha t) - \left(\frac{\alpha}{\omega_0}\right)\cos(\alpha t)}{1 + \left(\frac{\alpha}{\omega_0}\right)^2}.$$

Notice that as $\alpha \to 0$, $y(t) \to \sin(\alpha t)$, i.e., for low frequencies the output tends to become equal to the input. This behavior is, of course, precisely what we mean by the term *low pass*. We can write the expression for $y(t)$ for the simple LPF in different form by recalling the trigonometric identity

$$B_1 \sin(\alpha t) + B_2 \cos(\alpha t) = \sqrt{B_1^2 + B_2^2} \sin\left(\alpha t + \tan^{-1}\left\{\frac{B_2}{B_1}\right\}\right),$$

where B_1 and B_2 can be functions of α (but not of t). Then,

$$y(t) = \frac{1}{\sqrt{1 + \left(\frac{\alpha}{\omega_0}\right)^2}} \sin\left(\alpha t - \tan^{-1}\left\{\frac{\alpha}{\omega_0}\right\}\right) = A(\alpha) \sin[\alpha t - \varphi(\alpha)].$$

That is, the output signal is a sinusoid with the same frequency as the input signal, but it is reduced in amplitude (by a factor that depends on the frequency), as well as phase shifted (by an *angle* that is frequency dependent). For the LPF of Figure C.8 these factors are

$$A(\alpha) = \frac{1}{\sqrt{1 + \left(\frac{\alpha}{\omega_0}\right)^2}},$$

$$\varphi(\alpha) = -\tan^{-1}\left(\frac{\alpha}{\omega_0}\right).$$

$A(\alpha)$ is a dimensionless factor, and $\varphi(\alpha)$ is measured in radians (recall that one radian $= 180°/\pi = 57.3°$). In particularly, we see that for $\alpha = \omega_0$ the LPF factors are

$$A(\omega_0) = \frac{1}{\sqrt{2}} = 0.707,$$

$$\varphi(\omega_0) = -\tan^{-1}(1) = -\frac{\pi}{4} \text{ radians} = -45°.$$

Looking back at the transfer function for the LPF (where now α is not necessarily a specific frequency α, but is in general a variable) we notice something interesting:

$$|H(j\omega)| = \frac{1}{\sqrt{1 + \left(\frac{\omega}{\omega_0}\right)^2}} = A(\omega),$$

$$\angle H(j\omega) = -\tan^{-1}\left(\frac{\omega}{\omega_0}\right) = \varphi(\omega).$$

This result [that the amplitude A and phase φ of $y(t)$ can be found directly from $H(j\omega)$] is actually true, in general for *any* filter (not just for the LPF), as we can show by returning to the expression for $y(t)$ just before I plugged in the explicit form of $H(j\omega)$:

$$y(t) = \frac{1}{2j}[e^{j\omega t}H(j\omega) - e^{-j\omega t}H(-j\omega)] = \text{Im}\left\{e^{j\omega t}H(j\omega)\right\}.$$

Remember, the last step follows because the expression inside the brackets is the difference of conjugates. Writing $H(j\omega)$ explicitly as a complex function in rectangular form [in *polar* form, of course, $H(j\omega) = A(\omega)e^{j\varphi(\omega)}$],

$$H(j\omega) = X(\omega) + jY(\omega),$$

and so

$$y(t) = \text{Im}\{[\cos(\omega t) + j\sin(\omega t)][X(\omega) + jY(\omega)]\},$$

or

$$y(t) = X(\omega)\sin(\omega t) + Y(\omega)\cos(\omega t),$$

or

$$y(t) = \sqrt{X^2 + Y^2}\sin\left(\omega t + \tan^{-1}\left\{\frac{Y}{X}\right\}\right).$$

But

$$|H(j\omega)| = \sqrt{X^2 + Y^2} = A(\omega)$$

and

$$\angle H(j\omega) = \tan^{-1}\left\{\frac{Y}{X}\right\} = \varphi(\omega),$$

and so

$$y(t) = |H(j\omega)|\sin[\omega t + \angle H(j\omega)] = A(\omega)\sin[\omega t + \varphi(\omega)].$$

More sophisticated circuits (filters) can be analyzed for their ac frequency-dependent behavior by systematically applying Kirchhoff's laws to them to find their transfer functions. For example, consider a filter of Figure C.9, made of two *cascaded* stages of simple RC filters. By convention,

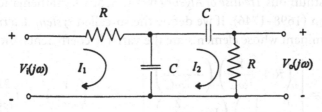

Figure C.9   A two-stage *RC* filter.

when we do an ac analysis of a filter we use capital letters for the voltage and current variables as a notational way to indicate that we are *not* considering arbitrary time functions. We are specifically considering only sinusoidally varying time signals, and of course are using the ac impedances for the various circuit components. In the notation of Figure C.9, the input and output voltages are V_i and V_0, respectively, and the two *loop currents* are denoted by I_1 and I_2. Thus, the current in the left C is $I_1 - I_2$ downward, or $I_2 - I_1$ upward. Using Kirchhoff's voltage law on each loop, we have

$$-V_i + I_1 R + \frac{1}{j\omega C}(I_1 - I_2) = 0,$$

$$\frac{1}{j\omega C}I_2 + I_2 R + \frac{1}{j\omega C}(I_2 - I_1) = 0,$$

which can be written in the more systematic form of

$$I_1\left(R + \frac{1}{j\omega C}\right) + I_2\left(-\frac{1}{j\omega C}\right) = V_i,$$

$$I_1\left(-\frac{1}{j\omega C}\right) + I_1\left(R + \frac{2}{j\omega C}\right) = 0.$$

This is a pair of simultaneous equations, where we consider I_1 and I_2 as the unknowns, and V_i as given. Don't lose sight of what we are ultimately after. We want to find the transfer function of the filter, the ratio $H(j\omega) = V_0(j\omega)/V_i(j\omega)$. We will do this by solving the above pair of simultaneous equations for I_2, in terms of V_i. Then, by observing that $V_0 = I_2 R$, we can find $H(j\omega)$. For such a simple system of equations one could solve the first for I_2 in terms of I_1 (or vice versa) and then substitute into the second. This is *not* a good *general* approach, however. For a system of equations just the next level of complexity up (three equations in three unknowns) you can go crazy trying to get the algebra straight! Cramer's rule, from the theory of determinants, is the proper technique. It is named after the Swiss mathematician Gabriel Cramer (1704–1752) who published it in 1750 but, in fact, it had appeared in print two years before, in the posthumous *Treatise of Algebra* by the Scottish mathematician Colin MacLaurin (1698–1746). If we define the so-called *system determinant* as the determinant whose elements are the variable coefficients, then

$$D = \begin{vmatrix} \left(R + \frac{1}{j\omega C}\right) & \left(-\frac{1}{j\omega C}\right) \\ \left(-\frac{1}{j\omega C}\right) & \left(R + \frac{2}{j\omega C}\right) \end{vmatrix} = R^2 - \left(\frac{1}{\omega C}\right)^2 - j\frac{3R}{\omega C},$$

and then Cramer's rule says that

$$I_2 = \left| \begin{matrix} \left(R + \dfrac{1}{j\omega C}\right) & V_i \\ \left(-\dfrac{1}{j\omega C}\right) & 0 \end{matrix} \right| \div D = V_i \dfrac{\dfrac{1}{j\omega C}}{R^2 - (\dfrac{1}{\omega C})^2 - j\dfrac{3R}{\omega C}}.$$

Thus, as $V_0 = I_2 R$, we have

$$\frac{V_0}{V_i} = H(j\omega) = \frac{-j\dfrac{R}{\omega C}}{R^2 - (\dfrac{1}{\omega C})^2 - j\dfrac{3R}{\omega C}}.$$

From this we could clearly find expressions for $A(\omega)$ and $\varphi(\omega)$, and thus determine $v_0(t)$ explicitly in response to the input $v_i(t) = \sin(\omega t)$. Notice that while $H(j\omega)$ is generally complex, there is one special frequency at which the filter's transfer function is purely real, i.e.,

$$H(j\omega) = +\frac{1}{3}$$

when

$$\omega = \frac{1}{RC}.$$

This result is of no direct interest to us in this book, but you will find in more advanced electronics courses that this property, of a purely *real* filter transfer function *at a particular frequency*, can be exploited to make oscillators, which *are* of interest in radio. Oscillators are obviously of importance in radio transmitters. Less obvious, perhaps, is that oscillators are also vital in AM radio receivers (in that part of the receiver called the *local oscillator*). For our purposes here, however, we'll just assume the existence of circuits that generate signals like $\sin(\omega t)$, without worrying about the details of how such circuits are made (but see Chapter 8).

You'll notice that I did not need to solve for I_1 to find H. Knowledge of I_1 provides useful information, too, however. If we know I_1 in terms of V_i, then we can calculate V_i/I_1, a ratio that represents the *input impedance* that the input signal source for V_i "sees" connected across its terminals (denoted by Z_i). This is important because that impedance determines the current the signal source has to be able to provide to the input of the filter. Thus, using Cramer's rule again,

$$I_1 = \left| \begin{matrix} V_i & \left(-\dfrac{1}{j\omega C}\right) \\ 0 & \left(R + \dfrac{2}{j\omega C}\right) \end{matrix} \right| \div D = V_i \dfrac{R + \dfrac{2}{j\omega C}}{R^2 - (\dfrac{1}{\omega C})^2 - j\dfrac{3R}{\omega C}}.$$

or

$$Z_i = \frac{R^2 - (\frac{1}{\omega C})^2 - j\frac{3R}{\omega C}}{R + \frac{2}{j\omega C}}.$$

In particular, at the frequency at which $H(j\omega)$ is purely real, direct substitution shows

$$Z_i = R(1.2 - j0.6), \quad \text{at} \quad \omega = \frac{1}{RC}.$$

Thus, while the transfer function of the filter is purely real at $\omega = \frac{1}{RC}$, the input impedance of the filter is definitely complex. This is a quite interesting result, as it shows that Z_i is independent of the value of C at $\omega = \frac{1}{RC}$. Thus, we can vary C (actually, both of the equal-valued capacitors, simultaneously) to change the value of $\omega = \frac{1}{RC}$ and yet the input impedance seen by the input signal source will *not* change (and so the signal source sees a constant current demand, a property of great importance in designing the variable frequency oscillator circuits that occur in AM radio receivers).

To conclude this discussion of the two-stage filter in Figure C.9, notice that the phase shift from input to output is, at frequency $\omega = \frac{1}{RC}$, zero degrees. We can see this directly simply by recalling that at this particular frequency the transfer function $[H(j\omega)]$ is real and *positive*. (A filter with a transfer function that is real and *negative* at some particular frequency would have a 180° phase shift from input to output, at that frequency—see Problem C.2). But what about the phase shifts through each stage? You might suspect that because the R and C locations are reversed in the two stages of the filter that the individual stage phase shifts will have opposite signs (but of equal magnitudes, since the two phase shifts have to add to zero). In fact, for the second stage, we have the transfer function

$$H_2 = \frac{R}{R + \frac{1}{j\omega C}},$$

which, at $\omega = \frac{1}{RC}$, is

$$H_2\left(j\frac{1}{RC}\right) = \frac{1 + j}{2}.$$

That is, the phase shift through the second stage at $\omega = \frac{1}{RC}$ is +45°, and so the phase shift through the first stage *must* be −45°. We can *verify* this last statement by making a direct calculation.

Naively, and *incorrectly*, one might calculate the first stage phase shift by writing the transfer function of that stage as

$$H_1 = \frac{\frac{1}{j\omega C}}{R + \frac{1}{j\omega C}},$$

and so, at $\omega = \frac{1}{RC}$,

$$H_1\left(j\frac{1}{RC}\right) = \frac{1-j}{2},$$

which does, indeed, have an angle (phase shift) of $-45°$. But this calculation, while producing the correct numerical result, is wrong! It is wrong because it ignores the influence of the second stage on the output of the first stage. This influence is called the "loading effect." What we should correctly do is indicate in Figure C.10, where the loading impedance of the second stage is denoted by Z. Thus, we have

$$Z = \frac{1}{j\omega C} + R = \frac{(1+j\omega RC)}{j\omega C},$$

and so

$$H_1 = \frac{\frac{(\frac{1}{j\omega C})Z}{(\frac{1}{j\omega C}) + Z}}{R + \frac{(\frac{1}{j\omega C})Z}{(\frac{1}{j\omega C}) + Z}}.$$

Doing the substitution of Z into H_1, you can show that

$$H_1 = \frac{1+j\omega RC}{1-(\omega RC)^2 + j3\omega RC},$$

and so, at $\omega = \frac{1}{RC}$,

$$H_1\left(j\frac{1}{RC}\right) = \frac{1-j}{3},$$

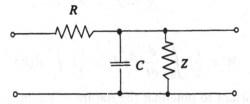

Figure C.10   "Loading" of the first stage by the second stage (Z).

which does indeed have an angle (phase shift) of $-45°$. Notice, however, that the denominator of H_1 is properly a 3, not the 2 given by the incorrect analysis. It was simply a *coincidence*, an *accident*, that the first, naive, incorrect analysis, which ignored the second-stage loading on the first stage, gave the correct answer.

Power (p) is the rate at which energy is delivered to a component, and is given by

$$p = vi,$$

where p has the units of watts (joules/second) when v (the voltage drop across the component) is in volts, and i (the current in the component) is in amperes. To see that this is dimensionally correct, first note that power is energy per unit time. Then, recall that voltage is energy per unit charge, and that current is charge per unit time. Thus, the product vi has units energy/charge times charge/time = energy/time, the units of power. If we integrate power over an interval of time, then the result is the total, net energy (W) delivered to the component during that interval. For example, for a resistor we have $v = iR$ and so

$$p = vi = (iR)i = i^2R,$$

or, in the time interval 0 to T, the net energy delivered to the resistor is

$$W = \int_0^T p(t)\, dt = \int_0^T (i^2)R\, dt = R \int_0^T i^2(t)\, dt.$$

Since the integrand is always nonnegative we conclude, *independent* of the time behavior of the current, that $W > 0$ if $i(t) \neq 0$. The electrical energy delivered to the resistor is totally converted to heat energy, i.e., the temperatures of resistors that carry currents increase.

For inductors and capacitors, however, the situation is remarkably different. For an inductor, for example,

$$p = vi = \left(L\frac{di}{dt}\right)i = \frac{1}{2}L\frac{d(i^2)}{dt},$$

and so

$$W = \frac{1}{2}L \int_0^T \frac{d}{dt}(i^2)\, dt = \frac{1}{2}L \int_0^T d(i^2).$$

Now, it is important to note that the nature of the integration limits on the last integral (0 and T) is *time*, while the variable of integration is i. It is perhaps clearer, then to write

$$W = \frac{1}{2}L \int_{i^2(0)}^{i^2(T)} d(i^2)$$

where $i(0)$ and $i(T)$ are the inductor currents at times $t = 0$ and $t = T$, respectively. Next, change variable to $u = i^2$. Then,

$$W = \frac{1}{2}L \int_{u(0)}^{u(T)} du = \frac{1}{2}L[u(t) - u(0)] = \frac{1}{2}L[i^2(T) - i^2(0)].$$

Thus, $i(t)$ can have any physically possible behavior over the interval $0 < t < T$ and yet, if $i(0) = i(T)$, then $W = 0$. In such a case, where the beginning and ending currents are equal, the total, net energy delivered to the inductor is zero. What has physically happened is that, as $i(t)$ has varied from its initial value at $t = 0$, energy is *stored* in a magnetic field around the inductor and then, as the current returns to its initial value at $t = T$, the stored energy is returned to the circuit (i.e., to the original source of the energy) as the field "collapses." Inductors do not dissipate electrical energy by turning it into heat energy and so, unlike resistors, ideal inductors don't get warm when conducting an electrical current. The same situation is true for ideal capacitors (see Problem C.3).

These power and energy concepts may appear to be quite elementary, but consider the following classic problem that may demonstrate that there *is* more than meets the quick glance here. Suppose we have a resistor R with current $i(t)$ in it. Then, as before,

$$W = \int_{-\infty}^{\infty} p(t)\,dt = R \int_{-\infty}^{\infty} i^2(t)\,dt$$

is the total energy dissipated as heat by the resistor. Also, as current is the time derivative of electric charge, we have

$$Q = \int_{-\infty}^{\infty} i(t)\,dt$$

as the total charge that passes through the resistor. Consider now a specific $i(t)$: for c a constant, let

$$i(t) = \begin{cases} 0, & t < 0 \\ c^{-4/5}, & 0 < t < c \\ 0, & t > c. \end{cases}$$

That is, $i(t)$ is a finite-valued pulse that is nonzero only for a finite length of time. The total charge transported through the resistor is

$$Q = \int_0^c c^{-4/5} \, dt = c^{1/5}.$$

Now, suppose we pick the constant c to be ever smaller, i.e., we let $c \to 0$. Then the pulselike current obviously does something a bit odd—it becomes ever briefer in duration but ever larger in amplitude. But notice that $\lim_{c \to 0} Q = 0$ which means that, even though the amplitude of the current pulse blows up, the pulse duration becomes shorter "even faster" so that the total charge transported through the resistor goes to zero. Now for the puzzle! What happens to W? Well, we have $i^2(t) = c^{-8/5}$ over the duration of the current pulse and so

$$W = R \int_0^c c^{-8/5} \, dt = Rc^{-3/5}.$$

So, $\lim_{c \to 0} W = \infty$, which means the resistor will instantly vaporize because all that infinite energy is delivered in zero time. But, how can *that* be, as in the limit of $c \to 0$ there is *no charge* transported through the resistor? Physics and electrical engineering students should ask their professors about what is going on here!

An important concept in many power calculations is that of the so-called "rms" value of a periodic (*not* necessarily sinusoidal) signal. Thus, suppose $v(t)$ denotes the voltage drop across a 1-Ω resistor. Then, the instantaneous power is

$$p(t) = \frac{v^2(t)}{R} = v^2(t), \text{ as } R = 1.$$

The total energy, W, dissipated by the resistor over one complete period of $v(t)$ is then

$$W = \int_0^T p(t) \, dt = \int_0^T v^2(t) \, dt.$$

Suppose that we define V_{rms} to be that constant, dc voltage that would dissipate the same energy in the same time interval. Then

$$V_{rms}^2 T = \int_0^T v^2(t) \, dt$$

or

$$V_{rms} = \sqrt{\frac{1}{T} \int_0^T v^2(t) \, dt}.$$

Now you can see where the name comes from; V_{rms} is the "(square) root of the mean (average over T) of the square" of $v(t)$.

Although this derivation of V_{rms} was done under the assumption that $v(t)$ is physically a voltage signal across a 1–Ω resistor (a *very* special situation), we can now simply extend the result and make it the *definition* of the rms value of *any* periodic signal $f(t)$, i.e.,

$$F_{rms} = \sqrt{\frac{1}{T} \int_0^T f^2(t)\, dt,}$$

whatever the physical nature of $f(t)$.

If we know the details of $f(t)$, then of course we can specifically calculate F_{rms}. For example, suppose $f(t) = F_M \sin(\omega t + \varphi)$. Then,

$$F_{rms} = \sqrt{\frac{\omega}{2\pi} \int_0^{2\pi/\omega} F_M^2 \sin^2(\omega t + \varphi)\, dt} = \frac{F_M}{\sqrt{2}}.$$

That is, the rms value of any sinusoidally time varying signal is simply the maximum value divided by $\sqrt{2}$ (multiplied by 0.707), independent of frequency and phase.

Now, let $i(t)$ be the current in an impedance $Z = R + jX$. Then, the energy dissipated by Z, over a period, is just the energy dissipated by R (as the reactive part of Z only temporarily *stores* energy in a field). The total energy, W, delivered to the impedance over a period is

$$W = \int_0^T i^2(t) R\, dt,$$

and so the *average* power is

$$P = \frac{W}{T} = \frac{1}{T} \int_0^T i^2(t) R\, dt = I_{rms}^2 R.$$

In particular, if $i(t) = I_m \sin(wt + \varphi)$, then the average power delivered to the impedance is

$$P = \frac{1}{2} I_m^2 R.$$

As shown earlier in this appendix, $V(s) = I(s)Z(s)$ or, for the ac case $(s = j\omega)$,

$$V(j\omega) = V_m e^{j\omega t}$$

and so

$$I(j\omega) = \frac{V(j\omega)}{Z(j\omega)} = \frac{V_m e^{j\omega t}}{Z(j\omega)}.$$

If we write the impedance in polar form,

$$Z = \sqrt{R^2 + X^2} e^{j \tan^{-1}(X/R)}$$

and then

$$I(j\omega) = \frac{V_m}{\sqrt{R^2 + X^2}} e^{j[\omega t - \tan^{-1}(X/R)]}$$

which immediately gives the maximum current as

$$I_m = \frac{V_m}{\sqrt{R^2 + X^2}}.$$

Thus, the average power in the impedance is

$$P = \frac{1}{2} \frac{V_m^2 R}{R^2 + X^2}.$$

Notice carefully the role of X. It plays no direct role in dissipating energy, but it plays a most important indirect role because it influences I_m in R.

An important theoretical property of inductors is that the current through them cannot change instantaneously. We can see this by writing the instantaneous power as

$$p = vi = Li\frac{di}{dt}.$$

If the current i could change instantaneously, then $di/dt = \infty$ at that instant and so $p = \infty$ at the same instant. But we reject as nonsense the idea that *any* physical quantity (such as power) in real circuit hardware can have an infinite value. By making a similar argument, one concludes too that the voltage drop across the capacitor cannot change instantly. Since the power in a resistor does not involve a derivative, however, then both the voltage drop across and the current through a resistor *can* change instantly.

As the final step in arriving at the ideal transformer, write the equations for the two magnetically coupled coils of Figure C.5 as

$$v_1 L_2 = L_1 L_2 \frac{di_1}{dt} + M L_2 \frac{di_2}{dt}$$

$$v_2 M = M^2 \frac{di_1}{dt} + L_2 M \frac{di_2}{dt}.$$

Now, imagine that the two coils are *totally coupled*, i.e., that $K = 1$. (In actual practice this condition can be closely approached, e.g., it is possible to have $K = 1 - 10^{-8}$). Then $M = \sqrt{L_1 L_2}$ and so the two equations become

$$v_1 L_2 = L_1 L_2 \frac{di_1}{dt} + L_2 \sqrt{L_1 L_2} \frac{di_2}{dt}$$

$$v_2 \sqrt{L_1 L_2} = L_1 L_2 \frac{di_1}{dt} + L_2 \sqrt{L_1 L_2} \frac{di_2}{dt}.$$

Thus, $v_1 L_2 = v_2 \sqrt{L_1 L_2}$ or,

$$v_1 = v_2 \sqrt{\frac{L_1}{L_2}}$$

or, recalling that the self-inductance of a coil depends on the *square* of the number of turns in the coil, we have

$$v_1 = v_2 \frac{n_1}{n_2}.$$

For unity coupling, then, we have $\frac{v_2}{v_1} = \frac{n_2}{n_1}$ and the ratio of the voltages across the two coils is equal to the so-called *turns-ratio* of the coils. This result is *one* part of the definition of the ideal transformer, but it is not the whole picture. To complete the evolution of coupled coils to the ideal transformer, I will now make one more assumption (in addition to the first one of unity coupling); I will also assume that the self-inductances of the two coils both become huge (i.e., that both n_1 and n_2 each become huge) all the while keeping a constant ratio (i.e., keeping the turns ratio fixed). Then, as the first equation above is

$$\frac{v_1}{L_1} = \frac{di_1}{dt} + \frac{M}{L_1} \frac{di_2}{dt}$$

and since $\frac{M}{L_2} = \frac{\sqrt{L_1 L_2}}{L_1} = \sqrt{\frac{L_2}{L_1}} = \frac{n_2}{n_1}$, then in the limit as $L_1 \to \infty$ we have

$$0 = \frac{di_1}{dt} + \frac{n_2}{n_1} \frac{di_2}{dt}.$$

With the physically reasonable initial conditions $i_1(0) = i_2(0) = 0$, this equation is easily integrated to give $i_1 + \frac{n_2}{n_1} i_2 = 0$ or,

$$\frac{i_2}{i_1} = -\frac{n_1}{n_2}$$

and so the ratio of the coil currents is the *negative of the reciprocal* of the turns ratio. This is the second condition that defines the ideal transformer. There is an astonishing implication of the two ideal transformer conditions that I think are not at all obvious from the physics: the ideal transformer neither dissipates energy (this *is* obvious because there is no loss mechanism present, i.e., no resistance) nor *stores* energy (this is *not* obvious since we know inductors *do* store energy and the ideal transformer is made from *two huge* inductances). Here's the proof: we have

$$i_1 = -\frac{n_2}{n_1} i_2$$

$$v_2 = \frac{n_2}{n_1} v_1$$

and so, calling v_1, i_1 the *input* and v_2, i_2 the *output*,

$$i_1 v_1 = -\frac{n_2}{n_1} i_2 v_1 = \text{instantaneous input power}$$

$$i_2 v_2 = i_2 \frac{n_2}{n_1} v_1 = \text{instantaneous output power.}$$

Thus,

$$i_1 v_1 + i_2 v_2 = -\frac{n_2}{n_1} i_2 v_1 + i_2 \frac{n_2}{n_1} v_1 = 0,$$

which says the *total instantaneous* absorbed power in the ideal transformer is zero. The ideal transformer only *transfers* energy from one pair of terminals (called the *primary*) to the other pair of terminals (called the *secondary*).

These properties are quite useful in determining what happens in circuits at the instant when a *switching* event occurs. For example, recall the solution for the battery current in the circuit of Figure C.4. I left the solution in the incomplete state, i.e., as

$$i(t) = \frac{1}{R} + I_1 e^{s_1 t} + I_2 e^{s_2 t},$$

where I_1 and I_2 are constants yet to be found. The two values of s are, as shown earlier, the solutions to the quadratic equation

$$s^2 + \frac{R^2 C + L}{2RLC} s + \frac{1}{2LC} = 0,$$

which are

$$s = \frac{1}{2} \left[-\frac{R^2 C + L}{2RCL} \pm \sqrt{\left(\frac{R^2 C + L}{2RLC} \right)^2 - \frac{2}{LC}} \right].$$

Since the complex roots to any algebraic equation with real coefficients always appear as conjugate pairs (this is a very deep theorem in algebra, and not trivial at all), then the two roots to a quadratic are either a complex conjugate pair *or* both roots are real (it is impossible for one root to be real and one root to be complex). It should be obvious *by inspection* of the formula for s that if both roots are real then both are negative, and that if both roots are complex then their real parts are negative. (See Problem C.4). That is, for this circuit we can *always* write $s_{1,2} = \sigma \pm j\omega$ where $\sigma < 0$ and $\omega > 0$. This means that, even without yet knowing I_1 and I_2, we can conclude

$$\lim_{t \to \infty} i(t) = \frac{1}{R}$$

as both exponential terms in $i(t)$ will vanish in the limit. That is, in the limit of $t \gg 0$ the battery current is strictly dc. This is, in fact, consistent with our previous deduction that, for dc, inductors are "shorts" and capacitors are "opens." That is, for $t \gg 0$ in the circuit of Figure C.4, replace the L with a short and the C with an open and, indeed, the dc battery current is obviously $\frac{1}{R}$. I'll now calculate I_1 and I_2.

Since there are two constants to determine, we will need to find two equations for I_1 and I_2. One equation is easy to find. Just *before* the switch is closed the voltage drop across the C was given as zero, and the current in the L was also given as zero. Since both of these quantities cannot change instantly, then both must still be zero just *after* the switch is closed. (If the switch is closed at time $t = 0$, it is standard in electrical engineering to write "just before" and "just after" as $t = 0^-$ and $t = 0^+$, respectively.)

Thus, at $t = 0^+$, the battery current flows entirely through the two resistors (which are in series) and we have

$$i(0^+) = \frac{1}{2R} = \frac{1}{R} + I_1 + I_2 \quad \text{or} \quad I_1 + I_2 = -\frac{1}{2R}.$$

For our second equation we need to do a bit more work. In the notation of Figure C.4, recall the set of equations that define the circuit for $t > 0$:

$$1 = iR + L\frac{di_1}{dt},$$
$$L\frac{di_1}{dt} = i_2 R + \frac{1}{C}\int_0^t i_2(u)\, du,$$
$$i = i_1 + i_2.$$

If we evaluate the first equation of this set for $t = 0^+$, we have

$$1 = i(0^+)R + L\frac{di_1}{dt}\bigg|_{t=0^+} = \frac{1}{2R}R + L\frac{di_1}{dt}\bigg|_{t=0^+}$$

or

$$\frac{di_1}{dt}\bigg|_{t=0^+} = \frac{1}{2L}.$$

From the third equation in the set we have

$$i(0^+) = i_1(0^+) + i_2(0^+),$$

which, because $i_1(0^+) = 0$, says

$$i_2(0^+) = i(0^+) = \frac{1}{2R}.$$

Also, if we differentiate the first and the second equations of the set, then we have

$$0 = R\frac{di}{dt} + L\frac{d^2i_2}{dt^2},$$

$$L\frac{d^2i_1}{dt^2} = R\frac{di_2}{dt} + \frac{1}{C}i_2$$

which, when combined, give

$$0 = R\frac{di}{dt} + R\frac{di_2}{dt} + \frac{1}{C}i_2.$$

But, since

$$\frac{di}{dt} = \frac{di_1}{dt} + \frac{di_2}{dt}$$

then

$$0 = 2R\frac{di}{dt} - R\frac{di_1}{dt} + \frac{1}{C}i_2.$$

Finally, evaluating this last result for $t = 0^+$ [and recalling our earlier results for $di_1/dt|_{t=0^+}$ and $i_2(0^+)$], we have

$$0 = 2R\frac{di}{dt}\bigg|_{t=0^+} - \frac{R}{2L} + \frac{1}{2RC}$$

or

$$\frac{di}{dt}\bigg|_{t=0^+} = \frac{1}{4}\left(\frac{1}{L} - \frac{1}{R^2C}\right).$$

That is, our second equation for I_1 and I_2 is

$$I_1 s_1 + I_2 s_2 = \frac{1}{4}\left(\frac{1}{L} - \frac{1}{R^2 C}\right).$$

With two equations for the two unknowns, I_1 and I_2, it is clear that we can solve for them in terms of the circuit parameters R, L, C, and the two values for s (which, in turn, are known functions of R, L, and C). We thus have the complete solution for the battery current, $i(t)$, for $t > 0$, in terms of the circuit parameters (and of the given initial conditions).

Problems

C.1. The gyrator—invented in 1948, *on paper*, by the Dutch electrical engineer B. D. H. Tellegen (1900–1991)—is a four-terminal component mathematically defined as follows (using the notation of Figure C.11):

$$v_1 = -Ki_2$$
$$v_2 = Ki_1$$

where K is the *gyration resistance*.

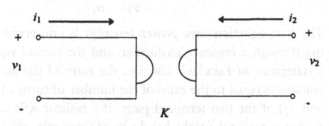

Figure C.11 A gyrator.

(a) Suppose a resistor R is connected across the right-hand terminals, and so $v_2 = -i_2 R$ (the minus sign simply indicates the current i_2 is drawn "opposite" to the direction consistent with the indicated polarity of v_2). What is the ratio v_1/i_1, the resistance that appears between the left-hand terminals (often called by electrical engineers as the resistance seen "looking into" the left-hand terminals)?

(b) Two gyrators, with $K_1 = K_2 = K$, are connected with a capacitor as shown in Figure C.12. Show that this arrangement behaves like

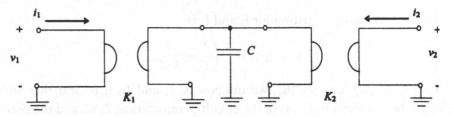

Figure C.12   Simulating a perfect inductor with a capacitor and two gyrators.

an inductor, i.e., that $i_2 = -i_1$ and that

$$v_1 - v_2 = L\frac{di_1}{dt}.$$

Find an expression for L as a function of K and C. (This is one way to electronically *simulate* an inductor—about which more is said in Appendix F—to a very high degree of accuracy; "real" inductors tend to deviate *considerably* from mathematical theory).

(c) A *lossless ideal* transformer is a four-terminal component mathematically defined as follows (using the notation of Figure C.13):

$$i_1v_1 + i_2v_2 = 0$$
$$\frac{v_2}{v_1} = \frac{n_2}{n_1}.$$

The first equation says power (energy) is conserved in passing through a *lossless* transformer, and the second equation is a statement of Faraday's law, i.e., the ratio of the ac terminal voltages is equal to the ratio of the number of turns of wire (n_1 and n_2) at the two terminal pairs. If a resistor R is connected as shown across the right-hand pair of terminals, what is v_1/i_1? This property of a transformer, of *impedance transformation*, is

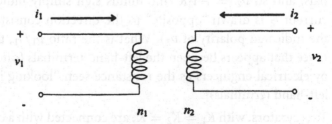

Figure C.13   Lossless transformer.

used in achieving the efficient coupling of ac energy from the high-resistance audio amplifier circuitry of an AM radio to the low resistance of a loudspeaker.

C.2. Consider the three-stage filter of Figure C.14.

(a) Find $H(j\omega) = \frac{V_0}{V_i}$. Hint: Your analysis should show that, at one particular frequency $\omega = \omega_0$, $H(j\omega_0) = -\frac{1}{29}$, a purely real value. Find an expression for ω_0.

(b) At $\omega = \omega_0$, what is Z_i, the input impedance? Hint: Your answer should be of the form $Z_i = NR$, where N is a particular complex number.

(c) At $\omega = \omega_0$ the phase shift through the entire filter is 180°. Find the phase shift of *each stage*, accurate to 0.001°, and verify that the three shifts add to 180°. *Don't forget to take loading into account!* Hint: the shifts through the first two stages are each less than 60°, and the third stage shift is greater than 60° (but less than 70°).

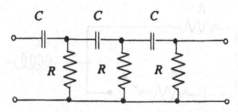

Figure C.14   Three-stage RC filter.

C.3. Derive an expression for the total net energy W delivered to a capacitor C over the time interval $0 < t < T$, and show that $W = 0$ if $v(0) = v(T)$. [The voltage drop across the capacitor, at any time t, is $v(t)$.]

C.4. In the circuit shown in Figure C.15, the capacitor is charged to V_0 volts. At time $t = 0$ the switch is closed and the capacitor begins to discharge. Show that the response of the circuit will be oscillatory only if the value R is at least some minimum value, which is a function of the values for L and C. Hint: With $v(t)$ as the voltage drop across the parallel R and C show that

$$LC\frac{d^2v}{dt^2} + \frac{L}{R}\frac{dv}{dt} + v = 0$$

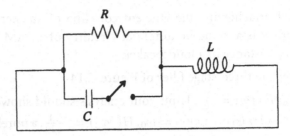

Figure C.15   A discharging capacitor.

and assume v is of the form e^{st}. Under what conditions does s have an imaginary part?

C.5. If the circuit in Figure C.15 is now modified slightly to look like Figure C.16, show that the discharge behavior of the circuit is oscillatory only if R is within an *interval*. That is, show the circuit oscillates only if R is neither too big nor too small. What is the interval that R must be in for oscillations to exist?

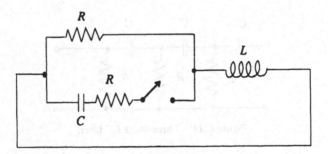

Figure C.16   A discharging capacitor.

C.6. In Figure C.17 the switch has been *closed* for a long time. At $t = 0$ it is opened.

(a) Just before the switch is opened, what are the currents in the two equal-valued inductors, and what is the voltage drop across the capacitor (at time $t = 0^-$)?

(b) For $t > 0$, determine under what condition the circuit will oscillate.

C.7. In Figure C.18 the switch has been closed for a long time. Then, at time $t = 0$, the switch is opened. Show that the differential equation for $v(t)$, the voltage drop across the parallel LR section, is

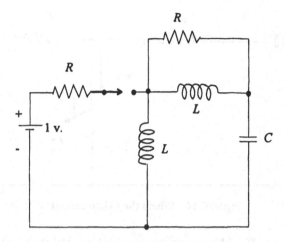

Figure C.17 An initial-value problem.

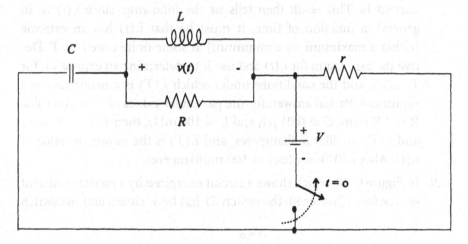

Figure C.18 Does this circuit oscillate?

$$\left(1 + \frac{r}{R}\right)\frac{d^2v}{dt^2} + \left(\frac{r}{L} + \frac{1}{RC}\right)\frac{dv}{dt} + \frac{1}{LC}v = 0,$$

and determine the condition on r, R, L, and C for which $v(t)$ oscillates. Partial answer: for the particular values of $R = r = 1$ K ohm, $C = 0.005$ μF, and $L = 1$ mH, $v(t)$ is an exponentially damped oscillation of frequency 15.915 kHz.

C.8. In Figure C.19 the switch has been open for a long time. Then, at time $t = 0$, the switch is closed. The problem is to calculate the current in the switch, $i_s(t)$, for $t > 0$. To start, you should be able to con-

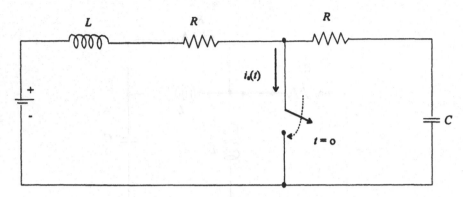

Figure C.19   What's the switch current?

vince yourself *without writing any mathematics* that $i_s(0^+) = i_s(\infty)$, and also to determine what this common initial and final switch current is. This result then tells us the following: since $i_s(t)$ is, in general, a function of time, it must be that $i_s(t)$ has an extreme (either a maximum on a minimum) at some finite time $t = T$. Derive the expression for $i_s(t)$ and use it to determine an expression for T, $i_s(T)$, and the conditions under which $i_s(T)$ is a maximum or a minimum. Partial answer: for the particular values of $V = 100$ volts, $R = 1$ K ohm, $C = 0.01\ \mu F$, and $L = 100$ mHz, then $T = 25.58\ \mu sec$ and $i_s(T) = 30.3$ milliamperes, and $i_s(T)$ is the *minimum* value of $i_s(t)$. Also, $i_s(0^+) = i_s(\infty) = 100$ milliamperes.

C.9. In Figure C.20, which shows a circuit energized by a constant current source (see Chapter 8), the switch S1 has been closed and the switch

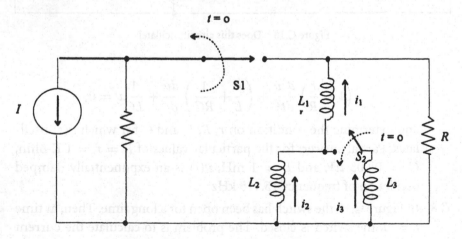

Figure C.20   A circuit with trapped energy.

S2 has been open for a long time. Then, at time $t = 0$, S1 is opened and S2 is closed (assume these two switching events occur both instantly and simultaneously).

(a) Explain why $i_1(0^-) = i_2(0^-) = I$, and $i_3(0^-) = 0$. What is the purpose of the shunt resistor (no value shown) across the constant current source?

(b) Find expressions for $i_2(t)$ and $i_3(t)$ for $t > 0$, and show that $\lim_{t \to \infty} i_2(t) = -\lim_{t \to \infty} i_3(t)$. Explain why this result means there is, after a long time, a steady current flowing in the inductor loop formed by L_2 and L_3, i.e., energy is trapped in this circuit, even though R appears to provide a shunt path! This is a particularly instructive example, because it shows how an *idealized* mathematical model can lead to nonphysical results, i.e., any real inductor would have some series resistance and there would actually not be a persistent current loop. (Try including series resistances R_2 and R_3 with the inductors L_2 and L_3 and observe how the mathematics complicates. This more realistic model should result in answers that reduce to the original ones as R_2 and $R_3 \to 0$.) Returning to the ideal model, you can check your answers by calculating the total energy dissipated by R in two different ways. First, simply find the difference between the initial and final energies stored in the magnetic fields of the inductors. Then, directly calculate the energy integral $R \int_0^\infty i_1^2(t)dt$. These two calculations must, of course, agree.

Thévenin's and Norton's Theorems

Kirchhoff's laws, and the constitutive laws of individual circuit components, are all that is needed to solve any circuit. But special circuit theorems can often transform a complicated solution into an easy one. The most famous of these theorems is *Thévenin's theorem*, named after the French government telegraph engineer Léon Charles Thévenin (1857–1926), who published it in 1883. The great German physicist Hermann Ludwig von Helmholtz (1821–1894)—Hertz, the discoverer of Maxwell's radio waves, was one of his students—had actually given a restricted form of the theorem in 1853, but it was Thévenin who showed its generality *and* brought it into widespread electrical engineering use.

Thévenin's theorem, which I'll state without proof, replaces any circuit made of just resistors and one or more sources (voltage and/or current) with a *single* voltage source and a *single* series resistor, as shown in Figure D.1. If we imagine the original circuit, and Thévenin's equivalent circuit, both sealed inside what

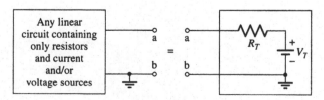

Figure D.1   A linear circuit and its Thévenin equivalent.

are called *black boxes* (we can't see inside a black box, just like we can't see inside a black hole!), then the two boxes are identical as far as their electrical behavior at terminals a/b is concerned.

An equivalent theorem, called *Norton's theorem* after the American electrical engineer Edward Lawry Norton (1898–1983), was published the year of Thévenin's death. Norton's theorem replaces the original circuit with an equivalent circuit consisting of just one current source and one parallel resistor, as shown in Figure D.2. It is important to understand that the equivalence of either Thévenin's or Norton's equivalent circuit is with respect to the observed conditions at the a, b terminals and *not* to conditions inside the various boxes. What goes on inside the three boxes almost surely is different, e.g., the *internal* power dissipated inside the original circuit box can be very different from the power dissipated inside the Thévenin box. Indeed, the powers dissipated in the Thévenin and Norton boxes can be different from each other (see Problem D.1).

As a prelude to stating Thévenin's theorem, I need to say just a bit more about the current and/or voltage sources that may be in the original circuit. There are two general kinds of such sources: *independent* and *dependent*. The independent voltage source is the one most students feel immediately comfortable with, probably because it is a good model of the common, ordinary battery. An independent ideal voltage source is a device that maintains a potential difference (either constant *or* time varying) across its terminals that is independent of the current through itself. By the same token, an independent ideal current source maintains the same current (either constant or time varying) through itself that is independent of the potential difference across its terminals. *Dependent* sources depend on some other circuit condition to determine what they do, and occur in theoretical models for what are called *active electronic* devices, e.g., transistors and vacuum tubes. See the end of this appendix and Chapter 8 for more on this.

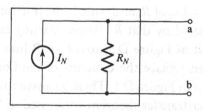

Figure D.2   The Norton (current source) equivalent.

A cautionary warning: **Never** short-circuit an ideal voltage source or open-circuit an ideal current source! In the first case the source will generate an infinitely large current in the short, and in the second case the source will generate an infinitely large potential difference across its terminals. These infinities are the mathematics telling you that you have attempted to build a physically unreasonable circuit.

To use Thévenin's theorem we need to only calculate the values of V_T and R_T in Figure D.1. From an operational point of view (i.e., the way we would actually *measure* V_T and R_T in the laboratory) it is immediately clear that if we first open-circuit terminals a and b (in Figure D.1) then $V_{ab} = V_T$; V_T is the *open-circuit* terminal voltage drop, often written as V_{oc}. Next, if we short-circuit terminals a and b, then the current in the short (written as I_{sc}) is given by

$$I_{sc} = \frac{V_T}{R_T} = \frac{V_{oc}}{R_T}.$$

Thus,

$$R_T = \frac{V_{oc}}{I_{sc}}.$$

An alternative (but of course equivalent) way to determine R_T is to replace all internal *independent* voltage sources with shorts, all internal *independent* current sources with opens, and then to calculate the resistance "seen" between terminals a and b. That resistance is R_T.

The Norton current source and resistance are just as easy to determine. First, it is immediately clear that $I_{sc} = I_N$. Also, $V_{oc} = R_N I_N = R_N I_{sc}$. Thus

$$R_N = \frac{V_{oc}}{I_{sc}} = R_T,$$

Here's an example of the use of Thévenin's theorem. In the circuit of Figure D.3 the resistor R is to be adjusted until it is dissipating maximum power. What is that value of R, and what is the fraction of the total battery power that is dissipated by that R? Before starting any calculations, let's put all of the circuit of Figure D.3 *except for R* into a box, as shown in Figure D.4, and then replace the contents of the box with its Thévenin equivalent, as shown in Figure D.1. Then, to answer our questions I'll use the maximum power transfer theorem (discussed in Chapter 5) that says maximum power will be delivered to R when $R = R_T$.

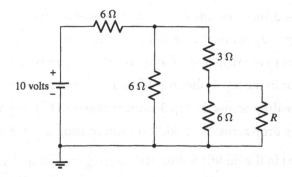

Figure D.3   Adjust R for maximum power.

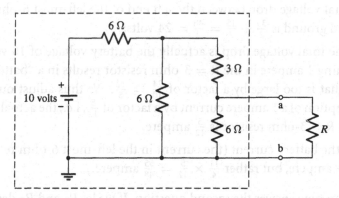

Figure D.4   The Thévenin equivalent of the circuit in Figure D.3.

Let's first find R_T "looking into" the terminals a and b of Figure D.4. With the independent 10-volt voltage source replaced with a short we see that (if necessary, see Appendix C for the ∥ notation),

$$R_T = [(6\|6) + 3]\|6 = (3 + 3)\|6 = 6\|6 = 3 \text{ ohms.}$$

That is, $R = R_T = 3$ ohms for maximum power in R, which is the first answer we are after.

To answer the second question, I'll next calculate the current in the $R = 3$-ohm resistor, and also the battery current, as we can then easily calculate the pertinent powers. There are a number of ways to do this, and what I'll show you now is a clever trick that works easily. Let's *assume* that the current in the $R = 3$-ohm resistor is 1 ampere. This is almost certainly *not* correct, but at the end of the analysis we'll see how to adjust our assumption (that's the trick!). Thus, we reason as follows from Figure D.4:

- the voltage drop across the $R = 3$-ohm resistor is 3 volts;
- the voltage drop across the right-most 6-ohm resistor is 3 volts;
- the current in the right-most 6-ohm resistor is $\frac{1}{2}$ ampere;
- the current in the top 3-ohm resistor is $1 + \frac{1}{2} = \frac{3}{2}$ ampere;
- the voltage drop across the top 3-ohm resistor is $3\left(\frac{3}{2}\right) = \frac{9}{2}$ volts;
- the voltage drop across the middle 6-ohm resistor is $\frac{9}{2} + 3 = \frac{15}{2}$ volts;
- the current in the middle 6-ohm resistor is $\frac{\frac{15}{2}}{6} = \frac{5}{4}$ ampere;
- the current in the left-most 6-ohm resistor is $\frac{5}{4} + \frac{3}{2} = \frac{11}{4}$ ampere;
- the voltage drop across the left-most 6-ohm resistor is $6\left(\frac{11}{4}\right) = \frac{33}{2}$ volts;
- the total voltage drop between the left end of the left-most 6-ohm resistor and ground is $\frac{33}{2} + \frac{15}{2} = \frac{48}{2} = 24$ volts;
- *But* the total voltage drop is actually the battery voltage of 10 volts. So, assuming 1 ampere in the $R = 3$-ohm resistor results in a "battery voltage" that is too large by a factor of $\frac{24}{10} = \frac{12}{5}$. We thus adjust our initial assumption of a 1 ampere current by a factor of $\frac{5}{12}$, i.e., the actual current in the $R = 3$-ohm resistor is $\frac{5}{12}$ ampere.
- Also, the battery current (the current in the left-most 6-ohm resistor) is not $\frac{11}{4}$ ampere, but rather $\frac{11}{4} \times \frac{5}{12} = \frac{55}{48}$ ampere.

We can now answer the second question. If we let P_R and P_B denote the power dissipated by R and the battery power, respectively, then

$$\frac{P_R}{P_B} = \frac{\left(\frac{5}{12}\right)^2 3}{(10)\left(\frac{55}{48}\right)} = \frac{\frac{75}{144}}{\frac{550}{48}} = \frac{(75)(48)}{(144)(550)} = 0.0454.$$

That is, when R is dissipating maximum power by being adjusted to be 3 ohms, it is dissipating only 4.54% of the total power provided by the battery. All the rest of the battery power is dissipated by the other resistors.

As a final example of finding a Thévenin equivalent circuit, consider Figure D.5 which shows a circuit with a *dependent* current source of $g_m v$. That is, the voltage drop v across resistor R *controls* the current source (g_m has the units of inverse-ohms or *siemens*, named after the German electrical engineer Werner Siemens (1816–1892), and is called a *transconductance*).

Our first calculation, for V_T, is easy. The current in R is $\frac{v}{R}$, which is directed downward. Thus,

$$\frac{v}{R} = -g_m v$$

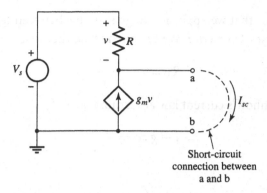

Figure D.5   A circuit with a *dependent* source.

which makes sense only if $v = 0$, i.e., if there is *no* voltage drop across R. Thus, the entire V_s voltage appears between terminals a and b and so $V_T = V_s$. To find R_T, we first find the short-circuit current I_{sc}. With terminals a and b connected together the bottom end of R is connected to ground, and so the entire voltage V_s appears across R ($v = V_s$). So, using Kirchhoff's current law at terminal a,

$$\frac{V_s}{R} + g_m V_s = I_{sc}$$

and so

$$R_T = \frac{V_T}{I_{sc}} = \frac{V_s}{\frac{V_s}{R} + g_m V_s} = \frac{1}{\frac{1}{R} + g_m}.$$

Alternatively, we could replace the independent voltage source with a short and arrive at Figure D.6 (we must retain the *dependent* current source):

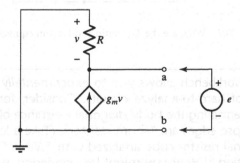

Figure D.6   Using Ohm's law to find R_T for the circuit of Figure D.5.

Imagine now that we apply a voltage source e between terminals a and b and that causes a current i. We can then write (because $v = -e$)

$$R_T = \frac{e}{i} = \frac{-v}{i}.$$

Applying Kirchhoff's current law at terminal a,

$$i + g_m e = -\frac{e}{R}$$

or,

$$i = e\left(\frac{1}{R} + g_m\right)$$

and so

$$R_T = \frac{e}{i} = \frac{1}{\frac{1}{R} + g_m}.$$

Problem

D.1. Find, for the circuit for Figure D.7, the Thévenin and Norton equivalent circuits. Then, with a 2-ohm resistor connected to terminals a and b, show that the *internal* dissipated powers are different in all three circuits but that the *external* power (i.e., in the 2-ohm resistor) is the same for all three circuits.

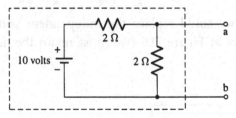

Figure D.7   What are the Thévenin and Norton equivalents?

Electronics Workbench allows you to *experimentally* find the internal resistance of hard-to-analyze circuits. Consider, for example, the problem of determining the body-diagonal resistance of a *four* dimensional cube whose edges are 1-ohm resistors. (Take a look back at the three-dimensional resistor cube analyzed with EWB in "What's New in the Second Edition.") How, you might be wondering, would you draw a four-dimensional resistor cube on a computer screen? Actually, it's easy. We assign to each vertex of the cube a quadruple of binary co-ordinates, from (0,0,0,0) to (1,1,1,1), i.e., a four-dimensional cube has sixteen vertices. Two vertices are adjacent (connected by a single edge) when their coordinates differ in just one position. This lets us determine which vertices connect to each other, and the result is the schematic of Figure D.8. Taking vertex 0 and vertex 15 as the two ends of a body-diagonal, EWB tells us that the resistance is $\frac{2}{3}$ ohm.

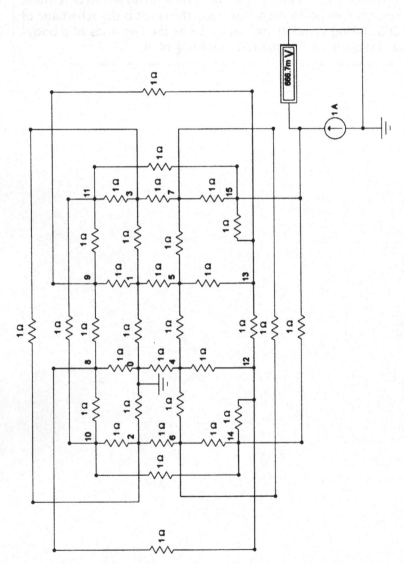

Figure D.8   A four-dimensional resistor cube (a "tesseract") collapsed onto a plane. Notice that some of the wire connectors cross, however, and so the circuit is not planar. The simulation shows that the body-diagonal resistance is 0.666 ohms ($\frac{2}{3}$ ohm).

Resonance in Electrical Circuits

Electrical circuits constructed from components with frequency-dependent impedances can possess a particularly useful property called *resonance*. A circuit is said to be operating at its resonant frequency when, for a fixed amplitude input signal, it exhibits its maximum response. Radio receivers use a "front-end" resonant circuit (the inductance of the antenna in parallel with a capacitance) called the *preselector* to help select one radio signal from all others, i.e., as I'll show you next, a parallel LC circuit has a maximum response at a certain frequency. This frequency is a function of the circuit values, and so the resonant or maximum response frequency of the receiver front end can be varied by varying, for example, the capacitor. That is one of the things that actually occurs when you turn the dial on your radio.

In Figure E.1 we have a parallel LC, with a fixed amplitude sinusoidal current, I, at frequency $\omega = 2\pi f$. If we define the response of this circuit to be the magnitude of the voltage V that is developed across the circuit by the current I, then it is easy to show that there is a particular frequency at which $|V|$ is maximum. Thus, from Ohm's law

$$V = IZ = I\frac{Z_c Z_L}{Z_c + Z_L} = I\frac{\left(\frac{1}{j\omega C}\right)j\omega L}{\left(\frac{1}{j\omega C}\right) + j\omega L}$$

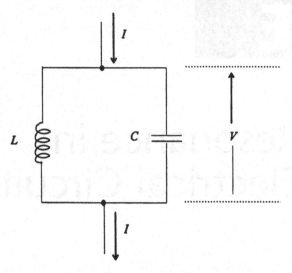

Figure E.1   A parallel LC circuit.

or

$$V = I\frac{j\omega L}{1 - \omega^2 LC}, \quad |V| = |I|\frac{\omega L}{|1 - \omega^2 LC|}.$$

Thus, $|V|$ is maximum (theoretically it goes to infinity) when

$$\omega = \omega_0 = \frac{1}{\sqrt{LC}}.$$

We say the parallel LC circuit is resonant at (or "tuned to") frequency

$$f_0 = \frac{1}{2\pi\sqrt{LC}} \text{ Hz.}$$

You can form a physical picture of what happens in the front end of a radio receiver by looking at Figure E.2. Suppose we have set the variable capacitor to some value, and so have tuned the antenna circuit (which contributes the L) to a particular resonant frequency, ω_0. Radio signals arriving at the antenna with much higher frequencies than ω_0 will see a (relatively) low impedance path to ground through the C, while signals at frequencies much lower than ω_0 will see a (relatively) low impedance path to ground through the L. Such signals are, quite literally, routed away from the interior of the receiver. It is only a signal arriving with frequency near ω_0 that can (pardon the pun!) "develop its full potential across the antenna."

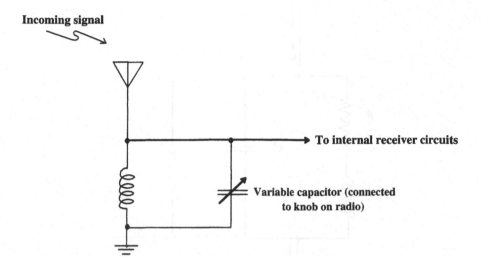

Figure E.2 A resonant antenna circuit (idealized).

A more realistic model of a parallel LC antenna circuit would include a resistance in series with the L, e.g., an antenna is often constructed in the form of a coil wrapped in many turns of fine wire, and while as discussed in Appendix C this is a good geometry for making an L, it also inherently brings with it a not insignificant ohmic resistance from the wire, itself. Now, from Figure E.3, we have

$$V = IZ, |V| = |I||Z|,$$

where

$$Z = \frac{(Z_R + Z_L)Z_c}{Z_R + Z_L + Z_c} = \frac{(R + j\omega L)\left(\frac{1}{j\omega C}\right)}{R + j\omega L + \frac{1}{j\omega C}} = \frac{R + j\omega L}{1 - \omega^2 LC + j\omega RC}.$$

Thus,

$$|Z| = \sqrt{\frac{R^2 + (\omega L)^2}{(1 - \omega^2 LC)^2 + (\omega RC)^2}}$$

To maximize $|V|$, the response of the antenna circuit, we must maximize $|Z|$. Mathematically it is easier to maximize $|Z|^2$ (which avoids the square-root operation), and it should be clear that the frequency at which $|Z|$ is maximum is the same frequency at which $|Z|^2$ is maximum. Thus, we set

$$\frac{d|Z|^2}{d\omega} = 0,$$

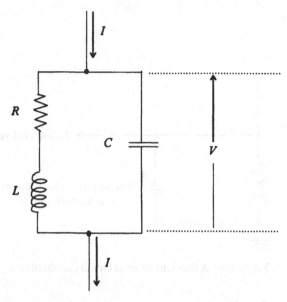

Figure E.3 A resonant antenna circuit (realistic).

and while formally this condition tells us only at what frequency (let's call it ω_0') $|Z|^2$ is an extreme (not necessarily a maximum), we know *physically* that it must be a maximum because the minimum $|Z|^2$ occurs when $\omega = \infty$ (because at high frequencies the C shorts the entire circuit to ground). For those who prefer mathematical analyses, calculate $\frac{d^2|Z|^2}{d\omega^2}$ (if you have the patience!) and check its sign at the resonant frequency ω_0'. You will find it is negative, showing that the extreme is indeed a maximum.

If you perform the detailed calculations of maximizing $|Z|^2$ you will find you get a quartic which is quadratic in $\omega_0'^2$. Using the quadratic equation on that gives

$$\omega_0' = \sqrt{-\frac{R^2}{L^2} + \frac{1}{LC}\sqrt{1 + \frac{2R^2 C}{L}}}$$

as the resonant frequency. Notice that if $R = 0$, then Figure E.3 reduces to Figure E.1 (and ω_0' reduces to ω_0). You can show, in fact, that $\omega_0' < \omega_0$ (see Problem E.1). Finally, notice that the expression for ω_0' makes physical sense only if

$$-\frac{R^2}{L^2} + \frac{1}{LC}\sqrt{1 + \frac{2R^2 C}{L}} > 0,$$

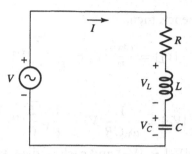

Figure E.4   Series R, L, C circuit.

as ω_0' must be real. If this condition is violated, then the circuit will no longer have a response that increases to a peak, and then decreases, as frequency increases.

Finally, take a look at Figure E.4 which shows a series R, L, C circuit. If we imagine that the voltage V is kept fixed, and that the response of the circuit is the current I, then as $Z = R + j(\omega L - \frac{1}{\omega C})$ is the impedance of the circuit, we have

$$I = \frac{V}{Z} = \frac{V}{R + j\left(\omega L - \frac{1}{\omega C}\right)},$$

and so

$$|I| = \frac{|V|}{\sqrt{R^2 + \left(\omega L - \frac{1}{\omega C}\right)^2}}.$$

It is clear that $|I|$ is maximum when $\omega = \omega_0 = \frac{1}{\sqrt{LC}}$ (independent of R) and so, *at resonance,*

$$I = \frac{V}{R} \text{ and so } |I| = \frac{|V|}{R}.$$

It is particularly interesting—you'll see why soon—to calculate V_L and V_C, the individual voltage drops across the L and the C, respectively. Thus,

$$V_L = Ij\omega L \text{ and so } |V_L| = \omega L|I|$$

and

$$V_C = \frac{I}{j\omega C} \text{ and so } |V_C| = \frac{1}{\omega C}|I|.$$

At the *resonant* frequency, then,

$$|V_L| = \frac{\omega_0 L}{R}|V| = \frac{1}{R}\sqrt{\frac{L}{C}}|V|$$

and

$$|V_C| = \frac{1}{\omega_0 CR}|V| = \frac{1}{R}\sqrt{\frac{L}{C}}|V|.$$

That is, *at resonance* $|V_L| = |V_C|$ and each of these two voltage drops can be far *greater* than V itself (for example, imagine $R \to 0$).

This last result may be surprising since $V = V_L + V_C + V_R$, but don't forget that these are complex quantities and so have phase, as well as magnitude. What appears to be an addition can, in fact, actually be a subtraction. Indeed, *at resonance*

$$V_L = Ij\frac{1}{\sqrt{LC}}L = jI\sqrt{\frac{L}{C}}$$

and

$$V_C = I\frac{1}{j\frac{1}{\sqrt{LC}}C} = -jI\sqrt{\frac{L}{C}}.$$

So, again, we see that $|V_L| = |V_C|$ but now also that, because of phase, $V_L + V_C = 0$! That is, at resonance the entire voltage drop across the R, L, and C is due only to the R, and not because the individual drops across the L and C are zero but rather because those two drops are equal in magnitude and 180° out of phase—and so they *cancel* each other.

Those individual V_L and V_C voltage drops *are* there, however. In fact, sticking your hands into the interior of a high-power-series resonant circuit (such as commonly occurs in radio transmitter circuitry) can be lethal. While the total voltage drop across the R, L, and C may be just a few volts, the individual voltage drops across the L and C may each be hundreds, even *thousands* of volts.

The phenomenon of resonance is associated with the physical process of *energy exchange*. That is, electrical energy in a resonant circuit is continually being swapped back and forth between the electric and magnetic fields of one or more capacitors and one or more inductors, respectively. The related phenomenon of voltage boost in a passive circuit (which we've seen can occur to an astonishing degree in a resonant circuit) can also occur even when an energy exchange mechanism is not present, e.g., in circuits with capacitors but no inductors. This is

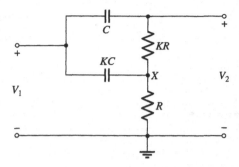

Figure E.5   A passive circuit with a voltage gain greater than unity.

often surprising to even longtime electrical engineers; when I've asked
electrical engineering professors and graduate students if it is possible
to construct a circuit from just resistors and capacitors that has, over an
interval of frequency (perhaps infinitely wide), a voltage gain greater
than unity, the response is almost always *no*. But that is incorrect.

Consider, for example, the circuit of Figure E.5, with KCL equations
at the V_2 and X nodes, respectively, of

$$(V_1 - V_2)j\omega C = \frac{V_2 - X}{KR}$$

$$(V_1 - X)j\omega KC = \frac{V_2 - X}{KR} = \frac{X}{R}.$$

In the circuit (and the equations) K is an arbitrary positive constant.

If we write $RC = \tau$, and the voltage gain $\frac{V_2}{V_1} = H$, then a little algebra
(see Problem E.5) results in

$$H = \frac{\omega^2 K^3 \tau^2 - j\omega(2K^2 + K)\tau}{\omega^2 K^3 \tau^2 - K - j\omega(2K^2 + K)\tau}.$$

From this it is clear that $|H| > 1$ for all $\omega \geq \omega_u$ where ω_u is the frequency
at which the magnitudes of the denominator and numerator of H are
equal (ω_u is the unity-gain frequency). A little algebra (see Problem E.5)
shows that

$$\omega_u = \frac{1}{K\tau\sqrt{2}} \text{ or, } f_u = \frac{1}{2\pi K\tau\sqrt{2}}.$$

In addition, a bit more algebra shows that the phase shift at the output
(relative to the input), at $\omega = \omega_u$, is given by (see Problem E.5)

$$180° - 2\tan^{-1}\left(\sqrt{2}\frac{2K + 1}{K}\right).$$

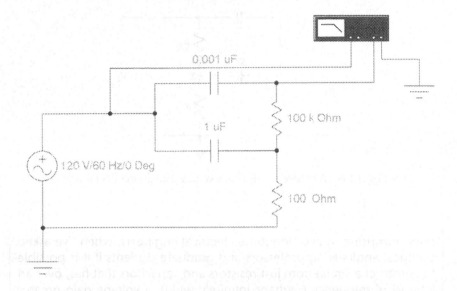

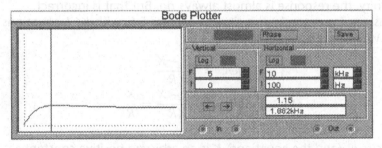

Figure E.6   This passive circuit, constructed from just two capacitors and two resistors, has a voltage gain greater than unity (i.e., greater than 0 db) for frequencies greater than about 1.129 kHz.

> For example, if $R = 100$ ohms, $C = 0.001$ μfd, and $K = 1000$, then $f_u = 1.125$ kHz and at that frequency the phase shift is 38.92°. An Electronics Workbench simulation of the circuit in Figure E.5, using these values, gave a gain of unity at 1.125 kHz with a phase shift of 38.94°, in excellent agreement with theory. In addition, the simulation shows the circuit has a maximum voltage gain of 1.15 at 1.882 kHz (as shown in Figure E.6). It is a broad maximum, too, extending out to 2.774 kHz.

The behavior of circuits constructed from just reactive components (no resistors) can have quite astounding frequency dependent properties. So, to conclude this appendix let me show you one example of this claim. In general, an impedance $Z = R + jX$ with a positive-real part ($R > 0$) *dissi-*

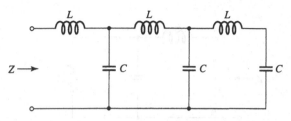

Figure E.7   A three-section LC-circuit.

pates energy. That is, energy put into such an impedance is lost. So, here's a snap question for you: If a circuit is constructed from only inductors and capacitors (such as is shown in Figure E.7), can it present an impedance that has a positive-real part? The answer, of course is, *no*, because there simply is no *physical* mechanism in such a circuit to dissipate energy, i.e, no resistors.

Well, that seems easy, but *no* is the correct answer only if the number of LC-sections (there are three in Figure E.7) is *finite*. If the number of sections is *infinite*, then $R > 0$ *is* possible! There are, of course, two puzzles associated with this claim (which I will soon demonstrate is true). First, how could one actually, really, build a circuit with an infinite of components in it? If the components are *lumped* (discrete, individual inductor coils and capacitors that you'd buy in an electronics shop) then of course you couldn't. But if they are *differential* components, such as the inductance and capacitance *per unit length* of a continuous transmission line (e.g., the coaxial cable that brings video signals to a television set, or the Web to a personal computer connected to an Internet service provider's server) then we can indeed imagine such a thing as a circuit with an infinite number of vanishingly small inductors and capacitors.

But still, even if we agree to that what could $R > 0$ mean? After all, even with an infinity of components a purely reactive circuit has not one, single resistor in it. So, why is there "lost" energy ($R > 0$)? Before answering that question, let me first show that such an impedance is possible. My demonstration depends on an enormously clever observation, one that I have been able to trace back to 1887, in an obscure paper by the English electrical engineer Oliver Heaviside (1850–1925).[1]

Imagine that Figure E.7 has an infinite number of LC sections, and then remove the first (left-most) section and ask what then is the input impedance of the remaining circuit? Since the circuit has an infinite number of sections, then the removal of one *makes no difference* and we will

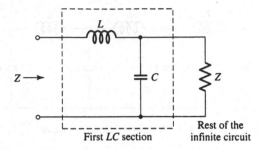

Figure E.8   Equivalent circuit of an *infinite*-section *LC* circuit.

still see an impedance of Z. Thus, we can replace an infinite circuit with the finite one of Figure E.8.

As a function of frequency (ω), we have

$$Z = j\omega L + \frac{\frac{1}{j\omega C}Z}{\frac{1}{j\omega C} + Z} = j\omega L + \frac{Z}{1 + j\omega CZ}.$$

And so

$$Z(1 + j\omega CZ) = j\omega L(1 + j\omega CZ) + Z$$

or

$$Z + j\omega CZ^2 = j\omega L - \omega^2 LCZ + Z$$

or

$$j\omega CZ^2 + \omega^2 LCZ - j\omega L = 0$$

or

$$Z^2 - j\omega LZ - \frac{L}{C} = 0.$$

Thus,

$$Z = \frac{j\omega L \pm \sqrt{-\omega^2 L^2 + \frac{4L}{C}}}{2}.$$

Which sign, + or −, should we use for Z? It has to be one or the other; it can't be both! One way to reason is that we should use the + sign because, as $\omega \to \infty$, the C in Figure E.8 shorts the parallel Z and it is then clear that, physically,

$$\lim_{\omega \to \infty} Z = j\omega L.$$

And that is precisely how our expression for Z will behave if we select the $+$ sign. And so we write

$$Z = R + jX = \frac{1}{2}\sqrt{\frac{4L}{C} - \omega^2 L^2} + j\omega\frac{L}{2}.$$

That is,

$$R = \frac{1}{2}\sqrt{\frac{4L}{C} - \omega^2 L^2} > 0 \text{ if } \frac{4L}{C} - \omega^2 L^2 > 0.$$

The condition for $R > 0$ is equivalent to $\omega < \frac{2}{\sqrt{LC}}$, which says that energy at frequencies less than $\frac{2}{\sqrt{LC}}$ is "dissipated." For $\omega > \frac{2}{\sqrt{LC}}$ we have Z purely imaginary, i.e., $R = 0$, and there is no puzzle. To understand how a circuit with no physical resistance can still dissipate energy, we have to broaden what we mean by *dissipate*. We usually think of the word as meaning the transformation of electrical energy into heat energy, but in more general terms it means the *loss* of electrical energy. In the case of our infinite circuit, $R > 0$ means that energy at frequencies $\omega < \frac{2}{\sqrt{LC}}$ *propagates away* from the input, down the circuit and off to infinity. It isn't turned into heat energy, but it is lost to us (although a so-called *terminating resistor* at the other end—at "infinity"—*can* absorb that energy when it arrives there).

Note

1. See my book *Oliver Heaviside: Sage in Solitude*, IEEE Press, 1988, pp. 230–232.

Problems

E.1. Show, for Figure E.3, that the stated condition for ω_0' to be real is equivalent to the condition $\frac{R^2 C}{L} < 1 + \sqrt{2}$, and so too large a value for R will cause resonance to disappear.

E.2. Show, for Figure E.3, that $\omega_0' < \omega_0$.

E.3. Consider the circuit of Figure E.1 again, but now include a parallel R as well. Define the current in the capacitor as I_c and (with $|I|$

constant) find the frequency ω_0' at which $|I_c|$ is maximum. State the conditions under which your answer makes physical sense, and determine the relationship between this frequency and ω_0 (the resonant frequency for the case $R = \infty$, the original circuit).

Answer:

$$\omega_0' = \frac{1}{\sqrt{LC - \left(\frac{1}{2}\right)\left(\frac{L}{R}\right)^2}}.$$

From this, $\omega_0' \geq \omega_0$.

E.4. Show that the circuit of Figure E.9 is a resonant filter, i.e., it has a maximum response at frequency $\omega_0 = \frac{1}{\sqrt{LC}}$. More specifically, derive an expression for $|\frac{V_2}{V_1}|$ and show it is infinite at $\omega = \omega_0$. Hint: Label the two currents as shown in the figure, and argue why $I_1 = I_2$. Then, with V_a and V_b as indicated, calculate each with respect to a *common reference point* in a circuit (it doesn't matter where, but the negative terminal for V_1 is a convenient choice). Finally, write $V_2 = V_a - V_b$. For obvious reasons, this circuit is often called a *lattice* filter.

Answer:

$$\left|\frac{V_2}{V_1}\right| = \left|\frac{(1 + \omega^2 LC)}{(1 - \omega^2 LC)}\right|.$$

Can you make a *physical* argument, directly from the circuit diagram and without writing any mathematics, for why

$$\lim_{\omega \to 0} \frac{V_2}{V_1} = 1$$

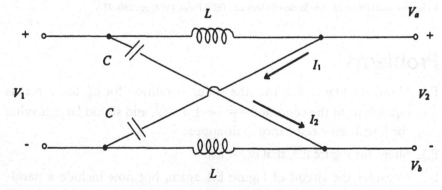

Figure E.9   A resonant filter circuit.

and

$$\lim_{\omega \to \infty} \frac{V_2}{V_1} = -1?$$

E.5. Verify all of the equations given in the shaded box concerning the circuit of Figure E.5.

E.6. Show that the magnetically coupled circuit of Figure E.10 has *two* resonant frequencies (see Appendix C for a discussion of the coupling inductance, M).

Answer: $\dfrac{\omega_0}{\sqrt{1 \pm k}}, k = \dfrac{M}{L}$ and $\omega_0 = \dfrac{1}{\sqrt{LC}}$.

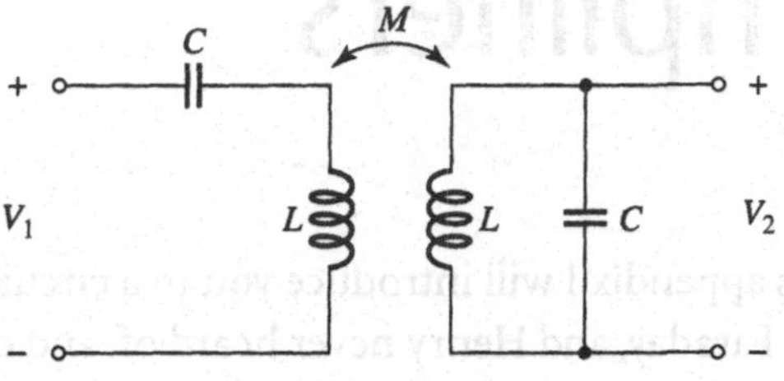

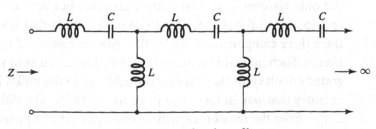

Figure E.10 A circuit with two resonant frequencies.

E.7. Consider the function

$$g(x) = x^{x^{x^{\cdot^{\cdot^{\cdot}}}}}, x > 0$$

where the stack of exponents is infinitely "high," and is evaluated "top-down." For example, a finite stack three levels high would be evaluated as

$$x^{x^x} = x^{(x^x)}.$$

Find the largest value of $x(= x_m)$ such that $g(x)$ has a finite value. Hint: Use the fact that, as long as $g(x)$ makes sense, $g(x)$ *increases* as x *increases*. Also, recall Heaviside's trick for the infinitely long circuit.

E.8. Show that the circuit of Figure E.11 is a bandpass filter.

Figure E.11 A bandpass filter.

Differential and Operational Amplifiers

In this appendix I will introduce you to a circuit component that Ohm, Faraday, and Henry never heard of, and could not possibly have even imagined in their wildest fantasies; the ideal differential amplifier (or *diff-amp*). The diff-amp played no role in the development of radio because it came into existence only in the 1940s, first as a laboratory curiosity and as a specialized analog computer device, and later as a circuit designer's "standard part" in the 1960s with the invention of integrated-circuit technology.

The differential amplifier is worth discussing in this book, however, because it can be used (in ways that the early radio pioneers would have considered to be magic) to build many of the circuits used in radio. For example, inductors figured largely in early radio circuitry, but it is very difficult to actually make a device that behaves like an ideal inductor. It is not difficult, however, to make integrated circuit devices that behave like nearly ideal resistors, capacitors, and diff-amps (which, in turn, use only resistors, capacitors, and transistors but *no inductors*), and so electrical engineers have invented circuits that use only these three components. That is, they have invented *inductorless circuits*. Such inductor-free circuits can be fabricated as tiny integrated circuit chips that bear no resemblance to the bulky, heavy circuitry that was all there was prior to the 1960s. The diff-amp differs from the resistor, capacitor, and inductor of Appendix C in three fundamental ways.

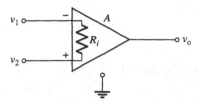

Figure F.1   Symbol for the differential amplifier ("diff-amp")

- They are passive (store or dissipate energy) while the diff-amp is active (contains internal energy sources and so it requires a power supply);
- They are two-terminal devices, while the diff-amp has three terminals;
- They are bilateral (work as well in one direction as the other) while the diff-amp has distinct input and output terminals and thus is unilateral.

The diff-amp is very easy to define. Its symbol is shown in Figure F.1 as a triangle pointing from its inputs (v_1 and v_2) to its output (v_o), where all three of these voltages are measured with respect to a common reference (shown as the ground symbol). The positive gain of the diff-amp is denoted by A, and so we have

$$v_o = A(v_2 - v_1), \quad A > 0.$$

This equation explains the diff-amp's name, in fact, as the output v_o is simply the gain times the *difference* between the plus-terminal input and the minus-terminal input. In a real differential amplifier A is very large, say 100,000 or so, while for an ideal diff-amp A is taken to be *infinity*. (Fortunately, $A = 10^5$ is nearly always sufficiently large to be a good "approximation" to infinity!)

The R_i in Figure F.1 denotes the resistance between the input terminals of the diff-amp. R_i can be quite large, on the order of hundreds of megohms—and the input current i shown in Figure F.1 is usually very small. Typically, i is in the nano-ampere range for a real diff-amp. In the case of an ideal diff-amp the input current is *always* zero (and this is so even if R_i is finite—if this seems paradoxical, keep reading).

That's it! All of the preceding probably seems pretty simple, perhaps even too simple to explain our interest in the diff-amp. So, let me now show you how that common first impression is wrong. To start, consider the circuit of Figure F.2, where $v_2 = 0$ because the +/input terminal is connected directly to ground. The voltage on the −/input terminal is denoted

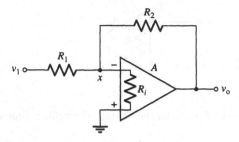

Figure F.2   A simple (but useful) diff-amp circuit.

by x, and so applying Kirchhoff's current law at that node we get

$$\frac{v_1 - x}{R_1} = \frac{x - v_o}{R_2} + \frac{x}{R_i},$$

along with

$$v_o = A(0 - x) = -Ax.$$

These two equations are easily manipulated to give the obvious

$$x = -\frac{v_o}{A}$$

and

$$v_1 = -\frac{v_o}{A}\left[1 + \frac{R_1}{R_i} + \frac{R_1}{R_2}(1 + A)\right].$$

Now we let $A \to \infty$ and assume that the output voltage v_o remains finite. After all, if v_o does not remain finite then we have a useless circuit (in fact, this assumption is experimentally observed to be the case *when there is a feedback path* from the output to the $-/$terminal). So, for the ideal differential amplifier we have

$$v_1 = -\frac{R_1}{R_2}v_o.$$

Defining the gain of the total circuit as $A_f = \frac{v_o}{v_1}$ gives

$$A_f = -\frac{R_2}{R_1}.$$

(The subscript of f is for *feedback*, as you'll notice that there is a connection in Figure F.2 between the output and the input via R_2.) That is, the effective gain of the total circuit has nothing to do with either A or with R_i as long as A is "big." The effective gain is completely determined by the resistors R_1 and R_2. The circuit of Figure F.2 is called an inverting amplifier because of the minus sign in the expression for A_f. The magnitude of this gain can be made anything we want by simply adjusting one or both of the two resistors. We can even get a positive gain by connecting two copies of the circuit in series (see also Problem F.1).

Now, looking back at the equation for x (the voltage at the $-$/terminal of the diff-amp), we see that if v_o remains finite as $A \to \infty$, then $x \to 0$ as $A \to \infty$. That is, the voltage at the $-$/input approaches ground potential as $A \to \infty$. But that terminal is *not* really ground as it is separated from the $+$/terminal (which *is* attached directly to ground) by R_i. For this reason the $-$/terminal is said, for an ideal diff-amp ($A \to \infty$) to be a *virtual ground*. This is a special case of the more general conclusion that, for v_o to remain finite as $A \to \infty$, it must be true that $v_1 \to v_2$, i.e., that the differential input signal $v_2 - v_1 \to 0$. Further, since this vanishing differential input is the voltage drop across R_i, then the input current to the diff-amp goes to zero as $A \to \infty$ *even though $R_i < \infty$.*

In summary, an ideal differential amplifier is characterized by having both input terminals at the *same* voltage (zero differential input voltage) and zero input current. These two conditions are all we need to write and solve the equations describing any circuit containing one or more ideal diff-amps. Let me show you a few interesting examples of how this is done (from now on, all diff-amps in this appendix will be taken as ideal).

First, consider the extraordinarily simple yet highly useful circuit of Figure F.3, which allows the amplified extraction of the tiny voltage (*very* small energy) signal present in a radio antenna. This circuit *isolates* the antenna voltage v_a from all other circuitry that may be connected to the right of the diff-amp because the antenna provides *zero* energy to the diff-amp (the

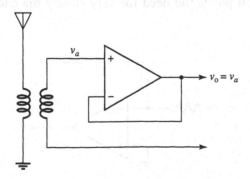

Figure F.3 The diff-amp *voltage follower* circuit.

diff-amp input current is zero). Still, the diff-amp output v_0 *must* equal v_a since $v_a - v_0 \ (= 0)$ is the differential input to the diff-amp. This circuit, with $v_0 = v_a$, is called a *voltage follower*, as the output reproduces (or *follows*) the input. The energy required by any circuitry to the right of the diff-amp is supplied by the diff-amp and its power supply and *not* by the antenna (which has none to spare!) As the final analysis of this appendix you'll find what I think a fascinating *theoretical* use of the voltage follower as a circuit isolator.

Next, consider the circuit of Figure F.4. Since the $-$/input terminal voltage is zero (because the $+$/input terminal is directly connected to ground), we have

$$\frac{v_i - 0}{R} = C\frac{d(0 - v_o)}{dt}$$

or

$$\frac{dv_o}{dt} = -\frac{1}{RC}v_i.$$

Thus, if $v_0(0) = 0$ (which is the case if the capacitor is initially uncharged and so at time $t = 0$ has zero voltage drop across its terminals, because v_o *is* the voltage drop across the C since the left terminal of the C is connected to virtual ground which is *always* at zero potential), then

$$v_o(t) = -\frac{1}{RC}\int_0^t v_i(u)\, du.$$

That is, the circuit of Figure F.4 is an *inverting integrator*, i.e., an integrator with negative gain of $-\frac{1}{RC}$ (e.g., if $R = 500K$ and $C = 1\ \mu fd$, the gain is -2).

It is possible to build an integrator circuit with positive gain using just one diff-amp (rather than putting two negative gain integrators in series), but the price you pay is the need for very closely matched resistors (see

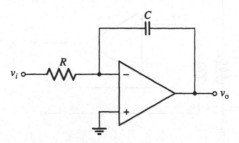

Figure F.4   The diff-amp integrator circuit with negative gain.

Problem F.2). If you swap the R and the C in Figure F.4 then you can show that v_o is proportional to the derivative of $v_i(t)$. Since such circuits are performing mathematical operations (integration or differentiation) the entire circuit of diff-amp *plus* the associated resistors and capacitors could logically be called an *operational amplifier* (or "op-amp," for short).

Now, let's consider a "magical" circuit, one that behaves like (or simu- lates) a *negative* capacitance. This may seem to be absurd, but the circuit of Figure F.5 does the trick. To write the equations that describe this circuit, let's call the diff-amp output voltage x and notice that, since the diff-amp has zero differential input voltage, then the voltage at the node connect- ing R_2 and C must be v (the voltage at the far left). So, using Kirchhoff's current law at the right-most and the left-most nodes (we can*not* sum cur- rents at the middle node because we do not know the diff-amp's output current which, unlike the diff-amp's input current, is not required to be zero—see Problem F.3), we can write

$$i = \frac{v - x}{R_1} \quad \text{and} \quad \frac{x - v}{R_2} = C\frac{dv}{dt}.$$

Or,

$$iR_1 = v - x \quad \text{and} \quad x - v = R_2C\frac{dv}{dt}.$$

Or,

$$-iR_1 = R_2C\frac{dv}{dt}.$$

Or,

$$i = -\frac{R_2C}{R_1}\frac{dv}{dt} = C_{eq}\frac{dv}{dt}.$$

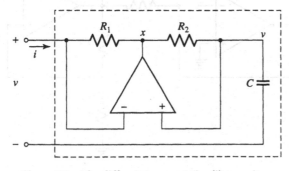

Figure F.5 The diff-amp as a *negative* (!) capacitor.

That is, the circuitry inside the dashed box of Figure F.5 "looks like" a negative capacitor with value

$$C_{eq} = -\frac{R_2}{R_1} C.$$

A similar calculation is the basis for Problem F.4.

As the final practical circuit example of this appendix, recall Problem C.1 in which *Tellegen's gyrator* was defined. There, without saying anything about how to actually make a gyrator, it was shown that if we had two of them we could make a circuit that would behave like an ideal floating inductor (i.e., like an inductor that can have *both* of its terminals connected with no constraints, as opposed to a grounded inductor which has only one terminal available for arbitrary connection). What I'll show you here is one way to make the gyrator itself, from just five resistors and two diff-amps. Recall that the gyrator has two pairs of terminals (each is called a *port*) which obey the relations

$$v_1 = -Ki_2$$
$$v_2 = Ki_1, \quad K > 0$$

where K is called the *gyration resistance*. Since the circuit shown in Figure F.6 satisfies these relations with $K = R$ (as I'll demonstrate next) then that circuit *is* the now not-so-mysterious gyrator.

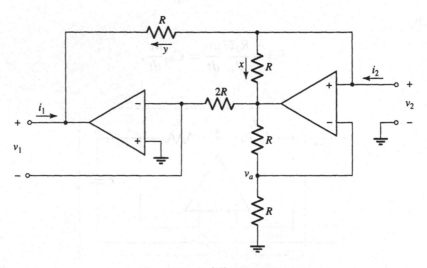

Figure F.6   A diff-amp gyrator.

Since v_a must be equal to v_2 (to have a zero differential input to the diff-amp on the right), then the output voltage of that diff-amp must be $2v_2$. We can then write the two currents x and y as

$$x = \frac{v_2 - 2v_2}{R} = -\frac{v_2}{R}$$

$$y = \frac{v_2 - v_1}{R}.$$

Since $i_2 = x + y$, then

$$i_2 = -\frac{v_2}{R} + \frac{v_2 - v_1}{R} = -\frac{v_1}{R}$$

or,

$$v_1 = -Ri_2,$$

which is the first of the two defining gyrator equations (with $K = R$).

To complete the analysis, simply note that the current flowing out of the virtual ground terminal on the left of Figure F.6 is i_1 (which is the current in the $2R$ resistor as no current flows *into* a diff-amp). Since the $2R$ resistor is connected to the output of the right diff-amp, with an output voltage of $2v_2$, then

$$i_1 = \frac{2v_2}{2R} = \frac{v_2}{R}$$

or

$$v_2 = i_1 R.$$

This is the second of the two defining gyrator equations, and so the circuit of Figure F.6 *is* a gyrator with one grounded port (v_2, i_2) and one floating port (v_1, i_1).

The circuit of Figure F.6, some of its practical problems (and their solution), and its use in constructing filters, is discussed in a famous paper by H. J. Orchard and Desmond F. Sheehan, "Inductorless Bandpass Filters," *IEEE Journal of Solid-State Circuits*, June 1970, pp. 108–118. That circuit is only *one* gyrator realization out of literally hundreds of circuits that have been published over the years. This book is not meant to be a treatise on gyrators, but perhaps just a couple more observations will add a lot more insight about gyrators for you.

If we write the defining equations in matrix form we get

$$\begin{bmatrix} v_1 \\ v_2 \end{bmatrix} = \begin{bmatrix} 0 & -K \\ K & 0 \end{bmatrix} \begin{bmatrix} i_1 \\ i_2 \end{bmatrix}.$$

The two zeros on the main diagonal of the 2×2 resistance matrix represent the gyrator's instantaneous losslessness (see Problem C.1 for more on this). All of the gyrator circuits that have been published can be separated into two categories; those in which the main diagonal zeros occur automatically without requiring any of the circuit components be "matched" in value (e.g., requiring two or more resistors to have precisely specified ratios), and those circuits which *do* require component matching. Gyrator circuits that are in the first category are called *structural* gyrators, and those in the second category are called *compensated* gyrators. The circuit of Figure F.6, which requires all of its resistors to be precisely the same value (R) except for one which must be $2R$, is a compensated gyrator. See Problem F.5 for an example of a structural gyrator.

Another way to mathematically describe a two-port circuit (as are gyrators and transformers) is with the so-called *transmission* matrix. That is, with the usual voltage and current symbols (v_1 and i_1 for the left port and v_2 and i_2 for the right port), we write

$$\begin{bmatrix} v_1 \\ i_1 \end{bmatrix} = \begin{bmatrix} A & B \\ C & D \end{bmatrix} \begin{bmatrix} v_2 \\ -i_2 \end{bmatrix}.$$

The use of $-i_2$ in the right-most column vector is motivated by what happens if we *cascade* (connect in series) two two-ports as shown in Figure F.7. That is, the current into the right port of the left box is the *negative* of the current into the left port of the right box.

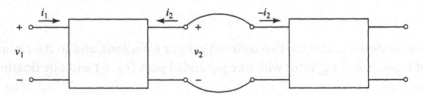

Figure F.7   Two two-ports in series (cascaded).

To see how this works, consider the transmission matrix descriptions for both the gyrator and the ideal transformer, respectively:

$$\text{gyrator:} \quad \begin{bmatrix} v_1 \\ i_1 \end{bmatrix} = \begin{bmatrix} 0 & R \\ \frac{1}{R} & 0 \end{bmatrix} \begin{bmatrix} v_2 \\ -i_2 \end{bmatrix}$$

$$\text{transformer:} \quad \begin{bmatrix} v_1 \\ i_1 \end{bmatrix} = \begin{bmatrix} \frac{1}{n} & 0 \\ 0 & n \end{bmatrix} \begin{bmatrix} v_2 \\ -i_2 \end{bmatrix}$$

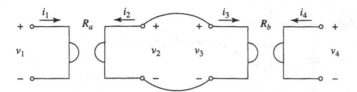

Figure F.8   Two gyrators in series.

where in the transformer transmission matrix the n is the turns ratio, i.e., the *secondary* port (v_2, i_2) has n times as many coil turns (see Appendix C) as does the *primary* port (v_1, i_1).

So, if we connect two gyrators in series, with gyration resistances R_a and R_b, as shown in Figure F.8, we have

$$\begin{bmatrix} v_1 \\ i_1 \end{bmatrix} = \begin{bmatrix} 0 & R_a \\ \frac{1}{R_a} & 0 \end{bmatrix} \begin{bmatrix} v_2 \\ -i_2 \end{bmatrix}$$

and

$$\begin{bmatrix} v_3 \\ i_3 \end{bmatrix} = \begin{bmatrix} v_2 \\ -i_2 \end{bmatrix} = \begin{bmatrix} 0 & R_b \\ \frac{1}{R_b} & 0 \end{bmatrix} \begin{bmatrix} v_4 \\ -i_4 \end{bmatrix}.$$

But this says

$$\begin{bmatrix} v_1 \\ i_1 \end{bmatrix} = \begin{bmatrix} 0 & R_a \\ \frac{1}{R_a} & 0 \end{bmatrix} \begin{bmatrix} 0 & R_b \\ \frac{1}{R_b} & 0 \end{bmatrix} \begin{bmatrix} v_4 \\ -i_4 \end{bmatrix} = \begin{bmatrix} \frac{R_a}{R_b} & 0 \\ 0 & \frac{R_b}{R_a} \end{bmatrix} \begin{bmatrix} v_4 \\ -i_4 \end{bmatrix}.$$

That is, the final, overall transmission matrix on the far right is the transmission matrix for an ideal transformer with a turns ratio of $n = \frac{R_b}{R_a}$. That is, two gyrators in series form an ideal transformer.

To conclude this appendix, let me show you an interesting theoretical circuit that uses the isolation property of the diff-amp voltage follower. Figure F.9 shows a *cascade* of identical RC-stages, each separated (isolated) from one another with a unity-gain buffer amplifier (see Figure F.3 again). The buffer amplifiers allow us to ignore any loading of an individual RC-stage by the one to its right, as the input current to a diff-amp is zero. Figure F.9 does not show a definite number of RC-stages because, in fact, we are soon going to let that number become arbitrarily large.

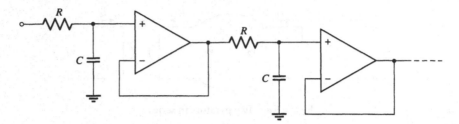

Figure F.9   Cascaded, isolated RC-sections.

The input-to-output transfer function for each stage is

$$\frac{\frac{1}{j\omega C}}{R + \frac{1}{j\omega C}} = \frac{1}{1 + j\omega RC}$$

and so (because of the absence of loading) we can write the total, overall transfer function of n-stages as

$$H(j\omega) = \frac{1}{(1 + j\omega RC)^n}.$$

The magnitude of $H(j\omega)$ is

$$|H(j\omega)| = \frac{1}{\{1 + (\omega RC)^2\}^{n/2}}.$$

At $\omega = 0$, $|H(j\omega)| = 1$, independent of the value of n. For a one-stage circuit ($n = 1$) the value of $|H(j\omega)|$ at $\omega = \omega_0 = \frac{1}{RC}$ is $|H(j\omega_0)| = \frac{1}{\sqrt{2}}$. Let's suppose that for any *prespecified* value of n we would like to retain this property, i.e.,

$$|H(j\omega_0)| = \frac{1}{\sqrt{2}}, \quad \omega_0 = \frac{1}{RC}.$$

This will obviously require us to recalculate the value of the RC product once we know n, and we can do that as follows. Let's call the value of the RC product $(RC)_n$, where $(RC)_1 = RC$. Then,

$$\frac{1}{\{1 + \omega_0^2 (RC)_n^2\}^{n/2}} = \frac{1}{\sqrt{2}} = \frac{1}{\{1 + \left(\frac{1}{RC}\right)^2 (RC)_n^2\}^{n/2}}.$$

Thus,

$$\left\{1 + \left(\frac{1}{RC}\right)^2 (RC)_n^2\right\}^n = 2$$

and so

$$(RC)_n = RC\sqrt{2^{1/n} - 1}.$$

For a given value of n we would choose our resistor and capacitor values with the aid of this equation, where $\omega_o = \frac{1}{RC}$ is the frequency at which we wish to have $|H(j\omega_o)| = \frac{1}{\sqrt{2}}$. Now, since

$$2^{1/n} = e^{\ln 2^{1/n}} = e^{\frac{1}{n}\ln 2},$$

and since (from Appendix A)

$$e^x = 1 + x + \frac{x^2}{2!} + \cdots,$$

then

$$2^{1/n} = 1 + \frac{1}{n}\ln(2) + \frac{\ln^2(2)}{2!n^2} + \cdots.$$

As n becomes large the terms in this expression decrease very rapidly and we have

$$2^{1/n} - 1 \approx \frac{1}{n}\ln(2), \quad n \gg 1.$$

So,

$$(RC)_n \approx RC\sqrt{\frac{\ln(2)}{n}}, \quad n \gg 1,$$

and thus

$$|H(j\omega)|^2 \approx \frac{1}{\left\{1 + \omega^2(RC)_n^2\right\}^n} = \frac{1}{\left\{1 + \frac{\omega^2(RC)^2\ln(2)}{n}\right\}^n}, \quad n \gg 1.$$

Our approximations *quickly* become better as we let $n \to \infty$, and in the limit a curious thing happens. Specifically, writing $a = (\omega RC)^2 \ln(2)$, we see that

$$\lim_{n\to\infty} |H(j\omega)|^2 = \lim_{n\to\infty} \frac{1}{\left\{1 + \frac{a}{n}\right\}^n} = \frac{1}{e^{an}} = e^{-an},$$

and so

$$\lim_{n\to\infty} |H(j\omega)| = e^{-\frac{an}{2}} = e^{-\frac{(\omega RC)^2}{2}\ln(2)}.$$

That is, a very long sequence of cascaded, identical isolated (by diff-amp voltage follower buffers) *RC* stages has a *Gaussian* amplitude frequency response. Who would have guessed it?

Problems

F.1. The circuit of Figure F.2 shows how to get a negative voltage gain with a single diff-amp. We can get any positive gain desired by simply connecting two such circuits in series, but that of course requires two diff-amps. Show that the circuit of Figure F.10 has positive gain with just one diff-amp, but that this gain does have the restriction $A_f = \frac{v_o}{v_i} \geq 1$.

(Answer: $A_f = 1 + \frac{R_2}{R_1}$.)

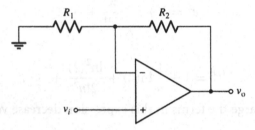

Figure F.10   Amplifier circuit with positive gain and *one* diff-amp.

F.2. Show that the single diff-amp circuit of Figure F.11 is an integrator with positive gain, i.e., show that $y(t) = G \int v(t)\, dt, G > 0$.

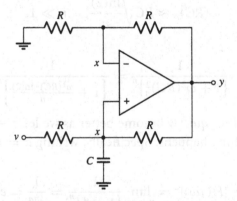

Figure F.11   An integrator circuit with positive gain and *one* diff-amp.

F.3. As mentioned in the text, you generally cannot use Kirchhoff's current law at any node to which the output of a diff-amp is connected. This isn't because the law is violated at such nodes, of course, as the law is fundamental and is *always* obeyed. Rather, we can't use it because a diff-amp's output current is usually simply not known. What we do, instead, is use the current law to *find* the diff-amp output current. For the circuit of Figure F.12 find the currents i_1 and i_2. Hint: you should find that one of the diff-amps has a *negative* output current, i.e., rather than being a current source that diff-amp is a current *sink*.

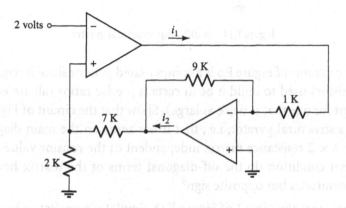

2 volts

9 K

1 K

7 K

i_2

2 K

Figure F.12   What are the diff-amp output currents?

F.4. Calculate the input resistance $R = \frac{v}{i}$ for the circuit shown in Figure F.13 and show that if $R_2 > R_1$, then $R < 0$.

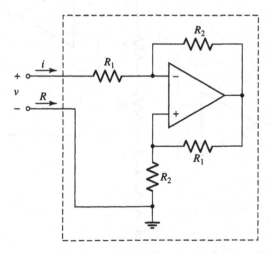

Figure F.13   A diff-amp circuit with a negative input resistance.

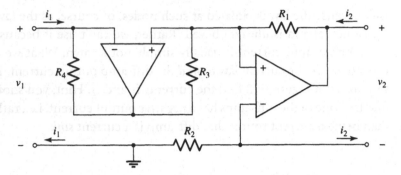

Figure F.14   A diff-amp structural gyrator.

F.5. The gyrator of Figure F.6 is a compensated gyrator since it requires the resistors used to build it be in certain precise ratios (all are equal except for one that is twice as large). Show that the circuit of Figure F.14 is a structural gyrator, i.e., that it has zeros on the main diagonal of its 2×2 resistance matrix independent of the resistor values. Under what condition do the off-diagonal terms of the matrix have equal magnitudes but opposite sign?

F.6. Show that the circuit of Figure F.15 simulates a perfect inductor with one end grounded, i.e., show that

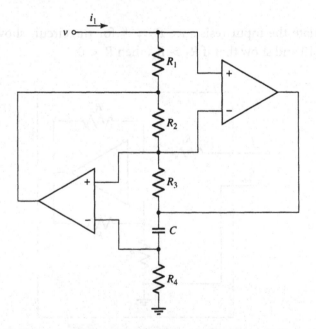

Figure F.15   A diff-amp circuit that behaves like a grounded inductor.

$$v = L_{eq}\frac{di}{dt}$$

and express L_{eq} in terms of R_1, R_2, R_3, R_4, and C. Hint: when $R_1 = R_2 = R_3 = R_4 = 10K$, and $C = 0.001 \,\mu fd$, then $L_{eq} = 100$ mH.

F.7. Calculate the phase shift of the n-stage cascaded, isolated RC circuit of Figure F.9, and in particular notice what happens to the phase shift as $n \to \infty$. What is the physical significance of your result?

Reversing the Order of Integration on Double Integrals, and Differentiating an Integral

In freshman calculus the physical interpretation of a one-dimensional integral is as the area under a curve in a plane, between lower and upper limits of the independent variable. Consider, for example, the integral

$$I_1 = \int_b^c f(x)\, dx,$$

where $f(x)$ is the curve sketched in Figure G.1. If the lower limit is changed to a, then we have

$$I_2 = \int_a^c f(x)\, dx$$

and clearly $I_2 < I_1$ because

$$I_2 = \int_a^c f(x)\, dx = \int_a^b f(x)\, dx + \int_b^c f(x)\, dx$$

$$= \int_a^b f(x)\, dx + I_1$$

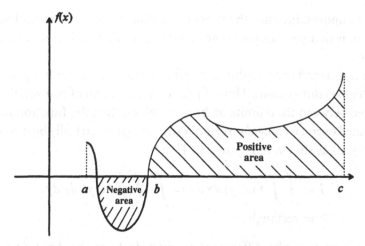

Figure G.1   The one-dimensional integral as the area under a curve.

and because

$$\int_a^b f(x)\, dx < 0.$$

The last statement follows because $f(x)$ is below the x-axis (and so bounds *negative* area) in the interval $a < x < b$.

We can have integrals with variable limits, e.g., if $i(t)$, $t > 0$, is the current into a capacitor C which at time $t = 0$ has zero electrical charge, then as discussed in Appendix C the voltage across the capacitor at any $t > 0$ is

$$v(t) = \frac{1}{C} \int_0^t i(u)\, du,$$

where the symbol "u" is a dummy variable of integration. Don't *ever* write the nonsense statement

$$v(t) = \frac{1}{C} \int_0^t i(t)\, dt,$$

which says t varies from 0 to t. One can occasionally find textbook writers who do such things, and it is *worse* than simply wrong. It shows a complete failure to appreciate the very concept of integration. One can even find authors, who make it clear that they know better, making this error. For example, in the preface of a recent book on magnetic recording, the author brushed aside his now and then sloppy mathematics (even more casual than mine!) by making the astonishing claim "integration is the important thing to understand, not the proper handling of 'dummy variables.'" My position is that you simply do not understand the first if you can't do the second.

We can extend the one-dimensional interpretation of an integral to two (and higher) dimensions. Thus, if $f(x, y)$ is a function of two variables, integrated between fixed limits on both variables, then the function is being integrated over a *rectangular* region R (defined geometrically by $a \leq x \leq b$, $c \leq y \leq d$) as shown in Figure G.2. That is,

$$I = \int \int f(x, y) \, dx \, dy = \int_c^d \int_a^b f(x, y) \, dx \, dy,$$

$$R = \text{rectangle}.$$

If we recognize the differential product $dx \, dy$ as the differential "area patch" in the region R, i.e., if we write $dA = dx \, dy$, then

$$I = \int_c^d \int_a^b f(x, y) \, dA.$$

This last statement says the differential area patch dA, at location (x, y) in R, is multiplied by the value of the function f at location (x, y), and the integral I is the sum of all these products for the set of area patches that covers R. If we introduce a third variable z, where $z = f(x, y)$, then z is the height of a surface above the x-y plane at point (x, y). Thus, the double integral is the *volume* under this surface, over the rectangular region R in

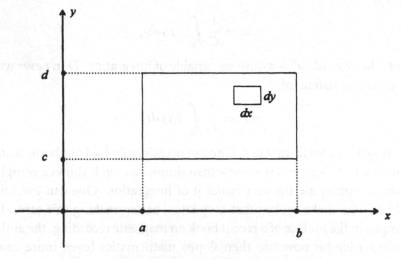

Figure G.2 A rectangular region of integration for a double integral.

the x-y plane. In those parts of R where $f(x, y) < 0$, we must think of *negative* volume just as we needed to imagine negative area as a possibility for a one-dimensional integral.

From all this it is physically clear that if the order of integration is reversed, then

$$\int_a^b \int_c^d f(x, y) \, dy \, dx = I.$$

That is, since the differential area patch $dA = dx \, dy$ also equals $dy \, dx$, then the order of integration on a double integral can be reversed when the region of integration R is a rectangle.

What if R is not a rectangle? This is not an abstract question, as non-rectangular regions of integration often occur. For example, consider the problem of evaluating the integral

$$S = \int_{-\infty}^{\infty} e^{-x^2} \, dx.$$

This integral is very important in the statistical theory of radio when random noise is present (an advanced topic *not* discussed in this book). It is an astonishing fact that this improper definite integral can be evaluated even though it is not possible to do the indefinite integral. To do this, we take the seemingly backward step of considering the apparently even more complicated two-dimensional integral

$$I_a = \int \int e^{-(x^2+y^2)} \, dx \, dy$$
$$R = \text{circle of radius } a$$

where R is now the circular region shown in Figure G.3. The limits of integration are not simply numbers (or infinity, which by a suitable stretch of imagination can be thought of as a number), but more generally are variables. In particular,

$$I_a = \int_{-a}^{a} \int_{-\sqrt{a^2-y^2}}^{\sqrt{a^2-y^2}} e^{-(x^2+y^2)} \, dx \, dy.$$

This is not an attractive problem! It becomes very attractive (i.e., easy), however, if we change to polar coordinates. That is, from (x, y) coordinates to (r, θ) coordinates where

$$x = r \cos(\theta)$$

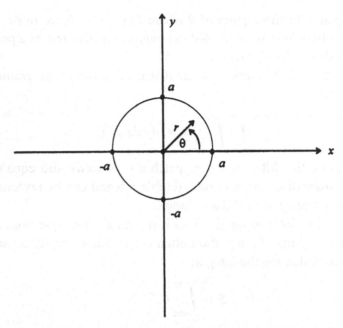

Figure G.3 A circular region of integration, of radius a, centered on the origin. R is the interior of the circle $x^2 + y^2 = a^2$.

$$y = r\sin(\theta)$$
$$x^2 + y^2 = r^2, \quad 0 \le r \le a, \quad 0 \le \theta < 2\pi,$$

as shown in Figure G.3. Then, physically, we clearly arrive at the same value for the integral if we replace the differential area patch in rectangular coordinates ($dx\,dy$) with the differential area patch in polar coordinates ($r\,dr\,d\theta$). In both cases, we cover the same circular region R. Thus,

$$I_a = \int_R \int e^{-(x^2+y^2)}\,dx\,dy = \int_R \int e^{-r^2} r\,dr\,d\theta = \int_0^a \int_0^{2\pi} e^{-r^2} r\,dr\,d\theta.$$

Notice that for any fixed value of r, the factor e^{-r^2} in the integrand is a constant for all θ, i.e., e^{-r^2} is the same at every point in the annular ring of width dr and inner (or outer) radius r (see Figure G.4).

Thus, if we divide the circular region R into concentric annular rings of width dr, then the differential contribution of one such ring to the integral is just e^{-r^2} multiplied by the differential area of the ring. But that differential area is just $2\pi r\,dr$ (the circumference of the ring, times the width of the ring). Thus, $dI_a = e^{-r^2} 2\pi r\,dr$. You may have already noticed that this

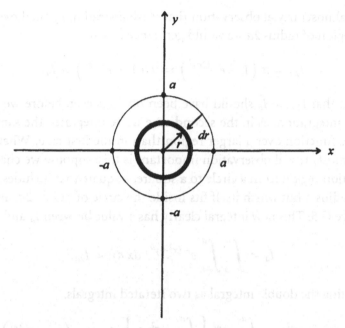

Figure G.4   A circular region of integration can be thought of as a nesting of annular rings.

is precisely what we get if we formally write the two-dimensional integral as two *iterated* one-dimensional integrals, i.e., as

$$I_a = \int_0^a \int_0^{2\pi} e^{-r^2} r \, dr \, d\theta = \int_0^a e^{-r^2} r \left\{ \int_0^{2\pi} d\theta \right\} dr = \int_0^a e^{-r^2} 2\pi r \, dr.$$

That is, first we integrate out one variable, then do a second one-dimensional integral on the remaining variable. Indeed, this is no mere coincidence, as it can be shown that *any* two-dimensional integral is equal to two iterated, one-dimensional integrals, if the region of integration is finite [and if $f(x, y)$ is what mathematicians call "reasonably well-behaved"]. That is,

$$\int \int f(x, y) \, dx \, dy = \int \varphi(y) \, dy,$$

where $\varphi(y) = \int f(x, y) \, dx$. Any book[1] on advanced calculus will have a proof of this, which while not difficult is somewhat lengthy.

To find I_a, then, we simply integrate over all r, i.e.,

$$I_a = \int_0^a e^{-r^2} 2\pi r \, dr = 2\pi \left(-\frac{1}{2} e^{-r^2} \Big|_0^a \right) = \pi \left(1 - e^{-a^2} \right).$$

It is a (almost) trivial observation that if we instead integrated over a circular region of radius $2a$ we would get (since $a > 0$),

$$I_{2a} = \pi \left(1 - e^{-(2a)^2}\right) = \pi \left(1 - e^{-4a^2}\right) > I_a.$$

The fact that $I_{2a} > I_a$ should have been obvious even before we did the explicit integration, as in the second case we've integrated the same non-negative function over a larger region than in the first case. What makes this (almost) trivial observation important is this: suppose we change the integration region from a circle to a square, a square that includes the circle of radius a but which itself fits inside the circle of radius $2a$, as shown in Figure G.5. This new integral clearly has a value between I_a and I_{2a}, i.e.,

$$I_a < \int_{-a}^{a} \int_{-a}^{a} e^{-(x^2+y^2)} \, dx \, dy < I_{2a}.$$

Or, writing the double integral as two iterated integrals.

$$\pi(1 - e^{-a^2}) < \int_{-a}^{a} e^{-y^2} \left\{ \int_{-a}^{a} e^{-x^2} \, dx \right\} dy < \pi \left(1 - e^{-4a^2}\right).$$

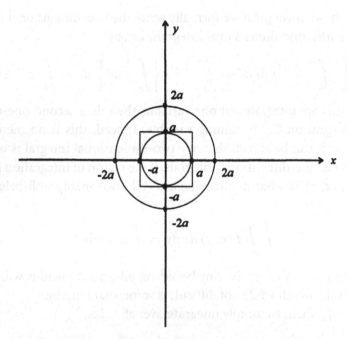

Figure G.5 Bounding the area of a square region of integration from above and below with two circular regions.

Now, as $\alpha \to \infty$, the upper and lower bounds on the double (or iterated) integral both *become equal* to π and so the value of the double integral (with infinite limits) is π:

$$\int_{-\infty}^{\infty} e^{-y^2} \left\{ \int_{-\infty}^{\infty} e^{-x^2} \, dx \right\} \, dy = \pi.$$

Recalling the original integral that started this discussion, we have

$$S = \int_{-\infty}^{\infty} e^{-x^2} \, dx = \int_{-\infty}^{\infty} e^{-y^2} \, dy$$

and so

$$S^2 = \pi \quad \text{or} \quad S = \sqrt{\pi}.$$

That is, we have

$$\int_{-\infty}^{\infty} e^{-x^2} \, dx = \sqrt{\pi},$$

as well as (because the integrand is even)

$$\int_{0}^{\infty} e^{-x^2} \, dx = \frac{1}{2}\sqrt{\pi}.$$

For yet another way to evaluate this integral, see the shaded box at the end of this appendix (and Problem G.9 for yet *another* approach!)

To summarize, in the process of arriving at these results I have argued that the order in which we integrate variables over a finite region is not important. For example, we first integrated out θ, then r, when we thought of the circular region as nested annular rings (see Figure G.4 again). But we could have equally well have integrated out r first, *then* θ, if we had thought of the circular region as composed of narrow pie wedges as shown in Figure G.6. You are generally on safe ground doing this if the integrand is continuous over the finite region of integration, a requirement satisfied in any real, physical problem.

If the integration region is infinite in extent (i.e., if at least one of the integration limits is $\pm\infty$), however, then it may *not* be okay to reverse the order of integration (see Problem G.3). It can be shown that for such situations the concept of *uniform convergence* is required. This is getting a bit far afield for this book, so let me simply refer you to any good book on advanced calculus. What works for finite processes does not always work for infinite ones. For example, is the sum of two rational numbers always rational? Obviously, yes. Is the sum of any finite number of rational num-

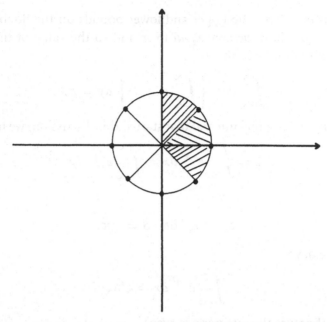

Figure G.6   A circular region of integration can be thought of as a ring of pie wedges.

bers always rational? The answer is immediately yes, as a consequence of the first conclusion. Is the sum of an *infinite* number of rational numbers always rational? *No*—as a counter-example, consider Euler's famous summation of the sum of the reciprocals squared (see Problem 10.1). When we are faced with reversing the order of improper integrals in this book, as in Section 2, we will do what most engineers and physicists do: *assume* uniform convergence until we get some indication that this isn't a good assumption (such as arriving at an obviously nonphysical result).

Reversing the order of integration is an idea that often is crucial to making headway on a problem that may otherwise be extraordinarily difficult or even intractable. For example, suppose $f(x)$ is *any* function with a derivative, i.e., $f'(x)$ exists. We can use reversal of the order of integration to evaluate the following general integral

$$I = \int_0^\infty \frac{f(ax) - f(bx)}{x}\, dx,$$

where a and b are arbitrary constants. Since $f'(x)$ exists, we can write the integrand as

$$\int_b^a f'(xy)\, dy$$

and so

$$I = \int_0^\infty \int_b^a f(xy)\, dy\, dx.$$

This follows because

$$\int_b^a f'(xy)\, dy = \left.\frac{f(xy)}{x}\right|_b^a = \frac{f(ax) - f(bx)}{x}.$$

Reversing the order of integration, we have

$$I = \int_b^a \int_0^\infty f'(xy)\, dx\, dy = \int_b^a \left\{ \left.\frac{f(xy)}{y}\right|_0^\infty dy \right\} = \int_b^a \frac{f(\infty) - f(0)}{y}\, dy$$
$$= \{f(\infty) - f(0)\}\{\ln(y)|\}_b^a$$

or finally,

$$\int_0^\infty \frac{f(ax) - f(bx)}{x}\, dx = \{f(\infty) - f(0)\} \ln\left(\frac{a}{b}\right).$$

This is, I think, an amazing result. It is called *Frullani's integral*, after the Italian mathematician Giuliano Frullani (1795–1834). The result *is* amazing, but not unfailing so. It assumes that not only does $f'(x)$ exist, but also that both $f(\infty)$ and $f(0)$ exist, too. Thus, the general formula *fails* to apply to a situation such as $\int_0^\infty \frac{\{\cos(ax) - \cos(bx)\}}{x}\, dx$, as $f(x) = \cos(x)$ and so $f(\infty)$ is *not* defined. As a special case that does work, if $f(x) = e^{-x}$ then $f(\infty) = 0$ and $f(0) = 1$ and so

$$\int_0^\infty \frac{e^{-ax} - e^{-bx}}{x}\, dx = \ln\left(\frac{b}{a}\right).$$

This is a good example of how a single integral can be done as a *double* integral, with the general trick being that of first introducing a dummy parameter (y) as a variable of integration in an inner integral, and then reversing the order of integration (see Problem G.2).

Looking at integrals now from a somewhat different perspective than that of the first half of this appendix, not infrequently there occurs a need to differentiate an integral. That is, suppose $g(y)$ is defined by the integral

$$g(y) = \int_0^\pi f(x, y)\, dx.$$

What is the derivative of $g(y)$? Is it, perhaps, correct to write

$$\frac{dg}{dy} = \frac{d}{dy} \int_0^\pi f(x, y)\, dx = \int_0^\pi \frac{\partial f}{\partial y}\, dx?$$

That is, is the derivative of the integral simply the integral of the (partial) derivative? In fact, in this case (with constants as the limits of integration) the answer is *yes*. But what if one (or both) of the limits on the integral is (are) dependent on the independent variable (y), as well? That is, if

$$g(y) = \int_{v(y)}^{u(y)} f(x, y)\, dx,$$

then is it still correct to write

$$\frac{dg}{dy} = \int_{v(y)}^{u(y)} \frac{\partial f}{\partial y}\, dx?$$

In fact, now the answer is *no*! Knowing how to correctly differentiate an integral with variable limits is important (I'll do an electrical example at the end of this appendix), and it is really quite easy to determine how to do so, using just freshman calculus.

Formally, from the definition of the derivative,

$$\frac{dg}{dy} = \lim_{h \to 0} \frac{g(y+h) - g(y)}{h}.$$

Now,

$$g(y+h) - g(y) = \int_{v(y+h)}^{u(y+h)} f(x, y+h)\, dx - \int_{v(y)}^{u(y)} f(x, y)\, dx.$$

If $u(y)$ and $v(y)$ are differentiable functions, then

$$\frac{du}{dy} = \lim_{h \to 0} \frac{u(y+h) - u(y)}{h}, \quad \frac{dv}{dy} = \lim_{h \to 0} \frac{v(y+h) - v(y)}{h},$$

and so, to a first approximation when h is very small,

$$u(y+h) = h\frac{du}{dy} + u(y)$$

$$v(y+h) = h\frac{dv}{dy} + v(y)$$

and so

$$\int_{v(y+h)}^{u(y+h)} f(x, y+h)\, dx = \int_{h\,dv/dy+v(y)}^{h\,du/dy+u(y)} f(x, y+h)\, dx.$$

Making the same sort of argument for $f(x, y)$ when h is very small, we can write as a first approximation that

$$f(x, y+h) = h\frac{\partial f}{\partial y} + f(x, y),$$

and so

$$
\int_{v(y+h)}^{u(y+h)} f(x, y+h)\, dx = \int_{v(y)+h\,dv/dy}^{u(y)+h\,du/dy} \left\{ f(x, y) + h\frac{\partial f}{\partial y} \right\} dx
$$
$$
= \int_{v+h\,dv/dy}^{u+h\,du/dy} f\, dx + h\int_{v+h\,dv/dy}^{u+h\,du/dy} \frac{\partial f}{\partial y}\, dx.
$$

This gives us

$$
g(y+h) - g(y) = \int_{v+h\,dv/dy}^{u+h\,du/dy} f\, dx + h\int_{v+h\,dv/dy}^{u+h\,du/dy} \frac{\partial f}{\partial y}\, dx - \int_{v}^{u} f\, dx.
$$

Using Figure G.7 as a guide, this last expression can be written as

$$
g(y+h) - g(y) = \int_{u}^{u+h\,du/dy} f(x, y)\, dx - \int_{v}^{v+h\,dv/dy} f(x, y)\, dx
$$
$$
+ h\int_{v+h\,dv/dy}^{u+h\,du/dy} \frac{\partial f}{\partial y}\, dx.
$$

Now, as $h \to 0$, our approximations become better and better. For the first two integrals, their integrands are essentially constant over their entire vanishingly narrow intervals of integration (which are of length $h\frac{du}{dy}$ and

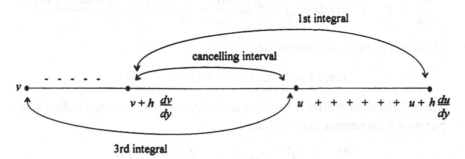

Figure G.7   The overlapping of intervals of integration.

$h \frac{dv}{dy}$, respectively), i.e., as $h \to 0$, then

$$\int_u^{u+h\,du/dy} f(x, y)\, dx = f[u(y), y]h\frac{du}{dy}$$

and

$$\int_v^{v+h\,dv/dy} f(x, y)\, dx = f[v(y), y]h\frac{dv}{dy}.$$

Thus,

$$\frac{dg}{dy} = \lim_{h \to 0} \frac{f[u(y), y]h\frac{du}{dy} - f[v(y), y]h\frac{dv}{dy} + h\int_{v+h\,dv/dy}^{u+h\,du/dy} \frac{\partial f}{\partial y}\, dx}{h}$$

or, at last, we have our result:

$$\frac{dg}{dy} = \int_{v(y)}^{u(y)} \frac{\partial f}{\partial y}\, dx + f[u(y), y]\frac{du}{dy} - f[v(y), y]\frac{dv}{dy}.$$

This expression is called *Leibniz's formula*, after the German mathematician Gottfried Wilhelm Leibniz (1646–1716). It shows us that if the integration limits are not functions of the independent variable, then the derivative of an integral is simply the integral of the derivative. As an example of the formula's use in such a case, consider Frullani's integral once more. That is,

$$I(a, b) = \int_0^\infty \frac{f(ax) - f(bx)}{x}\, dx.$$

From Leibniz's formula we have (*a* is a constant, but the integral will vary as we vary the "constant" value of a and so we *can* differentiate with respect to *a*):

$$\frac{\partial I}{\partial a} = \int_0^\infty f'(ax)\, dx = \int_0^\infty f'(u)\frac{du}{a} = \frac{1}{a}f(u)\Big|_0^\infty = \frac{f(\infty) - f(0)}{a}.$$

Then, integrating with respect to *a*,

$$I(a, b) = \{f(\infty) - f(0)\}\ln(a) + C(b),$$

where $C(b)$ is an arbitrary *function* of integration. Repeating the first step, but now with respect to *b*, we have

$$\frac{\partial I}{\partial b} = -\int_0^\infty f'(bx)\, dx = \frac{f(\infty) - f(0)}{b}.$$

But, since

$$\frac{\partial I}{\partial b} = \frac{dC}{db},$$

then an integration gives

$$C(b) = -\{f(\infty) - f(0)\}\ln(b) + D$$

where D is an arbitrary *constant* of integration. Thus,

$$I(a, b) = \{f(\infty) - f(0)\}\ln(a) - \{f(\infty) - f(0)\}\ln(b) + D$$

or

$$I(a, b) = \{f(\infty) - f(0)\}\ln\left(\frac{a}{b}\right) + D.$$

Finally, since $I(a, a) = 0$, then $D = 0$ and we again arrive at the result

$$\int_0^\infty \frac{f(ax) - f(bx)}{x}\, dx = \{f(\infty) - f(0)\}\ln\left(\frac{a}{b}\right).$$

Two different methods, the same answer.

If the integration limits are functions of the independent variable, then we must use the full Leibniz formula. As an elementary example of this, consider the series circuit in Figure G.8. At time $t = 0$ the switch is thrown to the right and the capacitor (charged to V_0) is allowed to discharge through the series inductor and resistor for $t > 0$. From Kirchhoff's voltage law around a closed loop, it follows that

$$L\frac{di}{dt} + iR - V_0 + \frac{1}{C}\int_0^t i(u)\, du = 0,$$

where $i(t)$ is the (clockwise) current. Differentiating,

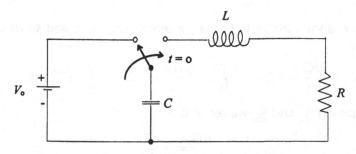

Figure G.8   A series RLC circuit, with the C charged.

$$\frac{d}{dt}\left\{L\frac{di}{dt}\right\} + \frac{d}{dt}\{iR\} - \frac{dV_0}{dt} + \frac{d}{dt}\left\{\frac{1}{C}\int_0^t i(u)\,du\right\} = 0.$$

Since V_0 is a constant, and using Leibniz's formula, we have

$$L\frac{d^2i}{dt^2} + R\frac{di}{dt} + \frac{1}{C}\left\{\int_0^t \frac{\partial i(u)}{\partial t}\,du + i(t)\frac{dt}{dt} - i(0)\frac{d(0)}{dt}\right\} = 0.$$

Since $i(u)$ is not a function of t, we have

$$\frac{\partial i(u)}{\partial t} = 0.$$

Thus,

$$L\frac{d^2i}{dt^2} + R\frac{di}{dt} + \frac{1}{C}i = 0,$$

a differential equation easily solved with the method of assuming an exponential solution.

Here's a way to evaluate $\int_0^\infty e^{-x^2}\,dx$ by differentiating integrals. Define the two functions

$$F(x) = \left\{\int_0^x e^{-t^2}\,dt\right\}^2 \quad \text{and} \quad G(x) = \int_0^1 \frac{e^{-x^2(1+t^2)}}{1+t^2}\,dt.$$

Then,

$$\frac{dF}{dx} = 2\left\{\int_0^x e^{-t^2}\,dt\right\}e^{-x^2} = 2e^{-x^2}\int_0^x e^{-t^2}\,dt$$

and

$$\frac{dG}{dx} = \int_0^1 \frac{-2x(1+t^2)e^{-x^2(1+t^2)}}{1+t^2}\,dt = -2\int_0^1 xe^{-x^2(1+t^2)}\,dt$$

$$= -2xe^{-x^2}\int_0^1 e^{-x^2t^2}\,dt.$$

Now, change variable in this last integral to $u = xt$ and so $dt = \frac{du}{x}$. Then,

$$\frac{dG}{dx} = -2xe^{-x^2}\int_0^x e^{-u^2}\frac{du}{x} = -2e^{-x^2}\int_0^x e^{-u^2}\,du.$$

Comparing $\frac{dF}{dx}$ and $\frac{dG}{dx}$ we see that

$$\frac{dG}{dx} = -\frac{dF}{dx}$$

or,

$$\frac{d}{dx}(G+F) = 0.$$

This says $G(x) + F(x) = $ constant for all $x > 0$. To find this constant, notice that $F(0) = 0$ and that

$$G(0) = \int_0^1 \frac{dt}{1+t^2} = \tan^{-1}(t)|_0^1 = \frac{\pi}{4}.$$

Thus, $G(0) + F(0) = \frac{\pi}{4} = G(x) + F(x)$, $x \geq 0$. Now, clearly

$$\lim_{x \to \infty} G(x) = 0.$$

So,

$$\lim_{x \to \infty} F(x) = \frac{\pi}{4} = \left\{ \int_0^\infty e^{-t^2} \, dt \right\}^2,$$

which is equivalent to

$$\int_0^\infty e^{-t^2} \, dt = \frac{1}{2}\sqrt{\pi}.$$

Note

1. Even though they are older works, I particularly recommend the two volumes by Richard Courant, *Differential and Integral Calculus*, Nordeman 1945. These beautifully written books are masterpieces of clarity, insight, and (even though Courant was a pure mathematician) they project a strong sense of physical reality.

Problems

G.1. Using the area interpretation of the integral, and Euler's identity, give a plausibility argument for the Riemann–Lebesgue lemma: for any two limits a and b,

$$\lim_{\omega \to \pm\infty} \int_a^b e^{-j\omega t} f(t) \, dt = 0$$

for any "well-behaved" function $f(t)$. (Take "well behaved" to mean "bounded.") For the special case $a = -\infty$ and $b = \infty$ the integral becomes the *Fourier transform* of $f(t)$, denoted by $F(j\omega)$, which plays a big role in this book and in radio. That is,

$$F(j\omega) = \int_{-\infty}^{\infty} f(t)e^{-j\omega t}\,dt,$$

and the Riemann–Lebesque lemma says $\lim_{\omega \to \pm\infty} F(j\omega) = 0$, which has very important *physical implications*. See the end of Chapter 13 for a formal proof of this special case of the lemma, and the physical interpretation of the result.

G.2. Show that

$$\int_0^{\infty} \frac{\{\cos(ax) - \cos(bx)\}}{x^2}\,dx = \left(\frac{\pi}{2}\right)(b - a),$$

where a and b are both positive constants. Hint: your first thought may be to write $f(x) = \frac{\cos(x)}{x}$ and then to attempt to use Frullani's integral. (Try this, and see that it does *not* work.) Rather, use the same idea behind the first *derivation* of Frullani's integral, i.e., introduce a new parameter, y, to make a double integral, and then reverse the order of integration. You will find it useful to know that

$$\int_0^{\infty} \frac{\sin(yx)}{x}\,dx = \frac{\pi}{2} \quad \text{for all } y > 0,$$

a result derived in Appendix H.

G.3. As a counterexample to demonstrate that reversing the order of integration of improper integrals is *not* necessarily okay, consider the function $f(x, y) = (2 - xy)xye^{-xy}$. Notice that

$$f(x, y) = \frac{\partial}{\partial x}(x^2 y e^{-xy}).$$

and that

$$f(x, y) = \frac{\partial}{\partial y}(xy^2 e^{-xy}).$$

Use this to show that

$$\int_0^1 dx \int_0^{\infty} f(x, y)\,dy = 0,$$

while

$$\int_0^{\infty} dy \int_0^1 f(x, y)\,dx = 1.$$

G.4. Consider the formidable appearing integral

$$I(a) = \int_0^{\infty} e^{-x^2 - (a^2/x^2)}\,dx, \quad a \geq 0.$$

Evaluate $I(a)$ by showing it satisfies the differential equation $\frac{dI}{da} = -2I$ [use the result derived in the first part of this appendix, $I(0) = \int_0^\infty e^{-x^2}\, dx = (\frac{1}{2})\sqrt{\pi}$]. Hint: make the change of variable $t = \frac{1}{x}$, followed by the second change of variable $u = at$.

G.5. Show that

$$\int_{-\infty}^{\infty} e^{-x^2} \cos(2\lambda x)\, dx = \sqrt{\pi}\, e^{-\lambda^2}$$

and that

$$\int_{-\infty}^{\infty} e^{-x^2} \sin(2\lambda x)\, dx = 0.$$

Hint: write the first integral as $I(\lambda)$ and show that it satisfies the differential equation $\frac{dI}{d\lambda} = -2\lambda I$, with $I(0) = \sqrt{\pi}$. You will find it helpful to remember integration by parts to do the integral that results from calculating $\frac{dI}{d\lambda}$. The second integral can be done in the same way [with $I(0) = 0$], but actually you should see the integral is *zero by inspection*. Why?

G.6. In Problem 10.1 the value of Euler's famous sum $\sum_{n=1}^{\infty} \frac{1}{n^2}$ is found using Fourier series. Another way to do it is with a double integral. Thus, notice that

$$\int_0^1 \int_0^1 \frac{dx\, dy}{1 - xy} = \int_0^1 \int_0^1 (1 + xy + x^2 y^2 + \cdots)\, dx\, dy$$

because, over the entire region of integration (the unit square), we can expand the left integrand as the convergent geometric series that is the right integrand. Then, doing the easy integrations on the right, you'll see that

$$\int_0^1 \int_0^1 \frac{dx\, dy}{1 - xy} = 1 + \frac{1}{2^2} + \frac{1}{3^2} + \cdots = \sum_{n=1}^{\infty} \frac{1}{n^2}.$$

So, if we can evaluate the double integral we have the answer to Euler's sum. For an incredibly clever evaluation of the integral, see George F. Simmons, *Calculus Gems*, McGraw-Hill, 1992, pp. 323–325. Professor Simmons ends the derivation with the comment that an obvious extension of this approach gives

$$\int_0^1 \int_0^1 \int_0^1 \frac{dx\, dy\, dz}{1 - xyz} = \sum_{n=1}^{\infty} \frac{1}{n^3},$$

and that if *you* can figure out how to do the triple integral then you'll become famous—because nobody yet has been able to do it! Summing the reciprocals cubed has baffled the best mathematicians for centuries and its solution will be a major achievement on the same level as establishing Fermat's Last Theorem (but probably *more* valuable to working analysts).

G.7. Two integrals that appear in the Fourier theory discussion of Chapter 12 are the *Fresnel integrals*, named after the French physicist Augustin Jean Fresnel (1788–1827), who in 1818 encountered them in his theoretical studies in optics. Even earlier, in 1743, Euler ran into them in the course of studying the physics of a coiled spring. These integrals are

$$F_1 = \int_0^\infty \sin(x^2)\,dx \quad \text{and} \quad F_2 = \int_0^\infty \cos(x^2)\,dx.$$

It is common to see authors of books on complex variable theory asserting that the evaluations of F_1 and F_2 require doing contour integrals in the complex plane, a sophisticated method far beyond the level of this book. In fact, however, it is possible to evaluate the Fresnel integrals using just freshman calculus. Here's an outline (*you* fill in the missing steps) of how Euler did it.

a. Consider the so-called *gamma function*, defined by Euler (in two letters to a friend in 1729 and 1730) as

$$\Gamma(n) = \int_0^\infty e^{-x} x^{n-1}\,dx$$

where $n > 0$. Show, by integrating by parts, that $\Gamma(n + 1) = n\Gamma(n)$, as well as that $\Gamma(1) = 1$. Thus, conclude that *if n is a positive integer, then* $\Gamma(n + 1) = n!$ (Since the defining integral exists for *all* positive n, not just for the positive integers, then $\Gamma(n)$ is a generalization of the factorial function, which is just what Euler was after.)

b. Now, setting $n = \frac{1}{2}$, we have

$$\Gamma\left(\frac{1}{2}\right) = \int_0^\infty \frac{e^{-x}}{\sqrt{x}}\,dx.$$

Change variable to $x = t^2$ and thus show that

$$\Gamma\left(\frac{1}{2}\right) = 2\int_0^\infty e^{-t^2}\,dt.$$

As shown in this appendix, however, the integral equals $\frac{1}{2}\sqrt{\pi}$ and so conclude that

$$\Gamma\left(\frac{1}{2}\right) = \sqrt{\pi}.$$

c. Next, change variable in the gamma integral as follows: with p and q as real, positive constants, write $x = u(p + jq)$ and show that

$$\int_0^\infty x^{n-1} e^{-px} e^{-jqx}\, dx = \frac{\Gamma(n)}{(p + jq)^n}.$$

d. Write $p + jq$ in polar form, i.e., as the complex number with magnitude r and angle α. So,

$$p + jq = re^{j\alpha} = r\{\cos(\alpha) + j\sin(\alpha)\},$$

substitute into the result from part c, equate real and imaginary parts on each side of the resulting expression, and so conclude that

$$\int_0^\infty x^{n-1} e^{-px} \cos(qx)\, dx = \frac{\Gamma(n)}{r^n}\cos(n\alpha)$$

$$\int_0^\infty x^{n-1} e^{-px} \sin(qx)\, dx = \frac{\Gamma(n)}{r^n}\sin(n\alpha).$$

e. Set $p = 0$ and $q = 1$ (which gives $r = 1$ and $\alpha = \frac{\pi}{2}$, of course), and also set $n = \frac{1}{2}$, and so conclude (using part b) that

$$\int_0^\infty \frac{\cos(x)}{\sqrt{x}}\, dx = \frac{\Gamma\left(\frac{1}{2}\right)}{\sqrt{2}} = \sqrt{\frac{\pi}{2}}$$

$$\int_0^\infty \frac{\sin(x)}{\sqrt{x}}\, dx = \frac{\Gamma\left(\frac{1}{2}\right)}{\sqrt{2}} = \sqrt{\frac{\pi}{2}}.$$

f. In the F_1 and F_2 integrals change variable to $u = x^2$ and show that

$$F_1 = \frac{1}{2}\int_0^\infty \frac{\sin(u)}{\sqrt{u}}\, du$$

$$F_2 = \frac{1}{2}\int_0^\infty \frac{\cos(u)}{\sqrt{u}}\, du$$

and so, using the results of part e, conclude (as did Euler) that

$$F_1 = \int_0^\infty \sin(x^2)\, dx = \frac{1}{2}\sqrt{\frac{\pi}{2}}$$

$$F_2 = \int_0^\infty \cos(x^2)\, dx = \frac{1}{2}\sqrt{\frac{\pi}{2}}.$$

G.8. If $C(x) = \int_x^\infty \cos(t^2)\, dt$ and $S(x) = \int_x^\infty \sin(t^2)\, dt$, then evaluate $J = \int_0^\infty \{C^2(x) + S^2(x)\}\, dx$. Historical note: this problem occurs in a 1925 paper on physical optics published in the *Proceedings of the Royal Society A*, whose author—the physicist Sir Arthur Schuster (1851–1934)—couldn't evaluate J. The following year the great English mathematician G. H. Hardy (1877–1947) wrote a brief note in the *Proceedings of the London Mathematical Society* in which he gives the following terse solution, to which you are to fill in the missing details. Hardy begins by writing that "it is plain that"

$$C^2(x) + S^2(x) = \int_x^\infty \int_x^\infty \cos(t^2 - u^2)\, dt\, du.$$

a. Show why this is "plain." (Hint: write

$$C^2(x) = \left\{ \int_x^\infty \cos(t^2)\, dt \right\} \left\{ \int_x^\infty \cos(u^2)\, du \right\},$$

do the same for $S^2(x)$, and recall the trigonometric identity for the cosine of the difference of two angles).

b. Hardy next forms J (and sounds just like an engineer when he writes he will "ignore any difficulties in changing the order of integrations"), and concludes that

$$J = \int_0^\infty \left\{ \int_x^\infty \int_x^\infty \cos(t^2 - u^2)\, dt\, du \right\} dx$$

$$= \int_0^\infty \left\{ \int_0^t u \cos(t^2 - u^2)\, du \right\} dt$$

$$+ \int_0^\infty \left\{ \int_0^u t \cos(t^2 - u^2)\, dt \right\} du.$$

Hardy then says the last two double integrals are obviously equal, and so

$$J = 2 \int_0^\infty \left\{ \int_0^t u \cos(t^2 - u^2)\, du \right\} dt.$$

Explain why all the above is so. (Hint: Start by interpretating the triple integral as a volume integral and then, by examining the limits on the integrals, determine the shape of the volume. Finally, write the volume integral in a different way.) More formally, you should show that

$$\int_0^\infty \left\{ \int_x^\infty \int_x^\infty f(t, u) \, dt \, du \right\} dx$$

$$= \int_0^\infty \left\{ \int_0^t uf(t, u) \, du \right\} dt + \int_0^\infty \left\{ \int_0^u tf(t, u) \, dt \right\} du.$$

For the Hardy–Schuster problem, of course,

$$f(t, u) = \cos(t^2 - u^2).$$

c. Hardy completes his derivation by stating that

$$\int_0^\infty \left\{ \int_0^t u \cos(t^2 - u^2) \, du \right\} dt = \frac{1}{2} \int_0^\infty \sin(t^2) \, dt = \frac{1}{4} \sqrt{\frac{\pi}{2}}$$

and so $J = \frac{1}{2}\sqrt{\frac{\pi}{2}}$. Show that this is so. Hint: Look back at the Fresnel integral F_1 discussed in Problem G.7.

G.9. Consider, one last time, the evaluation of $\int_0^\infty e^{-x^2} \, dx$. In 1780 the French genius Pierre-Simon Laplace (1749–1827) did it this way (you are to fill in the details):

a. Starting with the double integral

$$I = \int_0^\infty \left\{ \int_0^\infty e^{-s(1+x^2)} \, ds \right\} dx,$$

actually do the integrals (they are easy) and show that $I = \frac{\pi}{2}$.

b. Next, reverse the order of integration to write I as

$$I = \int_0^\infty e^{-s} \left\{ \int_0^\infty e^{-sx^2} \, dx \right\} ds$$

and make the change of variable $u = x\sqrt{s}$ in the inner integral. Show that this gives

$$I = \left(\int_0^\infty s^{-1/2} e^{-s} \, dx \right) \left(\int_0^\infty e^{-u^2} \, du \right).$$

c. Then, for the s integral, make the change of variable $u = \sqrt{s}$ and show that this results in

$$I = 2 \left(\int_0^\infty e^{-u^2} \, du \right)^2.$$

d. Combine the results of parts a and c to show that, indeed,

$$\int_0^\infty e^{-x^2} \, dx = \frac{1}{2}\sqrt{\pi}.$$

The Fourier Integral Theorem (How Mathematicians Do It)

Unlike the intuitive engineering approach of the text (Chapter 12) for arriving at the Fourier transform integral pair (by letting the period of a periodic function increase to infinity), a pure mathematician *starts* with the exotic mathematical identity

$$\int_{-\infty}^{\infty} \frac{e^{j\omega x}}{\omega} \, d\omega = j\pi \, \mathrm{sgn}(x) = \begin{cases} j\pi, & x > 0 \\ -j\pi, & x < 0 \end{cases}$$

where the *sign function*, $\mathrm{sgn}(x)$, is shown in Figure H.1. In this identity there is no association of ω with frequency; it is simply an arbitrary symbol, an arbitrary dummy variable of integration. This theoretical result in pure mathematics can be easily derived by doing a contour integral in the complex plane.[1] Such a *simple* derivation, ironically, is beyond the mathematical level of this book. Still, the above mathematical identity is just within the reach of freshman calculus (just like the Fresnel integrals in Problem G.7), and what I'll show you now is some very clever mathematics (I don't know who first did it). Going through this material is worthwhile, I think, just to see how much can be done if one is simply ingenious enough. For this appendix to make sense, you must have already read Chapter 14.

We begin by considering the function $g(y)$, defined by the following integral:

$$g(y) = \int_{0}^{\infty} e^{-xy} \frac{\sin(x)}{x} \, dx, \quad y \geq 0.$$

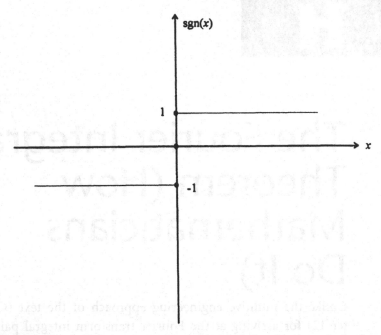

Figure H.1 The *sign* function (not to be confused with the trigonometric *sine* function!).

Pulling this integral out of thin air may seem arbitrary—and I guess it is! When I'm finished, however, you'll see that it is a good choice. We of course need the condition $y \geq 0$ to ensure the convergence of the integral, i.e., to keep the exponential factor from blowing up as $x \rightarrow \infty$ (which would happen if $y < 0$). Then, differentiating (and assuming we can interchange the order of differentiation and integration—see Appendix G), we have

$$\frac{dg}{dy} = \frac{d}{dy} \int_0^\infty e^{-xy} \frac{\sin(x)}{x} \, dx = \int_0^\infty -xe^{-xy} \frac{\sin(x)}{x} \, dx$$

or

$$\frac{dg}{dy} = -\int_0^\infty e^{-xy} \sin(x) \, dx.$$

This integral can be done with little difficulty with freshman calculus, by integrating by parts twice (try it). That will give

$$\frac{dg}{dy} = -\frac{1}{1 + y^2}.$$

Then, doing an indefinite integration [also quite straightforward with freshman calculus by changing variable to $y = \tan(\theta)$], we have

$$g(y) = C - \tan^{-1}(y)$$

where C is the constant of integration. To obtain the value of C, we need only to evaluate $g(y)$ for some particular value of y. In fact, it is easy to prove (see Problem H.1) that the integral vanishes in the limit of $y \to \infty$, i.e., that $\lim_{y\to\infty} g(y) = 0$. Thus,

$$\lim_{y\to\infty} g(y) = \lim_{y\to\infty} \{C - \tan^{-1}(y)\} = C - \tan^{-1}(\infty) = 0$$

or $C = \tan^{-1}(\infty) = \frac{\pi}{2}$.

We therefore now have

$$g(y) = \frac{\pi}{2} - \tan^{-1}(y) = \int_0^\infty e^{-xy} \frac{\sin(x)}{x} dx.$$

In particular, for the special case of $y = 0$ [and as $\tan^{-1}(0) = 0$], we arrive at an integral which occurs all through the mathematical theory of radio,[2]

$$\int_0^\infty \frac{\sin(x)}{x} dx = \frac{\pi}{2}.$$

If we observe that $\frac{\sin(x)}{x}$ is even, then we can also immediately write

$$\int_{-\infty}^0 \frac{\sin(x)}{x} dx = \frac{\pi}{2}.$$

We are now in the home stretch. Make the change of variable $x = Kv$, where K is an arbitrary real constant. Then.

$$\int_0^\infty \frac{\sin(x)}{x} dx = \begin{cases} \int_0^\infty \frac{\sin(Kv)}{Kv} K dv = \int_0^\infty \frac{\sin(Kv)}{v} dv \\ \qquad\qquad = \frac{\pi}{2} \qquad \text{if } K > 0 \\[2ex] \int_0^{-\infty} \frac{\sin(Kv)}{Kv} K dv = -\int_{-\infty}^0 \frac{\sin(Kv)}{v} dv \\ \qquad\qquad = -\frac{\pi}{2} \qquad \text{if } K < 0. \end{cases}$$

That is,

$$\int_0^\infty \frac{\sin(Kv)}{v} dv = \frac{\pi}{2} \operatorname{sgn}(K).$$

Finally, writing this last statement using the original symbols that opened this appendix, and extending the interval of integration to $-\infty$ to ∞, we have

$$\frac{1}{\pi} \int_{-\infty}^{\infty} \frac{\sin(\omega x)}{\omega} \, d\omega = \text{sgn}(x).$$

This remarkable result is called *Dirichlet's discontinuous integral* (the discontinuity is of course at $x = 0$), after the German mathematician Gustave Peter Lejeune Dirichlet (1805–1859). Now, recognizing that $\frac{\sin(\omega x)}{\omega}$ is the imaginary part of $\frac{e^{j\omega x}}{\omega}$, and that since $\frac{\cos(\omega x)}{\omega}$ is an odd function of ω we have $\int_{-\infty}^{\infty} \frac{\cos(\omega x)}{\omega} \, d\omega = 0$, then we conclude

$$\frac{1}{\pi} \int_{-\infty}^{\infty} \frac{e^{j\omega x}}{\omega} \, d\omega = \frac{1}{\pi} \int_{-\infty}^{\infty} j \frac{\sin(\omega x)}{\omega} \, d\omega = j \, \text{sgn}(x).$$

Or, *at last*, we arrive at the mathematical identity which opened this appendix,

$$\int_{-\infty}^{\infty} \frac{e^{j\omega x}}{\omega} \, d\omega = j\pi \, \text{sgn}(x).$$

Now, if we differentiate $\text{sgn}(x)$ we get zero everywhere except at $x = 0$, where $\text{sgn}(x)$ has a jump in value of 2. Thus,

$$\frac{d}{dx} \text{sgn}(x) = 2\delta(x).$$

But

$$\frac{d}{dx} \int_{-\infty}^{\infty} \frac{e^{j\omega x}}{\omega} \, d\omega = \int_{-\infty}^{\infty} \frac{1}{\omega} \frac{d}{dx} e^{j\omega x} \, d\omega$$

$$= \int_{-\infty}^{\infty} \frac{1}{\omega} j\omega e^{j\omega x} \, d\omega = j \int_{-\infty}^{\infty} e^{j\omega x} \, d\omega$$

Therefore,

$$j \int_{-\infty}^{\infty} e^{j\omega x} \, d\omega = j2\pi \, \delta(x),$$

or,

$$\delta(x) = \frac{1}{2\pi} \int_{-\infty}^{\infty} e^{j\omega x} \, d\omega.$$

From this it immediately follows that

$$\delta(x - y) = \frac{1}{2\pi} \int_{-\infty}^{\infty} e^{j\omega(x-y)} \, d\omega.$$

Suppose next that we take an arbitrary function, $h(y)$, and write

$$\int_{-\infty}^{\infty} \delta(x - y)h(y) \, dy.$$

From the sampling property of the impulse function (see Chapter 14), this is equal to $h(x)$. Thus,

$$h(x) = \int_{-\infty}^{\infty} h(y) \left[\frac{1}{2\pi} \int_{-\infty}^{\infty} e^{j\omega(x-y)} \, d\omega \right] dy$$

$$= \frac{1}{2\pi} \int_{-\infty}^{\infty} e^{j\omega x} \left[\int_{-\infty}^{\infty} h(y)e^{-j\omega y} \, dy \right] d\omega.$$

This statement, called the *Fourier integral theorem*, is generally written as a *pair* of integrals, i.e., the interior integral is written as

$$H(j\omega) = \int_{-\infty}^{\infty} h(y)e^{-j\omega y} \, dy,$$

and the exterior integral is written as

$$h(x) = \frac{1}{2\pi} \int_{-\infty}^{\infty} H(j\omega)e^{j\omega x} \, d\omega.$$

Note that the $\frac{1}{(2\pi)}$ factor could actually be split between the two integrals in an infinitude of ways; some authors, for example, put $\frac{1}{\sqrt{2\pi}}$ in front of each integral for the sake of symmetry. Electrical engineers usually write the Fourier integral pair as above, however, because if we *do* associate ω with frequency (and y with time—just replace all the y's with t's) then the pair already has symmetry. That is, with $\omega = 2\pi f$, we have

$$H(jf) = \int_{-\infty}^{\infty} h(t)e^{-j2\pi f t} \, dt,$$

$$h(t) = \int_{-\infty}^{\infty} H(jf)e^{j2\pi f t} \, df,$$

which is the Fourier transform pair (of integrals).

Notes

1. See, for example, Bernard Friedman, *Lectures on Applications-Oriented Mathematics* (V. Twersky, editor), Holden-Day, 1969, pp. 17–21.

2. This important integral was first evaluated by Euler sometime between 1776 and his death in 1783, and was published posthumously. He used a much different method (involving complex variables), rather than the elementary approach I've used here—see Problem H.3.

Problems

H.1. In the "freshman calculus" derivation of the mathematical identity that opens this appendix, I made the claim that

$$\lim_{y \to \infty} \int_0^\infty e^{-xy} \frac{\sin(x)}{x} \, dx = 0,$$

but said only that it is "easy" to show this is so. Prove my claim. Hint: Observe that

$$\left| \int_0^\infty \frac{e^{-xy} \sin(x)}{x} \, dx \right| \leq \int_0^\infty \left| \frac{e^{-xy} \sin(x)}{x} \right| \, dx$$

$$= \int_0^\infty |e^{-xy}| \left| \frac{\sin(x)}{x} \right| \, dx \leq \int_0^\infty e^{-xy} \, dx,$$

where the last statement follows because $|\frac{\sin(x)}{x}| \leq 1$ for all x and because $e^{-xy} > 0$ for all real x and y. Now simply evaluate the last integral and take the limit $y \to \infty$.

H.2. The exponential-integral function, written as $Ei(t)$, occurs in advanced engineering and physics. It is defined as

$$Ei(t) = \begin{cases} \displaystyle\int_t^\infty \frac{e^{-u}}{u} \, du & \text{for } t \geq 0 \\ 0 & \text{for } t < 0. \end{cases}$$

Show that its Fourier transform is $Ei(j\omega) = \frac{\ln(1+j\omega)}{(j\omega)}$. Hint: First, change variable to $x = \frac{u}{t}$ and show that $Ei(t) = \int_1^\infty \frac{e^{-xt}}{x} \, dx$, $t \geq 0$. Then write the Fourier transform integral (which will of course be a double integral) and reverse the order of integration.

H.3. Here's another way, using Euler's identity, to derive Dirichlet's discontinuous integral.

a. Evaluate $\int_0^\infty e^{(-p+jq)x}\, dx$ and thereby show that

$$\int_0^\infty e^{-px} \cos(qx)\, dx = \frac{p}{p^2 + q^2}, \quad p > 0.$$

b. Integrate the previous result with respect to q from a to b, reverse the order of integration, and thus show that

$$\int_0^\infty e^{-px} \frac{\sin(bx) - \sin(ax)}{x}\, dx = \tan^{-1}\left(\frac{b}{p}\right) - \tan^{-1}\left(\frac{a}{p}\right).$$

c. Finally, by setting $b = 0$ and letting $p \to 0$, show how the result of part b reduces to Dirichlet's integral. Comment on the relative merits of this derivation and the one given in the main text (e.g., which one makes the most "assumptions"?)

H.4. Consider the integral $I(k, h) = \int_0^h \frac{\sin(kx)}{x}\, dx$, where h is *any* positive constant. By making the obvious change of variable show that $\lim_{k\to\infty} I(k, h) = \frac{\pi}{2}$, *independent* of the particular value of h.

APPENDIX

I

The Hilbert
Integral Transform

Suppose we have a system with real $x(t)$ as the input; this particular system is to shift the phase of each frequency component of $x(t)$ by 90° while having no effect on the amplitudes of the components. Call the system output $\bar{x}(t)$. Such a system is called a quadrature filter, and $\bar{x}(t)$ is called the *Hilbert transform* [after the great German mathematician David Hilbert (1862–1943)] of $x(t)$. Notice that the Hilbert transform, unlike the Fourier, does not change domains, i.e., $x(t)$ and $\bar{x}(t)$ are both time functions. The quadrature filter (or Hilbert *transformer*) is an ideal phase shifter with infinite bandwidth.

Let $h(t)$ denote the impulse response of the Hilbert transformer, with Fourier transform $H(j\omega)$. $H(j\omega)$ is also, of course, the transfer function (see Appendix C) of the Hilbert transformer. If there is to be any possibility of actually building this system then, like $x(t)$, $h(t)$ must be real. Thus, $|H(j\omega)|$ must be even, and the phase angle of $H(j\omega)$ must be odd, as shown in Figure I.1. To mathematically express these magnitude and 90° phase-shift requirements, I'll write

$$H(j\omega) = \begin{cases} -j, & \omega > 0 \\ +j, & \omega < 0. \end{cases}$$

Now, from other work (see Chapters 14 and 15) we know

$$F[\cos(\omega_0 t)] = \pi[\delta(\omega - \omega_0) + \delta(\omega + \omega_0)]$$

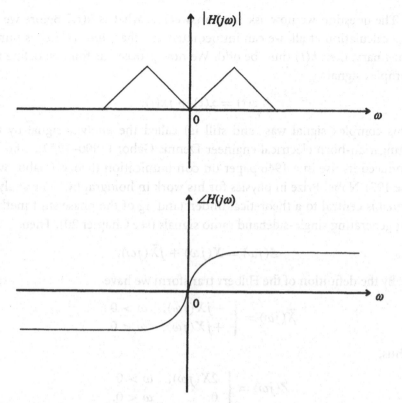

Figure I.1   The general forms of the magnitude and phase responses of a real system.

and

$$F[\sin(\omega_0 t)] = -j\pi[\delta(\omega - \omega_0) - \delta(\omega + \omega_0)].$$

Applying the $H(j\omega)$ of the Hilbert transformer to $F[\cos(\omega_0 t)]$, we can calculate the transformer output to be

$$H(j\omega)F[\cos(\omega_0 t)] = \pi[-j\delta(\omega - \omega_0) + j\delta(\omega + \omega_0)]$$

$$= -j\pi[\delta(\omega - \omega_0) - \delta(\omega + \omega_0)] = F[\sin(\omega_0 t)].$$

This tells us that the Hilbert transform of $\cos(\omega_0 t)$ is $\sin(\omega_0 t)$. In a similar manner,

$$H(j\omega)F[\sin(\omega_0 t)] = -j\pi[-j\delta(\omega - \omega_0) - j\delta(\omega + \omega_0)]$$

$$= -\pi[\delta(\omega - \omega_0) + \delta(\omega + \omega_0)] = -F[\cos(\omega_0 t)].$$

This tells us that the Hilbert transform of $\sin(\omega_0 t)$ is $-\cos(\omega_0 t)$. See part b of Problem I.2 for a generalization of these special cases.

The question we now ask (and answer) is: what is $h(t)$? Before we do *any* calculation at all, we can immediately say that since $H(j\omega)$ is purely imaginary, then $h(t)$ must be odd. We now proceed as follows: define the complex signal

$$z(t) = x(t) + j\overline{x}(t).$$

This complex signal was (and still is) called the *analytic* signal by the Hungarian-born electrical engineer Dennis Gabor (1900–1979), who introduced its use in a 1946 paper on communication theory. (Gabor won the 1971 Nobel Prize in physics for his work in holography.) The analytic signal is central to a theoretical understanding of the phase shift method for generating single-sideband radio signals (see Chapter 20). Then,

$$Z(j\omega) = X(j\omega) + j\overline{X}(j\omega).$$

By the definition of the Hilbert transform we have

$$\overline{X}(j\omega) = \begin{cases} -jX(j\omega), & \omega > 0 \\ +jX(j\omega), & \omega < 0. \end{cases}$$

Thus,

$$Z(j\omega) = \begin{cases} 2X(j\omega), & \omega > 0 \\ 0, & \omega < 0. \end{cases}$$

This can be compactly written as

$$Z(j\omega) = 2X(j\omega)S(j\omega),$$

where $S(j\omega)$ is the unit step in the frequency domain [I could also write this as $u(\omega)$ and be consistent with the notation used in the discussion in this book on step functions in time]. This is shown in Figure I.2. From the frequency convolution theorem (see Chapter 15) we have

$$z(t) = 2x(t) * s(t).$$

To find $s(t)$ I'll use an approach different from the one used in the text (so you'll see something new) to find the inverse Fourier transform of the unit step in the frequency domain (see Chapter 14). That is, write

$$S(j\omega) = \frac{1}{2}\,\mathrm{sgn}(\omega) + \frac{1}{2} = S_1(j\omega) + S_2(j\omega)$$

as shown in Figure I.3.

Since $S(j\omega) = S_1(j\omega) + S_2(j\omega)$, then $s(t) = s_1(t) + s_2(t)$. We have, for $s_2(t) \leftrightarrow S_2(j\omega) = \frac{1}{2}$,

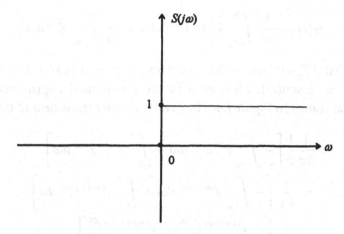

Figure I.2   The unit frequency step function.

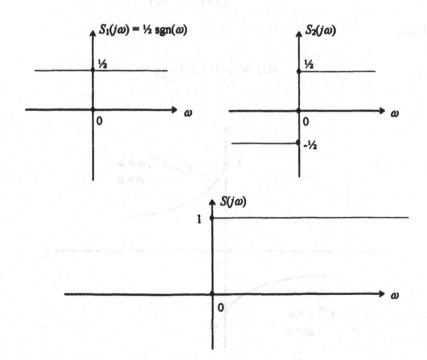

Figure I.3   The unit frequency step as the sum of a constant and the sign function.

$$s_2(t) = \frac{1}{2\pi} \int_{-\infty}^{\infty} S_2(j\omega)e^{j\omega t}\, d\omega = \frac{1}{4\pi} \int_{-\infty}^{\infty} e^{j\omega t}\, d\omega.$$

Recall that $\int_{-\infty}^{\infty} e^{j\omega t}\, d\omega = 2\pi\delta(t)$. Thus, $s_2(t) = (\frac{1}{2})\,\delta(t)$. For $s_1(t) \leftrightarrow S_1(j\omega) = \frac{1}{2}\,\mathrm{sgn}(\omega)$, I'll start with the exponential approximation to $S_1(j\omega)$ as shown in Figure I.4. The inverse Fourier transform of this is

$$\frac{1}{2\pi}\frac{1}{2}\left[-\int_{-\infty}^{0} e^{a\omega}e^{j\omega t}\, d\omega + \int_{0}^{\infty} e^{-a\omega}e^{j\omega t}\, d\omega\right]$$

$$= \frac{1}{4\pi}\left[-\int_{-\infty}^{0} e^{(a+jt)\omega}\, d\omega + \int_{0}^{\infty} e^{(-a+jt)\omega}\, d\omega\right]$$

$$= \frac{1}{4\pi}\left[-\frac{e^{(a+jt)\omega}}{a+jt}\bigg|_{0}^{-\infty} + \frac{e^{(-a+jt)\omega}}{-a+jt}\bigg|_{0}^{\infty}\right]$$

$$= \frac{1}{4\pi}\left[-\frac{1}{a+jt} - \frac{1}{-a+jt}\right]$$

or, as $a \to 0$, we have

$$s_1(t) = -\frac{1}{j2\pi t} = j\frac{1}{2\pi t}.$$

Thus,

$$s(t) = \frac{1}{2}\delta(t) + j\frac{1}{2\pi t}$$

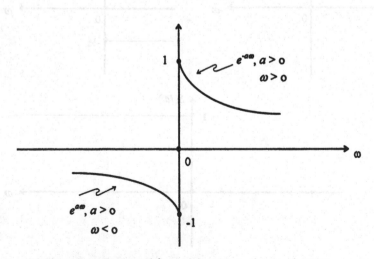

Figure I.4   Exponential approximation to the sign function.

and so

$$z(t) = 2x(t) * \left[\frac{1}{2}\delta(t) + j\frac{1}{2\pi t}\right] = x(t) * \delta(t) + jx(t) * \frac{1}{\pi t}.$$

But, $x(t) * \delta(t) = x(t)$, and so

$$z(t) = x(t) + jx(t) * \frac{1}{\pi t} = x(t) + j\bar{x}(t).$$

Thus, at last,

$$\bar{x}(t) = x(t) * \frac{1}{\pi t} = x(t) * h(t)$$

and so, $h(t) = \frac{1}{(\pi t)}$, $-\infty < t < \infty$.

Notice that $h(t)$ is odd, as argued at the start of this analysis. The quadrature filter is clearly not realizable, as its $h(t)$ is noncausal. Notice, too, that we now have the interesting Fourier transform pair (for $|t| < \infty$)

$$h(t) = \frac{1}{\pi t} \leftrightarrow H(j\omega) = -j\,\text{sgn}(\omega).$$

In integral form, we have

$$\bar{x}(t) = \int_{-\infty}^{\infty} x(\tau)h(t - \tau)\,d\tau$$

or,

$$\bar{x}(t) = \frac{1}{\pi}\int_{-\infty}^{\infty} \frac{x(\tau)}{t - \tau}\,d\tau.$$

This transform integral first appeared in an English-language scholarly journal in 1909, when G. H. Hardy—see Problem G.8—derived it using an approach different from the one I've used here. He thought it was original with him, but later learned that Hilbert had known of it since 1904, and so thereafter Hardy called it the Hilbert transform. But even Hilbert was not the first—the integral, in fact, appears in the very first sentence of the 1873 doctoral dissertation of the Russian mathematician Yulian-Karl Vasilievich Sokhotsky (1842–1927).

Direct evaluation of convolution integrals is almost always a tricky business and the Hilbert transform, with its discontinuous integrand, is certainly no exception. In particular, to evaluate this integral we must pay *very* careful attention to the location of the discontinuity (which is at $\tau = t$), as discussed in Chapter 15. To see how such an integration is done in detail, suppose $x(t)$ is a unit amplitude pulse over the interval $0 < t < T$, and zero at all other times. There are then three cases we must consider (keep in mind that the variable in the *integral* is τ, not t):

1. the discontinuity occurs before the start of the pulse, i.e., at $\tau = t < 0$;
2. the discontinuity occurs during the pulse, i.e., at $0 < \tau = t < T$;
3. the discontinuity occurs after the end of the pulse, at $\tau = t > T$.

Figure I.5 shows case 1, with $x(\tau)$ and $h(t - \tau)$ plotted together. From this plot you can see that, for $t < 0$,

$$\bar{x}(t) = \frac{1}{\pi} \int_0^T \frac{1}{t - \tau} \, d\tau = \frac{1}{\pi} \ln \left(\frac{t}{t - T} \right), \; t < 0.$$

Figure I.6 shows the situation for the much more interesting (mathematically speaking) case 2. Since the integrand blows up as τ approaches t, either from the left or the right (and this, of course, is the source of all the difficulty in doing this integral), I'll resort to the artifice of integrating from 0 to $t - \varepsilon$, and then from $t + \varepsilon$ to T, with ε very small but still with $\varepsilon > 0$. In this way we avoid the integrand explosion at $\varepsilon = t$. Then, we'll explore what happens as we let ε vanish. If we're lucky, the limit will exist. That is, I'll calculate

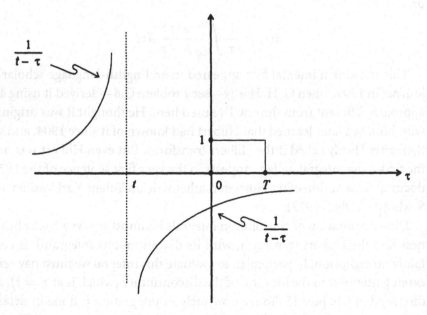

Figure I.5   Case 1 for the discontinuity before the pulse.

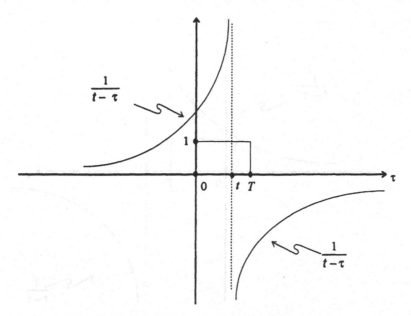

Figure I.6   Case 2 for the discontinuity during the pulse.

$$\bar{x}(t) = \lim_{\varepsilon \to 0} \left[\frac{1}{\pi} \int_0^{t-\varepsilon} \frac{1}{t-\tau} d\tau + \frac{1}{\pi} \int_{t+\varepsilon}^T \frac{1}{t-\tau} d\tau \right]$$
$$= -\frac{1}{\pi} \lim_{\varepsilon \to 0} \left[\ln\left(\frac{\varepsilon}{t}\right) + \ln\left(\frac{t-T}{-\varepsilon}\right) \right].$$

If we combine the two logarithms we see a wonderful thing happen—the εs cancel *even before* we have to consider actually taking the limit. Essentially what has happened is that the integrand explosions on each side of $\tau = t$ have equal magnitudes but opposite signs, and so cancel one another as we integrate across the discontinuity. (I use this same trick in Chapter 15.) That is,

$$\bar{x}(t) = -\frac{1}{\pi} \lim_{\varepsilon \to 0} \ln\left(\frac{\varepsilon}{t} \frac{t-T}{-\varepsilon}\right) = \frac{1}{\pi} \ln\left(\frac{t}{T-t}\right), \quad 0 < t < T.$$

And finally, for case 3, Figure I.7 shows the way things are. We have

$$\bar{x}(t) = \frac{1}{\pi} \int_0^T \frac{1}{t-\tau} d\tau = \frac{1}{\pi} \ln\left(\frac{t}{t-T}\right), \quad t > T.$$

If you look carefully at the expressions I have derived for $\bar{x}(t)$ in these three cases, you can see that they can actually be written in one all-purpose

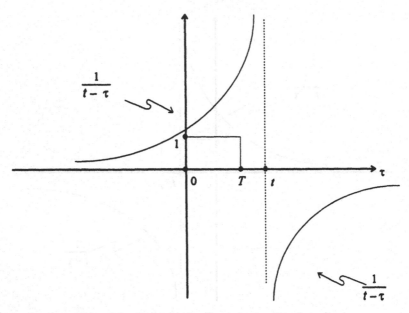

Figure I.7   Case 3 for the discontinuity after the pulse.

expression as

$$\bar{x}(t) = \frac{1}{\pi} \ln \left| \frac{t}{t - T} \right|, \quad |t| < \infty.$$

Figure I.8 shows $x(t)$ and $\bar{x}(t)$ plotted together. The complexity of this calculation, even for such a simple $x(t)$, is perhaps the reason for why tables of Hilbert transforms are not very long.

MATLAB can take much of the agony out of doing Hilbert transforms; the command hilbert(x) computes the Hilbert transform of the real part of the complex vector x and places the result in the imaginary part of x. To see how this works, take a look at the program htpulse.m which creates a unit amplitude pulse signal (with $T = 1$) in the form of the vector x, and then sets the vector xt equal to the imaginary part of hilbert(x). The result is displayed as Figure I.9, with the original pulse and its transform plotted together (it *does* look remarkably like Figure I.8, don't you think?)

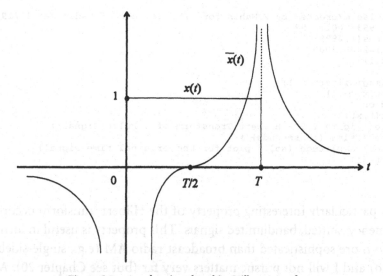

Figure I.8   A pulse signal and its Hilbert transform.

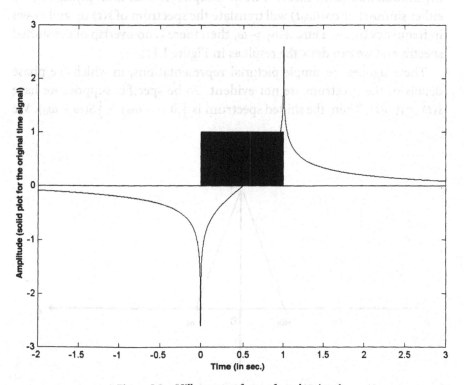

Figure I.9   Hilbert transform of a pulse signal.

```
%htpulse.m/created by PJNahin for "The Science of Radio 2e"(12/19/99)
t=-1.999:.001:2.999;
x=zeros(1,4999);
for i=2000:3000
    x(i)=1;
end
xt=imag(hilbert(x));
bar(t,x,'-.')
hold on
plot(t,xt)
title('Figure I.9 - Hilbert Transform of a Pulse Signal')
xlabel('time (in seconds)')
ylabel('amplitude (solid plot for the original time signal)')
figure(1)
```

A particularly interesting property of the Hilbert transform occurs for frequency-shifted, bandlimited signals. This property is useful in forms of radio more sophisticated than broadcast radio AM (e.g., single-sideband radio) and I will not pursue matters very far (but see Chapter 20). As an example of Fourier mathematics, however, it is appropriate for this book. So, suppose $s(t)$ is a bandlimited signal centered at dc, i.e., $s(t)$ is a so-called *baseband* signal as shown in Figure I.10. We know by the heterodyne (or modulation) shift theorem from Chapter 15 that multiplying $s(t)$ by either $\sin(\omega_0 t)$ or $\cos(\omega_0 t)$ will translate the spectrum of $s(t)$ up and down in frequency by ω_0. Thus, if $\omega_0 > \omega_s$ then there is no overlap of the shifted spectra and we can draw the result as in Figure I.11.

These figures are simple pictorial representations, in which the phase details of the spectrum are not evident. To be specific, suppose we have $s(t)\cos(\omega_0 t)$. Then, the shifted spectrum is $\frac{1}{2}S(\omega - \omega_0) + \frac{1}{2}S(\omega + \omega_0)$. We

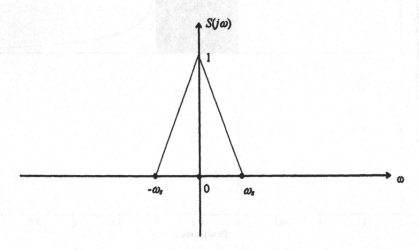

Figure I.10   The spectrum of a band-limited, baseband signal.

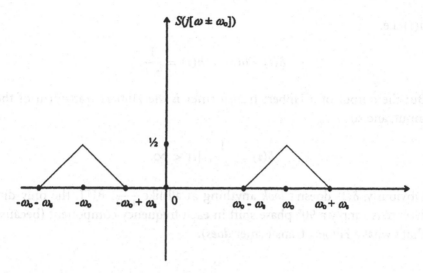

Figure I.11 The shifted spectrum of $s(t)$.

can then write the spectrum of the Hilbert transform of $s(t) \cos(\omega_0 t)$ as

$$-j\left(\frac{1}{2}\right) S(\omega - \omega_0) + j\left(\frac{1}{2}\right) S(\omega + \omega_0) = -j\left(\frac{1}{2}\right)[S(\omega - \omega_0) - S(\omega + \omega_0)].$$

Notice carefully that we can do this only because we know where *all* the spectral components at positive frequencies are located; at the location of $S(\omega - \omega_0)$. No part of $S(\omega + \omega_0)$ is at positive frequencies because of our assumption that $\omega_0 > \omega_s$. If $\omega_0 < \omega_s$ then there would be overlap of $S(\omega - \omega_0)$ and we would not know what to multiply by $-j$. Similarly, we know that *all* of $S(\omega + \omega_0)$ is at negative frequencies, and so it alone is what is multiplied by $+j$.

But, the spectrum of $s(t) \sin(\omega_0 t)$ is $-j\frac{1}{2}[S(\omega - \omega_0) - S(\omega + \omega_0)]$. Thus, we have the remarkable result

$$\overline{s(t) \cos(\omega_0 t)} = s(t)\overline{\cos(\omega_0 t)} = s(t) \sin(\omega_0 t).$$

if $s(t)$ is bandlimited at baseband, with its upper frequency cutoff (ω_s) less than ω_0. In a similar way you can show that

$$\overline{s(t) \sin(\omega_0 t)} = s(t)\overline{\sin(\omega_0 t)} = -s(t) \cos(\omega_0 t).$$

Finally, consider the following curious "puzzle." If we apply an impulse to the input of a Hilbert transformer, then the output is of course simply

$h(t)$, i.e.,

$$\delta(t) * h(t) = h(t) = \frac{1}{\pi t}.$$

But the output of a Hilbert transformer *is* the Hilbert transform of the input, and so

$$\overline{\delta}(t) = \frac{1}{\pi t}, \quad |t| < \infty.$$

Obviously, $\delta(t)$ doesn't *look* anything at all like $\frac{1}{(\pi t)}$. BUT, the only difference is simply a 90° phase shift in each frequency component (because that's what a Hilbert transformer *does*).

Problems

I.1. As mentioned at the beginning of this appendix, the requirement that the impulse response of the Hilbert transformer is real requires the phase angle of $H(j\omega)$ to be odd. Why did I use the formulation given, i.e., why not use instead

$$H(j\omega) = \begin{cases} +j, & \omega > 0 \\ -j, & \omega < 0? \end{cases}$$

Hint: take a look at Problem 13.6.

I.2. The Hilbert transform has several interesting properties. Here are three of them, in increasing order of difficulty, for you to try your hand at proving. The first two are actually pretty easy (you don't even need to write any mathematics), and there is a hint for the third.

a. $x(t)$ and $\overline{x}(t)$ have the same energy, and the same energy spectral density.

b. $\overline{\overline{x}}(t) = -x(t)$, i.e., the Hilbert transform of a Hilbert transform is the negative of the original time function.

c. $\int_{-\infty}^{\infty} x(t)\overline{x}(t)\, dt = 0$. To prove this, recall the result from Chapter 15 in which it is shown that if $m(t)$ and $g(t)$ are two time functions, then

$$\int_{-\infty}^{\infty} m(t)g(t)\, dt = \frac{1}{2\pi} \int_{-\infty}^{\infty} M(j\omega)G^*(j\omega)\, d\omega,$$

where M and G are the Fourier transforms of m and g, respectively. Thus, if $m(t) = x(t)$ and $g(t) = \bar{x}(t)$, then we have $M(j\omega) = X(j\omega)$, $G(j\omega) = \bar{X}(j\omega) = -j\,\text{sgn}(\omega)X(j\omega)$, and so $G^*(j\omega) = j\,\text{sgn}(\omega)X^*(j\omega)$. Therefore,

$$\int_{-\infty}^{\infty} x(t)\bar{x}(t)\,dt = \frac{1}{2\pi}\int_{-\infty}^{\infty} j\,\text{sgn}(\omega)|X(j\omega)|^2\,d\omega.$$

Now, to complete the proof, make an argument about why the last integral must vanish. Hint: think about the evenness and oddness of the various factors in the integrand.

I.3. By direct evaluation of the transform integral, prove that the Hilbert transform of any constant is zero.

I.4. Using the relationship between the Fourier transforms of $x(t)$ and $\bar{x}(t)$ show that, for $|t| < \infty$, the Hilbert transform of $\frac{1}{(t^2+1)}$ is $\frac{t}{(t^2+1)}$ (see Problem 12.3 for the appropriate Fourier transforms).

Table of Fourier Transform Pairs and Theorems

The notation of this table is that of the text, proper, i.e., $V(j\omega)$ is the Fourier transform of $v(t)$. Also, the lowercase u denotes the unit-step function (the domain is clear by context in each case), and the π symbol *when it has an argument* is the unit-gate function. These results are derived and/or discussed in the text on the pages in the brackets.

1. $v(at) \leftrightarrow \dfrac{1}{|a|}V(j\omega/a), a \neq 0$ [172]

2. $e^{-\sigma t}u(t) \leftrightarrow \dfrac{1}{\sigma + j\omega}, \sigma > 0$ [175]

3. $u(t) \leftrightarrow \pi\delta(\omega) + \dfrac{1}{j\omega}$ [203]

4. $\dfrac{u(t)}{\sqrt{t}} \leftrightarrow \sqrt{\dfrac{\pi}{2\omega}}(1-j)$ [176]

5. $\mathrm{sgn}(t) \leftrightarrow \dfrac{2}{j\omega}$ [206]

6. $|t| \leftrightarrow -\dfrac{2}{\omega^2}, \omega \neq 0$ [206]

7. $\dfrac{dv}{dt} \leftrightarrow j\omega V(j\omega)$ [178]

8. $tv(t) \leftrightarrow j\dfrac{dV}{d\omega}$ [178]

9. $\dfrac{1}{t^2+1} \leftrightarrow \pi e^{-|\omega|}$ [178]

10. $\dfrac{t}{t^2 + 1} \leftrightarrow -j\pi e^{-|\omega|}\,\text{sgn}(\omega)$ [178]

11. $e^{-at}\sin(\omega_c t)u(t) \leftrightarrow \dfrac{\omega_c}{(a + j\omega)^2 + \omega_c^2},\, a > 0$ [178]

12. $e^{-at^2} \leftrightarrow \sqrt{\dfrac{\pi}{a}}\,e^{-\frac{\omega^2}{4a}},\, a > 0$ [207]

13. $V(jt) \leftrightarrow 2\pi v(-\omega)$ [196]

14. $\pi(t) \leftrightarrow \dfrac{\sin\left(\frac{\omega}{2}\right)}{\left(\frac{\omega}{2}\right)}$ [171]

15. $\delta(t) \leftrightarrow 1$ [194]

16. $1 \leftrightarrow 2\pi\delta(\omega)$ [195]

17. $\dfrac{\sin(t)}{t}u(t) \leftrightarrow \dfrac{\pi}{2} \cdot \pi\left(\dfrac{\omega}{2}\right) + j\dfrac{1}{2}\ln\left|\dfrac{\omega - 1}{\omega + 1}\right|$ [229]

18. $\dfrac{1}{2}\delta(t) + j\dfrac{1}{2\pi t} \leftrightarrow u(\omega)$ [204]

19. $h(t) * x(t) \leftrightarrow H(j\omega)X(j\omega)$ [222]

20. $\cos(\omega_0 t) \leftrightarrow \pi[\delta(\omega - \omega_o) + \delta(\omega + \omega_o)]$ [220]

21. $h(t)x(t) \leftrightarrow \dfrac{1}{2\pi}H(j\omega)*X(j\omega)$ [222]

22. $\displaystyle\int_t^\infty \dfrac{e^{-u}}{u}\,du \leftrightarrow \dfrac{\ln(1 + j\omega)}{j\omega},\, t > 0$ [434]

23. $\displaystyle\int_0^t h(z)\,dz \leftrightarrow \pi H(0)\delta(\omega) + \dfrac{1}{j\omega}H(j\omega)$ [230]

Last Words

If you've *read* this far, as opposed to those who simply like to flip to the end of a book to "see how it all ends" without waiting, then you know that by current publishing fashions this is an eccentric book. Or, it is if I've succeeded in my goal of making it different from the lookalike sophomore circuits textbooks produced by most of the big commercial publishers (who react to mass-market, consensus-building surveys). Those aren't bad books, mind you—I'll even go so far as to admit that some of them are actually pretty good—but what the education world needs *least* is yet another one.

Writing a book is generally not an easy job, and the writing of this book reminds me of some words from an eighth-century poem, written by a long-forgotten Irish monk who nevertheless succeeded in immortalizing his beloved cat:

> I and Pangur Ban, my cat
> 'Tis a like task we are at;
> Hunting mice is his delight,
> Hunting words, I sit all night.

I hope that, in my many nightly hunts, I have found the right words for this book. If you think so, please let me know. I'll pass all such communications straight on to my editor, with no delay. But if you don't think so, and can suggest some improvement or addition (or even, God forbid, something I should *delete*), please let me know that, too. I may not pass it on to my editor, but I promise that I will consider what you write. I can be reached on the Internet at paul.nahin@unh.edu.

About the Author

Paul J. Nahin is Professor of Electrical Engineering at the University of New Hampshire, where he has often taught the undergraduate courses in circuits, networks, and electronics. During the academic year 1999–2000 he was Visiting Professor of Electrical Engineering at the University of Virginia, where he taught classes in undergraduate circuits and (in the Technology, Culture and Communications Department) the cultural history of AM broadcast radio.

NAME INDEX